A
HOUSE
BUILT
ON
SAND

A HOUSE BUILT ON SAND

Exposing

Postmodernist

Myths about

Science

Edited by

Noretta Koertge

New York   Oxford

Oxford University Press

1998

Oxford University Press

Oxford   New York
Athens   Auckland   Bangkok   Bogota   Bombay
Buenos Aires   Calcutta   Cape Town   Dar es Salaam
Delhi   Florence   Hong Kong   Istanbul   Karachi
Kuala Lumpur   Madras   Madrid   Melbourne
Mexico City   Nairobi   Paris   Singapore
Taipei   Tokyo   Toronto   Warsaw

and associated companies in
Berlin   Ibadan

Library of Congress Cataloging-in-Publication Data
A house built on sand : exposing postmodernist myths about science /
edited by Noretta Koertge.
p.   cm.
Includes bibliographical references and index.
ISBN 0-19-511725-5
1. Science.   2. Science—Social aspects.   3. Science and state.
4. Research—Philosophy.   I. Koertge, Noretta.
Q172.H68   1998
501—dc21   97-47506

1 3 5 7 9 8 6 4 2

Printed in the United States of America
on acid-free paper

Acknowledgments

Any collaborative volume depends not only on the talents of the contributors but also on their willingness to cooperate in choosing topics and meeting deadlines. The experience of working with the 16 scientists and humanists involved in this project was heartening in many ways. Despite widespread rumors about the incommensurability of the "Two Cultures," when we circulated drafts of our essays among ourselves there was an amazing consilience of basic intellectual values. People with backgrounds in science, engineering, and mathematics were very concerned with questions of historical accuracy. The historians and philosophers of science in our group placed a high premium on getting the scientific details right. And we all equally deplored the factual errors, weak arguments, and slippery rhetoric that permeated so much of the postmodernist science studies that we were critiquing.

Not that there weren't disagreements! Many of the essays in this volume were modified as a result, thanks to the good ideas of the readers and the good sense of the writers to be receptive to the suggestions of their colleagues.

All of this took time, but thanks to the technology of the internet and the willingness of the contributors to make this project a high priority, we were able to go from the initial prospectus to final manuscript in exactly one year. Many thanks to those who interrupted their preparation of tenure dossiers, grant proposals, and sabbatical leaves to meet the deadlines for this project.

A special acknowledgment also goes to Oxford University Press. We are grateful to those anonymous referees and to the press for their confidence in this project.

I would also like personally to thank various people who supported my editorial efforts: my wise E-mail buddies who served as informal editorial consultants; my patient HPS colleagues here at IU who listened to me over lunch and gave good advice; the spunky students in my "Science Wars" seminar for invigorating discussions; Karen Blaisdell for transforming electronic submissions in every conceivable format into a coherent manuscript; Anne Mylott for compiling the index; and my family—Deborah, Matt, and Emma—for helping me remember what is most important in life.

Bloomington, Indiana   N. K.
July 1997

Contents

Contributors

PAUL BOGHOSSIAN is professor of philosophy and chair of the department at New York University. His research interests lie mainly in the philosophy of mind, the philosophy of language and epistemology. He has published articles on color, rule-following, naturalism, eliminativism, self-knowledge, analytic truth, and a priori knowledge. E-mail: boghossn@is2.nyu.edu

ALLAN FRANKLIN is professor of physics and an active participant in the history and philosophy of science program at the University of Colorado. In addition to his work in high-energy physics, he has written extensively on the philosophy of experiment, notably *The Neglect of Experiment* (1986) and *Experiment, Right or Wrong* (1990). E-mail: franklin_a@gold.colorado.edu

PAUL GROSS is university professor emeritus at the University of Virginia. In addition to a long and distinguished research career in molecular biology, including a stint as director of the Woods Hole Marine Biological Laboratory, he served as director of the University of Virginia's Center for Advanced Studies. Both experiences contributed directly to his provocative and influential book (coauthored with Norman Levitt) *Higher Superstition: The Academic Left and Its Quarrels with Science* (1994). E-mail: prg@faraday.clas.virginia.edu

JOHN HUTH is professor of physics at Harvard University. He performs experiments in high energy particle physics. His current experiments using the Collider Detector at Fermilab produced recently published data demonstrating the discovery of the top quark. He is also engaged in research in advanced accelerator technology and experiments to test the mechanisms of spontaneous symmetry breaking between the weak and electromagnetic forces. E-mail: huth@huhepl.harvard.edu

MARGARET C. JACOB is professor of the history and sociology of science at the University of Pennsylvania and one of the pioneers of the social history of science.

Her most recent book, *Scientific Culture and the Making of the Industrial West*, was published by Oxford University Press in 1997. She is also coauthor of *Telling the Truth about History* (1994) with Lynn Hunt and Joyce Appleby. E-mail: mjacob@sas.upenn.edu

PHILIP KITCHER is professor of philosophy at the University of California at San Diego and former editor of the journal *Philosophy of Science*. His research interests range from the philosophy of mathematics to the ethics of biological research. His most recent books include *Vaulting Ambition: Sociobiology and the Quest for Human Nature* (1985), *The Advancement of Science: Science without Legend, Objectivity without Illusions* (Oxford University Press, 1993), and *The Lives to Come: The Genetic Revolution and Human Possibilities* (1996). E-mail: pkitcher@ucsd.edu

NORETTA KOERTGE is professor of the history and philosophy of science at Indiana University. Her research interests include both the historical and the normative aspects of scientific methodology. She edited *Nature and Causes of Homosexuality: A Philosophic and Scientific Inquiry* (1981) and cowrote (with Daphne Patai) *Professing Feminism: Cautionary Tales from the Strange World of Women's Studies* (1994). E-mail:koertge@indiana.edu

NORMAN LEVITT is professor of mathematics at Rutgers University and author of *Grassmannians and Gauss Maps in Piecewise-Linear Topology* (1989). His experience at the interdisciplinary seminars on science and culture at the Rutgers University Center for the Critical Analysis of Contemporary Culture provided motivation and material for the controversial book he coauthored with Paul Gross, *Higher Superstition: The Academic Left and Its Quarrels with Science* (1994). He is now writing about the relationship between science and democracy. E-mail: njlevitt@math.rutgers.edu

WILLIAM McKINNEY is professor of philosophy at Southeast Missouri State University. His research interests include theories of scientific discovery, philosophical analyses of experimental failures, and the philosophy of technology. E-mail: c430hup@semovm.semo.edu

MEERA NANDA first trained as a microbiologist and is now completing a second Ph.D. in the science and technology studies department at Rensselaer Polytechnic Institute. In addition to her work in science journalism, she has written scholarly articles on the implications of social constructionist and postmodern critiques of science and technology for the project of modernizing non-Western countries. E-mail: nandam@rpi.edu

WILLIAM NEWMAN is professor of history and philosophy of science at Indiana University. His current research draws connections between alchemical debates and the emergence of modern science. He is the author of *The Summa perfectionis of Pseudo-Geber: A Critical Edition, Translation and Study* (1991) and *Gehennical Fire: The Lives of George Starkey, an American Alchemist in the Scientific Revolution* (1994). E-mail: wnewman@indiana.edu

CASSANDRA PINNICK is professor of philosophy at Western Kentucky University. She has published articles on feminist epistemology, technology assessment, and the ethnographic approach to science studies. Her research interests focus on developing a historically based epistemology of science. E-mail: pinnick@wku.edu

MICHAEL RUSE is professor of philosophy and zoology at the University of Guelph. He is the founding editor of the *Journal of Philosophy and Biology* and the author of numerous books on history and the philosophy of biology, including *The Darwinian Revolution: Science Red in Tooth and Claw* (1979), *Sociobiology: Sense or Nonsense?* (1979), *Taking Darwin Seriously: A Naturalistic Approach to Philosophy* (1986), *The Darwinian Paradigm* (1989), and *Monad to Man: The Concept of Progress in Evolutionary Biology* (1996). E-mail: mruse@arts.uoguelph.ca

ALAN SOBLE is professor and research professor of philosophy at the University of New Orleans. In 1977 he founded the Society for the Philosophy of Sex and Love and served as its director until 1992. His recent books include *Pornography: Marxism, Feminism, and the Future of Sexuality* (1986), *The Structure of Love* (1990), and *Sexual Investigations* (1996). He is currently writing *The Limits of Feminist Scholarship*. E-mail: agspl@uno.edu

ALAN SOKAL is professor of physics at New York University and coauthor (with Roberto Fernandez and Jurg Frohlich) of *Random Walks, Critical Phenomena and Triviality in Quantum Field Theory* (1992). He recently wrote a series of critiques of cultural studies commentaries on the physical sciences, most notably his satirical article, "Transgressing the Boundaries: Toward a Transformative Hermeneutics of Quantum Gravity," which appeared in a recent issue of *Social Text*. E-mail: sokal@nyu.edu

PHILIP SULLIVAN is professor of aerospace engineering at the University of Toronto's Institute for Aerospace Studies. His research interests include hypersonic flow, air cushion vehicle dynamics, and biological fluid flows. He is currently writing a historically informed textbook on fluid mechanics. E-mail: sullivan@utias.utoronto.ca

A
HOUSE
BUILT
ON
SAND

Scrutinizing Science Studies

Noretta Koertge

T he "House" in our title refers to interdisciplinary endeavors called Science, Technology, and Society Studies (STS) or Science and Culture Studies. Within their veritable carnival of approaches and methodologies we find feminists and Marxists of every stripe, ethnomethodologists, deconstructionists, sociologists of knowledge and critical theorists—those who find significance in rhetoric and others who emphasize the role of patronage and the power of empire. One might expect to find irreconcilable differences between, for example, those who stress material conditions and those who focus on metaphors, and, indeed, there are lively debates on such matters. Yet we also find widely shared convictions and projects. Although it is always hazardous to try to summarize any group's principles and purposes, the following are some noteworthy precepts that appear to be widely shared:

- Every aspect of that complex set of enterprises that we call science, including, above all, its content and results, is shaped by and can be understood only in its local historical and cultural context.
- In particular, the products of scientific inquiry, the so-called laws of nature, must always be viewed as social constructions. Their validity depends on the consensus of "experts" in just the same way as the legitimacy of a pope depends on a council of cardinals.
- Although scientists typically succeed in arrogating special epistemic authority to themselves, scientific knowledge is just "one story among many." The more epistemological authority that science has in a given society, the more important it is to unmask its pretensions to be an enterprise dedicated to the pursuit of objective knowledge. Science must be "humbled."
- Since the quest for objective knowledge is a quixotic one, the best way to appraise scientific claims is through a process of political evaluation. Since the "evidence" for a scientific claim is never conclusive and is always open to negotiation, the best way to evaluate scientific results is to ask who stands to benefit if the claim is taken to be true. Thus, for the citizen the key question about a scientific result should not be how well tested the claim is but, rather, Cui bono?
- "Science is politics by other means": the results of scientific inquiry are profoundly and importantly shaped by the ideological agendas of powerful elites.

3

- There is no univocal sense in which the science of one society is better than that of another. In particular, Euroscience is not objectively superior to the various ethnosciences and shamanisms described by anthropologists or invented by Afrocentrists.
- Neither is there any clear sense in which we can talk about scientific progress within the European tradition. On the contrary, science is characterized chiefly by its complicity in all the most negative and oppressive aspects of modern history: increasingly destructive warfare, environmental disasters, racism, sexism, eugenics, exploitation, alienation, and imperialism.
- Given the impossibility of scientific objectivity, it is futile to exhort scientists and policymakers to try harder to remove ideological bias from the practice of science. Instead, what we need to do is deliberately introduce "corrective biases" and "progressive political values" into science. There is a call for "emancipatory science" and "advocacy research."

Postmodernists pride themselves on their reflexivity, so it is not surprising to find them defending their own approach by proclaiming it to be morally and politically superior to the traditional scientific emphases on disinterestedness, universalism, and empiricism. Convinced that scientists succeed in making their case by relying on rhetorical devices that include the deployment of judiciously selected "data" and artfully constructed "experimental demonstrations," advocates of the Cultural Studies approach have assembled a group of case studies that are intended to persuade us that their account of science is indeed exemplified by past and present scientific practice. Thus in the now notorious "Science Wars" issue of *Social Text*, Sandra Harding boasts that "the Right's objections [to science studies] virtually never get into the nitty-gritty of historical or ethnographic detail to contest the accuracy of social studies of science accounts. Such objections remain at the level of rhetorical flourishes and ridicule."[1] Harding's claim is exaggerated—there are already many book reviews and articles (some written by the contributors to this volume) cataloguing the factual errors, interpretative flaws, and shoddy scholarship that are all too pervasive in STS writings. (The Sokal "experiment" vividly illustrates what can easily happen in a field that repudiates all received scholarship, in which "text" is more important than "fact" and the political inspiration for a claim becomes the overriding evaluative criterion.)

But ironically, Harding's remark does reinforce the idea that it would be very useful to have a collection of hard-hitting critiques of the postmodernist case studies that are cited over and over again as evidence for the claim that the results of natural science tell us more about social context than they do about the natural world. Although the overall tone of this collection is critical of the dominant STS portrait of the development of science, we recognize the importance of many of the questions raised by social historians, sociologists, and students of science policy. But we believe that a prerequisite for dealing fruitfully and responsibly with these concerns is to be more careful in our descriptions of the history of science and current scientific practice. This volume is intended to expedite the winnowing process, for only then can we realize the positive potential of the science studies approach.

At the 1994 professional meeting of historians and philosophers of science in New Orleans (a meeting attended, coincidentally, by FBI agents looking for clues to the origins of the ideas in the Unabomber's *Manifesto*), Philip Kitcher read a paper entitled "How the Road to Relativism Is Paved—With the Best of Intentions but the Worst of Arguments." Because many of the political concerns discussed in science studies are legitimate and widely shared, many people in the academy have been loath to criticize publicly weak philosophical arguments or shaky history, feeling the science critics were well intentioned. However, more and more people (and not just on the right—the *Nation* was quite supportive of Sokal's intervention) are now realizing that short-run political solidarity is no substitute for scholarly integrity.

In this book we intend to provide a place where reason and good sense can be brought to bear on a field that has lost its mechanisms of scholarly self-control. Our analyses do not pull any punches in identifying errors. Some of the case studies we criticize have turned out to contain crude factual blunders (e.g., Latour's reading of relativity theory). Other cases involve egregious errors of omission, such as the attempt to explain the slow development of fluid dynamics solely in terms of gender bias. However, we will not pick nits, and when a flawed account contains positive aspects or something to be learned, even though it is mistaken, we will not hesitate to say so. Our only target is shoddy scholarship, but our concern stems from the conviction that intellectual values undergird progressive political values.

ON A MORE PERSONAL NOTE. As soon as the news of the Sokal intervention in the "Science Wars" hit the media, I was contacted by various people at my university—several physicists, a sociologist, a comparative literature student, and an art historian. They all were asking, What on earth is going on in science studies these days? Sometimes the question had an implicit (or explicit) follow-up: Is this the sort of thing you people in history and philosophy of science do?

The second question made me wince. I feel privileged to work in one of the oldest "science studies" departments in this country (although we have never used that label). From time to time my colleagues have team-taught or offered courses on their own in Indiana University's departments of physics, astronomy, biology, psychology, and economics. Our graduate students have had faculty from mathematics, computer science, linguistics, and cognitive science on their thesis committees. How could anyone, for a moment, believe that we had gone postmodern?!

But the price of respect is vigilance. Many serious scholars in history, philosophy, and sociology of science have been too busy doing good research and too sanguine about the good sense of their more distant colleagues. Some think that deconstruction deserves our "sympathetic indulgence." However, I believe it is high time for what sociologists call "boundary maintenance" but what I prefer to think of not as policing but as a form of Popperian respect. While doing graduate work at Chelsea College, I attended faithfully the famous "Popper Seminar" at the London School of Economics. I often heard Sir Karl (he had just been knighted) say (I quote from memory): "The greatest tribute you can pay someone is to criticize his work. Perhaps you will be able to improve on his ideas, but that is not necessary. We learn from our mistakes."

I believe this book responds well to the question about what is going on in science studies and helps raise the level of critical discussion. What it does not address is the question that kept resurfacing in the Science Wars seminar that I taught while the articles in this book were being written—namely, What is really going on in the Science Wars? Why has the postmodernist perspective become so popular?

The ten members of that seminar drew on a wide variety of disciplinary backgrounds and experience: fieldwork in anthropology and archeology; technical laboratory work in an astronomy field station and at Los Alamos; and Peace Corps service and teaching experience in Puerto Rico, Mexico, India, and Turkey. It quickly emerged that we had quite disparate attitudes toward, for example, technology, the philosophy of elementary science education, and governmental regulation! But we also had two immediate points of agreement. When we read *Higher Superstition,* we all found that it was more moderate and less polemical than we had anticipated (or remembered from earlier reading). Either we had been misled by reviews of the book, or the level of vitriol in the debate has risen so rapidly since the book was written that it no longer sounded so harsh (or both)![2]

In regard to our other point of agreement, it would provide a pleasing symmetry if I could report that the postmodernist pieces we read were also better modulated than we had anticipated or remembered. Unfortunately, that was not the case. I can easily visualize one participant gesticulating in an agitated fashion while saying something to the following effect: Why do they keep shooting themselves in the foot? I was prepared to be very sympathetic to the postmodernists, you know. I've seen the bad effects of scientism in my own field, and I can go along with them for quite a long ways. But then it gets so absurd I just have to put the book down. Does anyone really believe this stuff? What are they doing!

Some members of the seminar were inclined to bracket the parts in the readings they found gratuitously polemical or subversive to the authors' own projects, especially when they found the project a congenial one. But all of us found many of the interchanges puzzling, and at coffee breaks we kept stewing about what was really going on. Everyone had an ingredient to add: the Vietnam War and *Sputnik*; C. P. Snow, Alan Bloom, Kuhn, and Feyerabend; Three Mile Island, Bhopal, and the spotted owl; Reagan's defunding of social science and the demise of the supercollider; the explicit political mission of women's studies and black studies programs; anxiety about affirmative action and the bad job market for new Ph.D.s; Watergate and the end of the Cold War.

Neither our seminar nor this volume could fruitfully grapple with themes of such magnitude. A respectable account of the broader historical context of the Science Wars awaits the touch of a future Gibbon. It also does not make much sense to call for a new science studies paradigm or a polite division of scholarly labor. We cannot even "agree to disagree" in an authentic fashion until we better understand what divides us. Perhaps the best place to start talking is by focusing on particular scientific episodes and specific claims about the history and practice of science. That is what we do in this volume.

Notes

1. Sandra Harding (1996), Science Is "Good to Think With," *Social Text* 46–47:15.
2. For a synopsis of these later developments, see the new preface and supplementary notes in the paperback edition of *Higher Superstition* by Paul R. Gross and Norman Levitt (Baltimore: Johns Hopkins, 1998).

Part I

The Strange World
of Postmodernist
Science Studies

The three chapters in this introductory section both describe some of the major currents that constitute postmodernist studies of science and offer critical analyses of them. To the novice, the plethora of labels and acronyms is bound to be confusing, so in this volume we have tried, as far as possible, to follow Miss Manners's advice to use the forms of address most congenial to the individual. Thus although "postmodernism" or "POMO" can be used as a form of abuse, it is also the expression that many of the new commentators on science choose to describe their approach.

For example, the 1992 draft of the *National Science Education Standards* claimed to be based on a "contemporary approach, called postmodernism, [that] questions the objectivity of observation and the truth of scientific knowledge" (quoted in Holton 1997, 553). And in a recent book review in *Science*, Paul Forman, a historian of science at the Smithsonian, speaks approvingly of "our postmodern works" with its social constructionist epistemology and a "morality-based rather than truth-based *Weltgefuehl*" (1997, 750).[1]

Amongst postmodernist critiques of science, it is useful to distinguish between those people who describe what they do as Science and Technology Studies (STS) and the science commentators within the more loosely delineated approach called Cultural Studies. A comparison of the bibliographic items in the "Science Wars" issue of *Social Text*, a cultural studies magazine, and the 1995 *Handbook of Science and Technology Studies* (Jasanoff

et al.)[2] quickly reveals the significant overlap in these two approaches. Their disciplinary histories are quite different, however. STS has roots in the history and sociology of science as well as science and technology policy studies, whereas cultural studies draws more on recent trends in literary criticism and continental philosophy.

Any attempt at further classification would be counterproductive. As we will see, there are many varieties of feminist critiques of science. Some anthropologists of science think it is important to understand what the "natives" (i.e., the scientists) think they are doing; others believe it is necessary to de-familiarize scientific practices and to use only outsiders' categories to describe what is going on in the laboratory. Some case studies attempt to discuss the substance of the scientific debates and arguments; others focus on the rhetoric devices employed or the social networks connecting the actors.

Nevertheless, as the following chapters will show, there are enough shared elements in these writings to warrant treating them as a cluster. All of the writers we discuss clearly are committed to more or less extreme versions of constructionism and relativism. Their essays also seem to have a certain characteristic tone or attitude that is more difficult to pin down but that is in fact contained in the very term *postmodern*—namely, the conviction that we are now too sophisticated to be moved by the intellectual ideals or accomplishments of the last three centuries.

The following three chapters highlight different weaknesses in postmodernist science studies. Alan D. Sokal, a physicist, is particularly concerned about radical social constructionism and the conflation of ontology and epistemology. Paul Boghossian, a philosopher, emphasizes the incoherence of postmodernist variants of relativism. And Philip Kitcher, a philosopher working in a science studies program, is disturbed by the deleterious effects of various modern historiographies. All suggest that the central tenets of these postmodernist commentators on science are in fact antithetical to their professed progressive political goals.

Notes

1. The book reviewed is Gross et al., *The Flight from Science and Reason*. Forman agrees that postmodernism is found in most areas of scholarship, but he scoffs at "the organizer's aim of placing science back on its pre-postmodern pedestal" (p. 750) and the "hawking [of a] snake-oil cure for scientific illiteracy" (p. 752).

2. The volume was published with the cooperation of the Society for Social Studies of Science.

References

Paul Forman, Assailing the Seasons, *Science* 276 (1997), pp. 750–752.

Holton, Gerald, Science Education and the Sense of Self. In Paul R. Gross, Norman Levitt, and Martin W. Lewis (eds.). 1997. *The Flight from Science and Reason*. Baltimore: Johns Hopkins University Press, pp. 557–560.

Jasanoff, Shelia, Gerald E. Markle, James C. Petersen, and Trevor Pinch, eds. 1995. *Handbook of Science and Technology Studies*. Thousand Oaks: Sage Publications.

What the *Social Text* Affair
Does and Does Not Prove

Alan D. Sokal

> I did not write this work merely with the aim of setting the exegetical record straight. My larger target is those contemporaries who—in repeated acts of wish-fulfillment—have appropriated conclusions from the philosophy of science and put them to work in aid of a variety of social cum political causes for which those conclusions are ill adapted. Feminists, religious apologists (including "creation scientists"), counterculturalists, neoconservatives, and a host of other curious fellow-travelers have claimed to find crucial grist for their mills in, for instance, the avowed incommensurability and underdetermination of scientific theories. The displacement of the idea that facts and evidence matter by the idea that everything boils down to subjective interests and perspectives is—second only to American political campaigns—the most prominent and pernicious manifestation of anti-intellectualism in our time.
>
> Larry Laudan, *Science and Relativism* (1990, p. x)

I confess to some embarrassment at being asked to contribute an introductory essay to this collection of critical studies in the history, sociology, and philosophy of science. After all, I'm not a historian or a sociologist or a philosopher; I'm merely a theoretical physicist with an amateur interest in the philosophy of science and perhaps some modest skill at thinking clearly. *Social Text* cofounder Stanley Aronowitz was, alas, absolutely right when he called me "ill-read and half-educated."[1]

My own contribution to this field began, as the reader undoubtedly knows, with an unorthodox (and admittedly uncontrolled) experiment. I wrote a parody of postmodern science criticism, entitled "Transgressing the Boundaries: Toward a Transformative Hermeneutics of Quantum Gravity," and submitted it to the cultural-studies journal *Social Text* (of course without telling the editors that it was a parody). They published it as a serious scholarly article in their spring 1996 special issue devoted to what they call the "Science Wars."[2] Three weeks later, I revealed the hoax in an article in *Lingua Franca*,[3] and all hell broke loose.[4]

In this essay I'd like to discuss briefly what I think the "*Social Text* affair" does and does not prove. But first, to fend off the accusation that I'm an arrogant physicist who rejects all sociological intrusion on our "turf," I'd like to lay out some

9

positive things that I think social studies of science can accomplish. The following propositions are, I hope, noncontroversial:

1. Science is a human endeavor, and like any other human endeavor, it merits being subjected to rigorous social analysis. Which research problems count as important, how research funds are distributed, who gets prestige and power, what role scientific expertise plays in public-policy debates, in what form scientific knowledge becomes embodied in technology, and for whose benefit—all these issues are strongly affected by political, economic, and, to some extent, ideological considerations, as well as by the internal logic of scientific inquiry. They are thus fruitful subjects for empirical study by historians, sociologists, political scientists, and economists.

2. At a more subtle level, even the content of scientific debate—what types of theories can be conceived and entertained, what criteria are to be used for deciding among competing theories—is constrained in part by the prevailing attitudes of mind, which in turn arise in part from deep-seated historical factors. It is the task of historians and sociologists of science to sort out, in each instance, the roles played by "external" and "internal" factors in determining the course of scientific development. Not surprisingly, scientists tend to stress the "internal" factors while sociologists tend to stress the "external" factors, if only because each group tends to have a poor grasp of the other group's concepts. But these problems are perfectly amenable to rational debate.

3. There is nothing wrong with research informed by a political commitment as long as that commitment does not blind the researcher to inconvenient facts. Thus, there is a long and honorable tradition of sociopolitical critique of science,[5] including antiracist critiques of anthropological pseudoscience and eugenics[6] and feminist critiques of psychology and parts of medicine and biology.[7] These critiques typically follow a standard pattern: First, one shows, using conventional scientific arguments, why the research in question is flawed according to the ordinary canons of good science. Then—and only then—one attempts to explain how the researchers' social prejudices (which may well have been unconscious) led them to violate these canons. Of course, each such critique has to stand or fall on its own merits; having good political intentions doesn't guarantee that one's analysis will constitute good science, good sociology, or good history. But this general two-step approach is, I think, sound; and empirical studies of this kind, if conducted with due intellectual rigor, could shed useful light on the social conditions under which good science (defined normatively as the search for truths or at least approximate truths about the world) is fostered or hindered.[8]

I don't want to claim that these three points exhaust the field of fruitful inquiry for historians and sociologists of science, but they certainly do lay out a big and important area. And yet, some sociologists and literary intellectuals over the past two decades have gotten greedier: Roughly speaking, they want to attack the normative conception of scientific inquiry as a search for truths or approximate truths about the world; they want to see science as just another social practice, which produces "narrations" and "myths" that are no more valid than those produced by other social practices; and some of them want to argue further that these social practices encode a bourgeois and/or Eurocentric and/or masculinist worldview. Of course, like all brief summaries, this one is an oversimplification; and in any case, there is no canonical doctrine in the "new" sociology of science, just a bewildering variety of individuals and schools. More importantly, the task of summarization is here made more difficult by the fact that this literature is often ambiguous in crucial ways about its most fundamental claims (as I'll illustrate later using the cases of Latour and Barnes–Bloor). Still, I think most scientists and philosophers of science would be astonished to learn that "the natural world has a small or non-existent role in the construction of scientific knowledge," as the prominent sociologist of science Harry Collins claims,[9] or that "reality is the consequence rather than the cause" of the so-called "social construction of facts," as Bruno Latour and Steve Woolgar assert.[10]

The "*Social Text*" Affair

With this preamble out of the way, I'd now like to consider what (if anything) the "*Social Text* affair" proves—and also what it does not prove, because some of my overenthusiastic supporters have claimed too much. In this analysis, it's crucial to distinguish between what can be deduced from the fact of publication and what can be deduced from the content of the article.

From the mere fact of publication of my parody, I think that not much can be deduced. It doesn't prove that the whole field of cultural studies, or the cultural studies of science—much less the sociology of science—is nonsense. Nor does it prove that the intellectual standards in these fields are generally lax. (This might be the case, but it would have to be established on other grounds.) It proves only that the editors of one rather marginal journal were derelict in their intellectual duty, by publishing an article on quantum physics that they admit they could not understand, without bothering to get an opinion from anyone knowledgeable in quantum physics, solely because it came from a "conveniently credentialed ally" (as *Social Text* coeditor Bruce Robbins later candidly admitted),[11] flattered the editors' ideological preconceptions, and attacked their "enemies."[12]

To which, one might justifiably respond: So what?[13] The answer comes from examining the content of the parody. In this regard, one important point has been lost in much of the discussion of my article: Yes, the article is screamingly funny—I'm not modest; I'm proud of my work—but the most hilarious parts of my article were not written by me. Rather, they're direct quotations from the postmodern Masters, whom I shower with mock praise. In fact, the article is structured around the silliest quotations I could find about mathematics and physics (and the philosophy of mathematics and physics) from some of the most prominent French and American intellectuals; my only contribution was to invent a nonsensical argument linking these quotations together and praising them. This involved, of course, advocating an incoherent mishmash of trendy ideas—deconstructive literary theory, New Age ecology, so-called feminist epistemology,[14] extreme social–constructivist philosophy of science, even Lacanian psychoanalysis—but that just made the parody all the more fun. Indeed, in some cases I took the liberty of parodying extreme or ambiguously stated versions of views that I myself hold in a more moderate and precisely stated form.

Now, what precisely do I mean by "silliness"? Here's a very rough categorization: First of all, one has meaningless or absurd statements, name-dropping, and the display of false erudition. Second, one has sloppy thinking and poor philosophy, which come together notably (though not always) in the form of glib relativism.

The first of these categories wouldn't be so important, perhaps, if we were dealing with a few assistant professors of literature making fools of themselves holding forth on quantum mechanics or Gödel's theorem. It becomes more relevant because we're dealing with important intellectuals, at least as measured by shelf space in the cultural-studies section of university bookstores. Here, for instance, are Gilles Deleuze and Félix Guattari holding forth on chaos theory:

> To slow down is to set a limit in chaos to which all speeds are subject, so that they form a variable determined as abscissa, at the same time as the limit forms a universal constant that cannot be gone beyond (for example, a maximum degree of contraction). The first functives are therefore the limit and the variable, and reference is a relationship between values of the variable or, more profoundly, the relationship of the variable, as abscissa of speeds, with the limit.[15]

And there's much more—Jacques Lacan and Luce Irigaray on differential topology, Jean-François Lyotard on cosmology, Michel Serres on nonlinear time—but don't let me not spoil the fun.[16] (By the way, if you're worrying that I'm quoting out of context, just follow my footnotes, look up the originals, and decide for yourself. You'll find that these passages are even worse in context than out of context.)

Nor is all the nonsense of French origin. Connoisseurs of fashionable American work in the cultural studies of science will, I think, find ample food for thought.

Fine, the Science Studies contingent might now object: maybe some of our friends in the English department take Lacan or Deleuze seriously, but no one in our community does. True enough; but then take a look at Bruno Latour's semiotic analysis of the theory of relativity, published in *Social Studies of Science*, in which "Einstein's text is read as a contribution to the sociology of delegation."[17] Why is that? Because Latour finds Einstein's popular book on relativity full of situations in which the author delegates one observer to stand on the platform and make certain measurements and another observer to stand on the train and make certain measurements; and of course the results won't obey the Lorentz transformations unless the two observers do what they're told! You think I exaggerate? Latour emphasizes Einstein's

> obsession with transporting *in*formation through *trans*formations without *de*formation; his passion for the precise superimposition of readings; his panic at the idea that observers sent away might betray, might retain privileges, and send reports that could not be used to expand our knowledge; his desire to discipline the delegated observers and to turn them into dependent pieces of apparatus that do nothing but watch the coincidence of hands and notches.[18]

Furthermore, because Latour doesn't understand what the term *frame of reference* means in physics—he confuses it with *actor* in semiotics—he claims that relativity cannot deal with the transformation laws between two frames of reference but needs at least three:

> If there are only one, or even *two*, frames of reference, no solution can be found. . . . Einstein's solution is to consider *three* actors: one in the train, one on the embankment and a third one, the author [enunciator] or one of its representants, who tries to superimpose the coded observations sent back by the two others.[19]

Finally, Latour somehow got the idea that relativi ̣ / concerns the problems raised by the relative *location* (rather than the relative *motion*) of different observers. (Of course, even the word *observer* here is potentially misleading; it belongs to the pedagogy of relativity, not to the theory itself.) Here is Latour's summary of the meaning of relativity:

> Provided the two relativities [special and general] are accepted, more frames of reference with less privilege can be accessed, reduced, accumulated and combined, observers can be delegated to a few more places in the infinitely large (the cosmos) and the infinitely small (electrons), and the readings they send will be understandable. His [Einstein's] book could well be titled: "New Instructions for Bringing Back Long-Distance Scientific Travellers."[20]

I needn't pursue the point: Huth's essay in this volume provides a sober and detailed exegesis of Latour's confusions about relativity. The upshot is that Latour has produced 40 pages of comical misunderstandings of a theory that is nowadays routinely taught to intelligent college freshmen, and *Social Studies of Science* found it a worthy scholarly contribution.

OK, enough for examples of nonsense (although a lot more are available). More interesting intellectually, I think, are the sloppy thinking and glib relativism that have become prevalent in many parts of Science Studies (albeit not, by and large, among serious philosophers of science). When one analyzes these writings, one often finds radical-sounding assertions whose meaning is ambiguous and that can be given two alternative readings: one as interesting, radical, and grossly false; the other as boring and trivially true.

Let me start again with Latour, this time taken from his book *Science in Action*, in which he develops seven Rules of Method for the sociologist of science. Here is his Third Rule of Method: "Since the settlement of a controversy is the *cause* of Nature's representation, not the consequence, we can never use the outcome—Nature—to explain how and why a controversy has been settled."[21]

Note how Latour slips, without comment or argument, from "Nature's representation" in the first half of this sentence to "Nature" *tout court* in the second half. If we were to read "Nature's representation" in both halves, then we'd have the truism that scientists' representations of Nature (i.e., their theories) are arrived at by a social process and that the course and outcome of that social process can't be explained simply by its outcome. If, however, we take seriously "Nature" in the second half, linked as it is to the word *outcome*, then we would have the claim that the external world is created by scientists' negotiations: a claim that is bizarre, to say the least, given that the external world has been around for about 10 billion years longer than the human race.[22] Finally, if we take seriously "Nature" in the second half but expunge the word *outcome* preceding it, then we would have either (1) the weak (and trivially true) claim that the course and outcome of a scientific controversy cannot be explained solely by the nature of the external world (obviously some social factors play a role, if only in determining which experiments are technologically feasible at a given time, not to mention other, more subtle social influences) or (2) the strong (and manifestly false) claim that the nature of the external world plays no role in constraining the course and outcome of a scientific controversy.[23]

But if we apply the First Rule of Interpretation of Postmodern Academic Writing—"No sentence means what it says"—we can perhaps make sense of Latour's dictum. Let's read it not as a philosophical principle but, rather, as a methodological principle for a sociologist of science—more precisely, for a sociologist of science who does not have the scientific competence to make an independent assessment of whether the experimental/observational data do in fact warrant the conclusions the scientific community has drawn from them. (The principle applies with particular force when such a sociologist is studying contemporary science, because in this case there is no other scientific community besides the one under study that could provide such an independent assessment. By contrast, for studies of the distant past, one can always look at what subsequent scientists learned, including the results from experiments going beyond those originally performed.) In such a situation, the sociologist will be understandably reluctant to say that "the scientific community under study came to conclusion X because X is the way the world really is"—even if it is in fact the case that X is the way the world is and that is the reason the scientists came to believe it—because the sociologist has no independent grounds to believe that X is the way the world really is other than the fact that the scientific community under study came to believe it.

Of course, the sensible conclusion to draw from this dilemma is that sociologists of science ought not to study scientific controversies on which they lack the competence to make an independent assessment of the facts, if there is no other (e.g., historically later) scientific community on which they could justifiably rely for such an independent assessment. But it goes without saying that Latour and his colleagues would not enjoy this conclusion, because their goal, as Steve Fuller put it, is to "employ methods that enable them to fathom both the 'inner workings' and the 'outer character' of science without having to be expert in the fields they study."[24]

It seems to me that much sloppy thinking in Science Studies, like that in Latour's Third Rule of Method, involves conflating concepts that need to be distinguished. Most frequently this conflation is accomplished by terminological fiat: the author intentionally uses an old word or phrase in a radically new sense, thereby undermining any attempt to distinguish between the two meanings. The goal here is to achieve by definition what one could not achieve by logic. For example, we often find phrases like "the social construction of facts"[25] that intentionally elide the distinction between facts and our knowledge of them. Or to take another example, philosophers usually understand the word *knowledge* to mean "justified true belief" or some similar concept; but Barry Barnes and David Bloor redefine *knowledge* to mean "any collectively accepted system of belief."[26] Now, perhaps Barnes and Bloor are uninterested in inquiring whether a given belief is true or rationally justified;[27] but if they think these properties of beliefs are irrelevant to their purposes, then they should say so and explain why, without confusing the issue by redefining words.[28]

More generally, it seems to me that much sloppy thinking in Science Studies involves conflating two or more of the following levels of analysis:

1. *Ontology.* What objects *exist* in the world? What statements about these objects are *true*?
2. *Epistemology.* How can human beings obtain *knowledge* of truths about the world? How can they assess the *reliability* of that knowledge?

3. *Sociology of knowledge.* To what extent are the truths *known* (or *knowable*) by humans in any given society influenced (or determined) by social, economic, political, cultural, and ideological factors? Same question for the false statements erroneously believed to be true.
4. *Individual ethics.* What types of research *ought* a scientist (or technologist) to undertake (or refuse to undertake)?
5. *Social ethics.* What types of research *ought* society to encourage, subsidize, or publicly fund (or, alternatively, to discourage, tax, or forbid)?

These questions are obviously related—for example, if there are no objective truths about the world, then there isn't much point in asking how one can know those (nonexistent) truths—but they are conceptually distinct.

For example, Sandra Harding[29] (citing the work of Paul Forman)[30] points out that American research in the 1940s and 1950s on quantum electronics was motivated in large part by potential military applications. True enough. Now, quantum mechanics made possible solid-state physics, which in turn made possible quantum electronics (e.g., the transistor), which made possible nearly all modern technology (e.g., the computer). And the computer has had applications that are beneficial to society (e.g., by allowing the postmodern cultural critic to produce her articles more efficiently) as well as applications that are harmful (e.g., by allowing the U.S. military to kill human beings more efficiently). This raises a host of social and individual ethical questions: Ought society to forbid (or discourage) certain applications of computers? Forbid (or discourage) research on computers per se? Forbid (or discourage) research on quantum electronics? On solid-state physics? On quantum mechanics? And likewise for individual scientists and technologists. (Clearly, an affirmative answer to these questions becomes harder to justify as one goes down the list, but I do not want to declare any of these questions a priori illegitimate.)

Sociological questions also arise: To what extent is our (true) knowledge of computer science, quantum electronics, solid-state physics, and quantum mechanics—and our lack of knowledge about other scientific subjects, for example, the global climate—a result of public-policy choices that favor militarism? To what extent have the erroneous theories (if any) in computer science, quantum electronics, solid-state physics, and quantum mechanics been the result (in whole or in part) of social, economic, political, cultural, and ideological factors, in particular the culture of militarism?[31] These all are serious questions, which deserve careful investigation adhering to the highest standards of scientific and historical evidence. But they have no effect whatsoever on the underlying scientific questions: whether atoms (and silicon crystals, transistors, and computers) really do behave according to the laws of quantum mechanics (and solid-state physics, quantum electronics, and computer science). The militaristic orientation of American science has, quite simply, no bearing whatsoever on the ontological question, and only under a wildly implausible scenario could it have any bearing on the epistemological question (e.g., if the worldwide community of solid-state physicists, following what they believe to be the conventional standards of scientific evidence, were to hastily accept an erroneous theory of semiconductor behavior because of their enthusiasm for the breakthrough in military technology that this theory would make possible).

The extreme versions of social constructivism and relativism—such as the Edinburgh "Strong Programme"—are, I think, largely based on the same failure to distinguish clearly among ontology, epistemology, and the sociology of knowledge. This is how Barnes and Bloor describe the form of relativism that they defend:

> Our equivalence postulate is that all beliefs are on a par with one another with respect to the causes of their credibility. It is not that all beliefs are equally true or equally false, but that regardless of truth and falsity the fact of their credibility is to be seen as equally problematic. The position we shall defend is that the incidence of all beliefs without exception calls for empirical investigation and must be accounted for by finding the specific, local causes of this credibility. This means that regardless of whether the sociologist evaluates a belief as true or rational, or as false and irrational, he must search for the causes of its credibility. . . . All these questions can, and should, be answered without regard to the status of the belief as it is judged and evaluated by the sociologist's own standards.[32]

It seems clear from this passage, as well as from the paragraph that precedes it, that Barnes and Bloor are not advocating an ontological relativism; they recognize that "to say that all beliefs are equally true encounters the problem of how to handle beliefs which contradict one another" and that "to say that all beliefs are equally false poses the problem of the status of the relativist's own claims."[33] They might be advocating an epistemological relativism—that all beliefs are equally credible, or equally rational—and indeed, their attack on the universal validity of even the simplest rules of deductive inference (such as *modus ponens*) lends some support to this interpretation.[34] But more likely, what they are advocating is some form of methodological relativism for sociologists of knowledge. The problem is, what form?

If the claim were merely that we should use the same principles of sociology and psychology to explain the causation of all beliefs, irrespective of whether we evaluate them as true or false, rational or irrational, then I would have no particular objection (although I might have qualms about the hyperscientistic attitude that human beliefs are always to be explained causally through social science). But if the claim is that only *social* causes can enter into such an explanation—that the way the world *is* cannot enter—then I cannot disagree more strenuously.[35]

Let's take a concrete example: Why did the European scientific community become convinced of the truth of Newtonian mechanics somewhere between 1700 and 1750? Undoubtedly a variety of historical, sociological, ideological, and political factors must play a role in this explanation--one must explain, for example, why Newtonian mechanics was accepted quickly in England but more slowly in France[36]—but certainly some part of the explanation (and a rather important part at that) must be that the planets and comets really do move (to a very high degree of approximation, although not exactly) as predicted by Newtonian mechanics.[37] Or to take another example: Why did the majority view in the European and North American scientific communities shift from creationism to Darwinism over the course of the nineteenth century? Again, numerous historical, sociological, ideological, and political factors will play a role in this explanation; but can one plausibly explain this shift without any reference to the fossil record or to the Galápagos fauna?

In the unlikely event that the argument isn't already clear, here's a more homely example: Suppose we encounter a man running out of a lecture hall screaming at the top of his lungs that there's a stampeding herd of elephants in the room. What we are to make of this assertion and, in particular, how we are to evaluate its "causes" should, I think, depend heavily on whether or not there *is* in fact a stampeding herd of elephants in the room—or, more precisely, since I admit that we have no direct, unmediated access to external reality—whether when I and other people peek (cautiously!) into the room, *we* see or hear a stampeding herd of elephants (or the destruction that such a herd might have caused before exiting the room). If we do see such evidence of elephants, then the most plausible explanation of this set of observations is that there is (or was) in fact a stampeding herd of elephants in the lecture hall, that the man saw and/or heard it, and that his subsequent fright (which we might well share under the circumstances) led him to exit the room in a hurry and to scream the assertion that we overheard. And our reaction would be to call the police and the zookeepers. If, on the other hand, our own observations reveal no evidence of elephants in the lecture hall, then the most plausible explanation is that there was not in fact a stampeding herd of elephants in the room, that the man imagined the elephants as a result of some psychosis (whether internally or chemically induced), and that this led him to exit the room in a hurry and to scream the assertion that we overheard. And we'd call the police and the psychiatrists.[38] And I daresay that Barnes and Bloor, whatever they might write in journal articles for sociologists and philosophers, would do the same in real life.

The bottom line, it seems to me, is that there is no fundamental "metaphysical" difference between the epistemology of science and the epistemology of everyday life. Historians, detectives, and plumbers—indeed, all human beings—use the same basic methods of induction, deduction, and assessment of evidence as do physicists or biochemists. Modern science tries to carry out these operations in a more careful and systematic way—using controls and statistical tests, insisting on replication, and so forth—but nothing more.[39] Any philosophy of science—or methodology for sociologists—that is so blatantly wrong when applied to the epistemology of everyday life must be severely flawed at its core.

In summary, it seems to me that the "Strong Programme," like Latour's Third Rule of Method, is ambiguous in its intent; and, depending on how one resolves the ambiguity, it becomes either a valid and mildly interesting corrective to the most naive psychological and sociological notions—reminding us that "true beliefs have causes, too"—or else a gross and blatant error.

Against the "Science Wars"

Professor Kitcher concludes his contribution to this volume by saying, "I doubt that this essay will please anyone, for it attempts to occupy middle ground." In this he's certainly too pessimistic, for there's at least one counterexample: his essay pleases me. Indeed, I agree with nearly everything in it.

Now, perhaps this means only that I too—arrogant scientist though I may be—am one of those select few occupying the "middle ground." But I suspect that more of us occupy the "middle ground" in this debate than might at first appear. The point, of course, isn't to embrace "middle ground" (whatever that may be) abstractly and for its own sake, without regard to its content; that would be a grave dereliction of intellectual duty.[40] But here the middle ground as set forth in Kitcher's essay—based on a respect for both the "realist–rationalist cluster" and the "socio-historical cluster," even as we may debate their relative importance in specific cases—is so eminently sensible that nearly all scientists[41] and philosophers of science would give their assent, as would most (though apparently not all) sociologists of science. And this fact might give us some cause for reflection about the so-called—and I think grossly misnamed—"Science Wars."

The term was apparently first coined by *Social Text* coeditor Andrew Ross, who explained that

> the Science Wars [are] a second front opened up by conservatives cheered by the successes of their legions in the holy Culture Wars. Seeking explanations for their loss of standing in the public eye and the decline in funding from the public purse, conservatives in science have joined the backlash against the (new) usual suspects—pinkos, feminists, and multiculturalists.[42]

This theme was further elaborated in the now-famous special issue of *Social Text*.[43] But just as in the dreary "culture wars," the truth is rather more complicated than this Manichean portrayal allows. The alleged one-to-one correspondence between epistemological and political views is a gross misrepresentation.[44] So, too, is the idea that in this debate there are only two positions.

This conception of debate as combat is, in fact, probably the main reason that the *Social Text* editors fell for my parody. Acting not as intellectuals seeking the truth but as self-appointed generals in the "Science Wars," they apparently leaped at the chance to get a "real" scientist on their "side." Now, ruing their blunder, they must surely feel a kinship with the Trojans.

But the military metaphor is a mistake; the *Social Text* editors are not my enemies. Ross has legitimate concerns about new technologies and about the increasingly unequal distribution of scientific expertise. Aronowitz raises important questions about technological unemployment and the possibility of a "jobless future."[45] But, *pace* Ross, nothing is gained by denying the existence of objective scientific knowledge; it does exist, whether we like it or not. Political progressives should seek to have that knowledge distributed more democratically and to have it employed for socially useful ends. Indeed, the radical epistemological critique fatally undermines the needed political critique, by removing its factual basis. After all, the only reason that nuclear weapons are a danger to anyone is that the theories of nuclear physics on which their design is based are, at least to a very high degree of approximation, objectively true.[46]

Science Studies' epistemological conceits are a diversion from the important matters that motivated Science Studies in the first place: namely, the social, economic, and political roles of science and technology. To be sure, those conceits are not an accident; they have a history, which can be subjected to sociological study.[47] But Science Studies practitioners are not obliged to persist in a misguided epistemology; they can give it up, and go on to the serious task of studying science. Perhaps, from the perspective of a few years from now, today's "Science Wars" will turn out to have marked such a turning point.

Notes

1. Quoted in Janny Scott, "Postmodern Gravity Deconstructed, Slyly," *New York Times*, May 18, 1996, 1, 22.

2. Alan D. Sokal, "Transgressing the Boundaries: Toward a Transformative Hermeneutics of Quantum Gravity," *Social Text* 46/47 (Spring–Summer 1996): 217–52.

3. Alan Sokal, "A Physicist Experiments with Cultural Studies," *Lingua Franca* 6 (May–June 1996): 62–64.

4. The "official" reply from the editors of *Social Text* appears in "Mystery Science Theater," *Lingua Franca* 6 (July–August 1996): 54–64, along with a brief rejoinder from myself and letters from readers. For a more detailed explanation of my motivations in undertaking the parody, see Alan D. Sokal, "Transgressing the Boundaries: An Afterword," rejected for publication in *Social Text* but published in *Dissent* 43 (Fall 1996): 93–99, and, in slightly different form, in *Philosophy and Literature* 20 (October 1996): 338–46; and Alan Sokal, "A Plea for Reason, Evidence and Logic," *New Politics* 6 (Winter 1997): 126–29. For further commentary, see, for example, Tom Frank, "Textual Reckoning," *In These Times*, May 27, 1996, 22–24; Katha Pollitt, "Pomolotov Cocktail," *The Nation*, June 10, 1996, 9; Steven Weinberg, "Sokal's Hoax," *New York Review of Books*, August 8, 1996, 11–15; and Paul Boghossian, "What the Sokal Hoax Ought to Teach Us," *Times Literary Supplement*, December 13, 1996, 14–15.

5. I limit myself here to critiques challenging the substantive content of scientific theories or methodology. Other important types of critiques challenge the uses to which scientific knowledge is put (e.g., in technology) or the social structure of the scientific community.

6. See, for example, Stephen Jay Gould, *The Mismeasure of Man*, 2d ed. (New York: Norton, 1996).

7. See, for example, Anne Fausto-Sterling, *Myths of Gender: Biological Theories about Women and Men*, 2d ed. (New York: Basic Books, 1992); and Carol Tavris, *The Mismeasure of Woman* (New York: Simon & Schuster, 1992).

8. Of course, I don't mean to imply that the only (or even the principal) purpose of the history of science is to help working scientists. The history of science obviously has intrinsic value as a contribution to the history of human society and human thought. But it seems to me that history of science, when done well, can also help working scientists.

9. H. M. Collins, "Stages in the Empirical Programme of Relativism," *Social Studies of Science* 11 (1981): 3–10, quote at p. 3. Two qualifications need to be made: First, this statement is offered as part of Collins's introduction to a set of studies (edited by him) employing the relativist approach and constitutes his summary of that approach; he does not explicitly endorse this view, although an endorsement seems implied by the context. Second, while Collins appears to intend this assertion as an empirical claim about the history of science, it is possible that he intends it as neither an empirical claim nor a normative principle of epistemology but, rather, as a methodological injunction to sociologists of science: namely, to act *as if* "the natural world ha[d] a small or non-existent role in the construction of scientific knowledge" or, in other words, to ignore ("bracket") whatever role the natural world may in fact play in the construction of scientific knowledge. I shall argue later, in discussing Barnes and Bloor, that this approach is seriously deficient as methodology for sociologists of science.

10. Bruno Latour and Steve Woolgar, *Laboratory Life: The Social Construction of Scientific Facts* (London: Sage, 1979), 237.

11. Bruce Robbins, "*Social Text* and Reality," *In These Times*, July 8, 1996: 28–29, quote at p. 28.

12. The "Science Wars" special issue of *Social Text* was conceived primarily to attack Paul R. Gross and Norman Levitt's book *Higher Superstition: The Academic Left and Its Quarrels with Science* (Baltimore: Johns Hopkins University Press, 1994). See Andrew Ross, "Introduc-

tion," *Social Text* 46/47 (Spring–Summer 1996): 1–13; and also Andrew Ross and Stanley Aronowitz, unpublished letter to the author (and to other contributors to the "Science Wars" issue), March 8, 1995.

13. Indeed, a mainstream journal in the sociology of science would almost certainly not have fallen for my parody. (On the other hand, *Social Studies of Science* published a long article on the theory of relativity that, if it wasn't in fact a parody, might as well have been: see my discussion of Latour's article on page 12.) I chose *Social Text* because my primary motivation was political: see Sokal, "Afterword" and "A Plea."

14. I emphasize that this term is a misnomer, as these ideas are hotly debated among feminists, among whom I include myself. For incisive feminist critiques of "feminist epistemology," see Susan Haack, "Science 'from a Feminist Perspective,'" *Philosophy* 67 (1992): 5–18; Susan Haack, "Epistemological Reflections of an Old Feminist," *Reason Papers* 18 (Fall 1993): 31–43; Cassandra L. Pinnick, "Feminist Epistemology: Implications for Philosophy of Science," *Philosophy of Science* 61 (1994): 646–57; and Janet Radcliffe Richards, "Why Feminist Epistemology Isn't," in *The Flight from Science and Reason,* ed. Paul R. Gross, Norman Levitt, and Martin W. Lewis, *Annals of the New York Academy of Sciences* 775 (1996): 385–412.

15. Gilles Deleuze and Félix Guattari, *What Is Philosophy?* trans. Hugh Tomlinson and Graham Burchell (New York: Columbia University Press, 1994), 119.

16. For an extensive compilation of postmodern French philosophers' abuses of mathematics and physics, along with commentary for nonexperts, see Alan Sokal and Jean Bricmont, *Impostures intellectuelles* (Paris: Odile Jacob, 1997). In addition to Deleuze, Guattari, Lacan, and Irigaray, we also have chapters on Jean Baudrillard, Julia Kristeva, Bruno Latour, and Paul Virilio.

17. Bruno Latour, "A Relativistic Account of Einstein's Relativity," *Social Studies of Science* 18 (1988): 3–44, quote at p. 3.

18. Ibid., 22, italics in original.

19. Ibid., 10–11, italics in original.

20. Ibid., 22–23.

21. Bruno Latour, *Science in Action: How to Follow Scientists and Engineers through Society* (Cambridge, Mass.: Harvard University Press, 1987), 99, 258. This "rule" is the culmination of an argument (96–99) in which ontology, epistemology, and the sociology of knowledge are gradually conflated.

22. You might worry that here my argument is circular, in that it takes for granted the truth of the current scientific consensus in cosmology and paleontology. But this is not the case. First of all, my phrase "given that" is a rhetorical flourish that plays no essential role in the argument; the idea that the external world is created by scientists' negotiations is bizarre, irrespective of the details of cosmology and paleontology. Second, my phrase "the external world has been around" should, if one wants to be super-precise, be amended to read: "there is a vast body of extremely convincing (and diverse) evidence in support of the belief that the external world has been around . . . ; and *if* this belief is correct, *then* the claim that the external world is created by scientists' negotiations is bizarre to say the least." Indeed, all my assertions of fact—including "today in New York it's raining"—should be glossed in this way. Since I shall claim later that much contemporary work in Science Studies elides the distinction between ontology and epistemology, I don't want to leave myself open to the same accusation.

23. Re (2), the "homely example" in Gross and Levitt, *Higher Superstition,* 57–58, makes the point clearly.

24. Steve Fuller, *Philosophy, Rhetoric, and the End of Knowledge: The Coming of Science and Technology Studies* (Madison: University of Wisconsin Press, 1993), xii. For further analysis of *Science in Action,* see Olga Amsterdamska, "Surely You Are Joking, Monsieur Latour!" *Science, Technology, & Human Values* 15 (1990): 495–504.

25. Latour and Woolgar, *Laboratory Life*.

26. Barry Barnes and David Bloor, "Relativism, Rationalism and the Sociology of Knowledge," in *Rationality and Relativism*, ed. Martin Hollis and Steven Lukes (Oxford: Blackwell, 1981), 21–47, see n. 5 on p. 22. See also David Bloor, *Knowledge and Social Imagery*, 2d ed. (Chicago: University of Chicago Press, 1991), 5.

27. This is, in fact, the methodological import of their principles of "impartiality" and "symmetry" (Bloor, *Knowledge and Social Imagery*, 7). See below for further discussion and critique.

28. Note that Bloor, only nine pages after giving his nonstandard definition of "knowledge," reverts without comment to the standard definition of "knowledge," which he contrasts with "error": "It would be wrong to assume that the natural working of our animal resources always produces knowledge. They produce a mixture of knowledge and error with equal naturalness" (Bloor, *Knowledge and Social Imagery*, 14).

29. Sandra Harding, *Whose Science? Whose Knowledge? Thinking from Women's Lives* (Ithaca, N.Y.: Cornell University Press, 1991), chap. 4.

30. Paul Forman, "Behind Quantum Electronics: National Security as Basis for Physical Research in the United States, 1940–1960," *Historical Studies in the Physical and Biological Sciences* 18 (1987): 149–229.

31. I certainly don't exclude the possibility that current theories in any of these subjects might be erroneous. But critics wishing to make such a case would have to provide not only historical evidence of the claimed cultural influence but also scientific evidence that the theory in question is in fact erroneous. (The same evidentiary standards of course apply to past erroneous theories, but in this case the scientists may have already performed the second task, relieving the cultural critic of the need to do so from scratch.)

32. Barnes and Bloor, "Relativism, Rationalism," 23.

33. Ibid., 22.

34. Ibid., 35–47.

35. Bloor does state explicitly that "naturally there will be other types of causes apart from social ones which will cooperate in bringing about belief" (Bloor, *Knowledge and Social Imagery*, 7). The trouble is that he never makes explicit in what way natural causes will be allowed to enter into the explanation of belief or what precisely the symmetry principle will mean in this context. For a more detailed critique of Bloor's ambiguities (from a philosophical point of view slightly different from mine), see Larry Laudan, "The Pseudo-Science of Science?" *Philosophy of the Social Sciences* 11 (1981): 173–98. See also Peter Slezak, "A Second Look at David Bloor's *Knowledge and Social Imagery*," *Philosophy of the Social Sciences* 24 (1994): 336–61.

36. The consensus of historians appears to be that the slow acceptance of Newtonian mechanics in France arose from scholastic attachment to Cartesian theories as well as from certain theological considerations. See, for example, Pierre Brunet, *L'Introduction des théories de Newton en France au XVIIIe siècle* (Paris: A. Blanchard, 1931; reprint, Geneva: Slatkine, 1970); and Betty Jo Teeter Dobbs and Margaret C. Jacob, *Newton and the Culture of Newtonianism* (Atlantic Highlands, N.J.: Humanities Press, 1995).

37. Or more precisely: There is a vast body of extremely convincing astronomical evidence in support of the belief that the planets and comets do move (to a very high degree of approximation, although not exactly) as predicted by Newtonian mechanics; and if this belief is correct, then it is the fact of this motion (and not merely our belief in it) that forms part of the explanation of why the eighteenth-century European scientific community came to believe in the truth of Newtonian mechanics.

38. For what it's worth, these decisions can presumably be justified on Bayesian grounds, using our prior experience of the probability of finding elephants in lecture halls, of the incidence of psychosis, of the reliability of our own visual and auditory perceptions, and so forth.

39. Please note: I am not claiming that inference from scientific observations to scientific theories is as simple or unproblematic as inference from seeing elephants in front of me to the conclusion that elephants are in front of me. (In truth, even this latter inference is not so simple or unproblematic, because to fully ground it requires some knowledge of optics and the mechanisms of human vision.) As all practicing scientists and historians of science well know, the reasoning from scientific observations to scientific theories is far more indirect and typically involves a vast web of empirical evidence rather than a single observation. My point is simply that in all these cases—Newtonian mechanics, Darwinian evolution, and elephants—it is absurd to try to explain the "causes" of people's beliefs without including the natural (nonsocial) world as one of those causes, and a rather important one at that.

40. In American politics, the pernicious consequences of such a search for "middle ground" between two morally and intellectually bankrupt (and indeed barely distinguishable) positions hardly need further comment. "Middle ground" is, of course, meaningless until one specifies between what; and the corporate media's tacit definition of the outer limits of respectable opinion is, of course, a large part of the problem. On what grounds, for example, is single-payer health insurance—long in use in most industrialized countries—defined as "extreme" and "unrealistic" in the United States?

41. Including Gross and Levitt, as they make amply clear in *Higher Superstition*.

42. Andrew Ross, "Science Backlash on Technoskeptics," *The Nation*, October 2, 1995: 346–350, quote at p. 346. See also Ross, "Introduction," 6.

43. Five of the essay titles (Martin, Nelkin, Franklin, Kovel, Aronowitz) include the term *Science Wars*, and three more titles (Rose, Winner, Levidow) contain assorted martial metaphors.

44. My own leftist political views are a matter of record, as are those of many of my supporters (e.g., Michael Albert, Barbara Epstein, Meera Nanda, Ruth Rosen, and James Weinstein, among many others). Even Gross and Levitt, the original targets of Ross's wrath, make clear that their political views are basically left-liberal; they note that one of them (Levitt, as it turns out) is a member of Democratic Socialists of America (Gross and Levitt, *Higher Superstition*, 261, n, 7).

45. Stanley Aronowitz and William DiFazio. *The Jobless Future: Sci-Tech and the Dogma of Work* (Minneapolis: University of Minnesota Press, 1994).

46. This point was made almost a decade ago by Margarita Levin, "Caring New World: Feminism and Science," *American Scholar* 57 (1988): 100–6.

47. For an interesting conjecture, see Meera Nanda, "The Science Wars in India," *Dissent* 44 (Winter 1997): 78–83, at pp. 79–80. For a different (but not incompatible) conjecture, see Gross and Levitt, *Higher Superstition*, 74, 82–88, 217–33. Both these conjectures merit careful empirical investigation by intellectual historians.

2

What the Sokal Hoax
Ought to Teach Us

Paul A. Boghossian

In the autumn of 1994, the New York University theoretical physicist Alan Sokal submitted an essay to *Social Text*, the leading journal in the field of cultural studies. Entitled "Transgressing the Boundaries: Toward a Transformative Hermeneutics of Quantum Gravity," it purported to be a scholarly article about the "postmodern" philosophical and political implications of twentieth-century physical theories.[1] However, as the author himself later revealed in the journal *Lingua Franca*, his essay was merely a farrago of deliberate solecisms, howlers, and non sequiturs, stitched together so as to look good and to flatter the ideological preconceptions of the editors.[2] After review by five members of *Social Text*'s editorial board, Sokal's parody was accepted for publication as a serious piece of scholarship. It appeared in April 1996 in a special double issue of the journal devoted to rebutting the charge that cultural studies critiques of science tend to be riddled with incompetence.

Sokal's hoax is fast acquiring the status of a classic *succès de scandale*, with extensive press coverage in the United States and, to a growing extent, in Europe and Latin America. In the United States, more than 20 public forums devoted to the topic have either taken place or are scheduled, including packed sessions at Princeton, Duke, the University of Michigan, and New York University. But what exactly should it be taken to show?

I believe it shows three important things. First, that dubiously coherent relativistic views about the concepts of truth and evidence really have gained wide acceptance in the contemporary academy, just as it has often seemed. Second, that this onset of relativism has had precisely the sorts of pernicious consequence for standards of scholarship and intellectual responsibility that one would expect it to have. Finally, that neither of the preceding two claims need reflect a particular political point of view, least of all a conservative one.

It's impossible to do justice to the egregiousness of Sokal's essay without quoting it more or less in its entirety; what follows is a tiny sampling. Sokal starts off by establishing his postmodernist credentials: he derides scientists for continuing to cling to the "dogma imposed by the long post-Enlightenment hegemony over the Western intellectual outlook," that there exists an external world whose properties are independent of human beings, and that human beings can obtain reliable, if imperfect and tentative knowledge of these properties "by hewing to the 'objective' procedures and epistemological strictures prescribed by the (so-called) scientific method" (217). He asserts that this "dogma" has already been thoroughly undermined by the theories of general relativity and quantum mechanics and that physical reality has been shown to be "at bottom a social and linguistic construct" (217). In support of this, he adduces nothing more than a couple of pronouncements from physicists Niels Bohr and Werner Heisenberg, pronouncements that have been shown to be naive by discussions in the philosophy of science over the past 50 years.

Sokal then picks up steam, moving to his central thesis that recent developments in quantum gravity—an emerging and still-speculative physical theory—go much further, substantiating not only postmodern denials of the objectivity of truth but also the beginnings of a kind of physics that would be genuinely "liberatory," of real service to progressive political causes (226). Here the "reasoning" becomes truly venturesome, as he contrives to generate political and cultural conclusions from the physics of the very, very small. His inferences are mediated by nothing more than a hazy patchwork of puns (especially on the words *linear* and *discontinuous*), strained analogies, bald assertions, and what can be described only as non sequiturs of numbing grossness (to use a phrase that Peter Strawson applied to the far less culpable Immanuel Kant). For example, Sokal moves immediately from Bohr's observation that in quantum mechanics "a complete elucidation of one and the same object may require diverse points of view" to the following:

> In such a situation, how can a self-perpetuating secular priesthood of credentialed "scientists" purport to maintain a monopoly on the production of scientific knowledge? . . . The content and methodology of postmodern science thus provide powerful intellectual support for the progressive political project, understood in its broadest sense: the transgressing of boundaries, the breaking down of barriers, the radical democratization of all aspects of social, economic, political and cultural life. (229)

Sokal concludes by calling for the development of a correspondingly emancipated mathematics, one that, by not being based on standard (Zermelo–Fraenkel) set theory, would no longer constrain the progressive and postmodern ambitions of emerging physical science.

As if all this weren't enough, Sokal peppers his piece with as many smaller bits of transparent nonsense as could be made to fit on any given page. Some of these are of a purely mathematical or scientific nature—that the well-known geometrical constant π is a variable; that complex number theory, which dates from the nineteenth century and is taught to schoolchildren, is a new and speculative branch of mathematical physics; and that the crackpot New Age fantasy of a "morphogenetic field" constitutes a leading theory of quantum gravity. Others have to do with the alleged philosophical or political implications of basic science—that quantum field theory confirms Lacan's psychoanalytic speculations about the nature of the neurotic subject; that fuzzy logic is better suited to leftist political causes than classical logic is; and that Bell's theorem, a technical result in the foundations of quantum mechanics, supports a claimed linkage between quantum theory and "industrial discipline in the early bourgeois epoch." Throughout, Sokal quotes liberally and approvingly from the writings of leading postmodern theorists, including several editors of *Social Text*, passages that are often breathtaking in their combination of self-confidence and absurdity.

Commentators have made much of the scientific, mathematical, and philosophical illiteracy that an acceptance of Sokal's ingeniously contrived gibberish would appear to betray. But talk about illiteracy elides an important distinction between two different explanations of what might have led the editors to decide to publish Sokal's piece. One is that, although they understood perfectly well what the various sentences of his article actually mean, they found them plausible, whereas he, along with practically everybody else, doesn't. This might brand them as kooky, but wouldn't impugn their motives. The other hypothesis is that they actually had very little idea what many of the sentences mean, and so were not in a position to evaluate them for plausibility in the first place. The plausibility, or even the intelligibility, of Sokal's arguments just didn't enter into their deliberations.

I think it's very clear, and very important, that it's the second hypothesis that's true. To see why, consider, by way of example, the following passage from Sokal's essay:

> Just as liberal feminists are frequently content with a minimal agenda of legal and social equality for women and are "pro-choice," so liberal (and even some socialist) mathematicians are often content to work within the hegemonic Zermelo–Fraenkel framework (which, reflecting its nineteenth-century origins, already incorporates the axiom of equality) supplemented only by the axiom of choice. But this framework is grossly insufficient for a liberatory mathematics, as was proven long ago by Cohen 1966 (note 54, 242–243).

It's very hard to believe that an editor who knew what the various ingredient terms actually mean would not have raised an eyebrow at this passage. For the axiom of equality in set theory simply provides a definition of when it is that two sets are the same—namely, when they have the same members; obviously, this has nothing to do with liberalism or, indeed, with a political philosophy of any stripe. Similarly, the axiom of choice simply says that given any collection of mutually exclusive sets, there is always a set consisting of exactly one member from each of those sets. Again, this clearly has nothing to do with the issue of choice in the abortion debate. But even if one were somehow able to see one's way clear—I can't—to explaining this first quoted sentence in terms of the postmodern love for puns and wordplay, what would explain the subsequent sentence? Paul Cohen's 1966 publication proves that the question of whether or not there is a number between two other particular (transfinite cardinal) numbers isn't settled by the axioms of Zermelo–Fraenkel set theory. How could this conceivably count as proof that Zermelo–Fraenkel set theory is inadequate for the purposes of a "liberatory mathematics," whatever precisely that is supposed to be. Wouldn't any editor who knew what Paul Cohen had actually proved in 1966 have required just a little more by way of explanation here, in order to make the connection just a bit more perspicuous?

Since one could cite dozens of similar passages—Sokal goes out of his way to leave telltale clues to his true intent—the conclusion is inescapable that the editors of *Social Text* didn't know what many of the sentences in Sokal's essay actually meant;

and that they just didn't care. How could a group of scholars, editing what is supposed to be the leading journal in a given field, allow themselves such a sublime indifference to the content, truth, and plausibility of a scholarly submission accepted for publication?

By way of explanation, coeditors Andrew Ross and Bruce Robbins have said that as "a non-refereed journal of political opinion and cultural analysis produced by an editorial collective . . . *Social Text* has always seen itself in the 'little magazine' tradition of the independent left as much as in the academic domain."[3] But it's hard to see this as an adequate explanation; presumably, even a journal of political opinion should care whether what it publishes is intelligible.

What Ross and Co. should have said, it seems to me, is that *Social Text* is a political magazine in a deeper and more radical sense: under appropriate circumstances, it is prepared to let agreement with its ideological orientation trump every other criterion for publication, including something as basic as sheer intelligibility. The prospect of being able to display in their pages a natural scientist—a physicist, no less—throwing the full weight of his authority behind their cause was compelling enough for them to overlook the fact that they didn't have much of a clue to exactly what sort of support they were being offered. And this, it seems to me, is what's at the heart of the issue raised by Sokal's hoax: not the mere existence of incompetence in the academy, but, rather that specific form of it that arises from allowing ideological criteria to displace standards of scholarship so completely that not even considerations of intelligibility are seen as relevant to an argument's acceptability. How, given the recent and sorry history of ideologically motivated conceptions of knowledge—Lysenkoism in Stalin's Soviet Union, for example, or Nazi critiques of "Jewish science"—could it again have become acceptable to behave in this way?

The complete historical answer is a long story, but there can be little doubt that one of its crucial components is the brushfire spread, in vast sectors of the humanities and social sciences, of the cluster of simpleminded relativistic views about the concepts of truth and evidence that are commonly identified as "postmodernist." These views license, and typically insist upon, the substitution of political and ideological criteria for the historically more familiar assessment in terms of truth, evidence, and argument.

Most philosophers accept the claim that there is no such thing as a totally disinterested inquirer, one who approaches his or her topic utterly devoid of any prior assumptions, values, or biases. Postmodernism goes well beyond this historicist observation, as feminist scholar Linda Nicholson explains (without necessarily endorsing):

> The traditional historicist claim that all inquiry is inevitably influenced by the values of the inquirer provides a very weak counter to the norm of objectivity. . . . [T]he more radical move in the postmodern turn was to claim that the very criteria demarcating the true and the false, as well as such related distinctions as science and myth or fact and superstition, were internal to the traditions of modernity and could not be legitimized outside of those traditions. Moreover, it was argued that the very development and use of such criteria, as well as their extension to ever wider domains, had to be described as representing the growth and development of "specific regimes of power."[4]

As Nicholson sees, historicism, however broadly understood, doesn't entail that there is no such thing as objective truth. To concede that no one ever believes something solely because it's true is not to deny that anything is objectively true. Furthermore, the concession that no inquirer or inquiry is fully bias free doesn't entail that they can't be more or less bias free or that their biases can't be more or less damaging. To concede that the truth is never the only thing that someone is tracking isn't to deny that some people or methods are better than others at staying on its track.

Historicism leaves intact, then, both the claim that one's aim should be to arrive at conclusions that are objectively true and justified, independent of any particular perspective, and that science is the best idea anyone has had about how to satisfy it. Postmodernism, in seeking to demote science from the privileged epistemic position it has come to occupy, and thereby to blur the distinction between it and "other ways of knowing"—myth, and superstition, for example—needs to go much further than historicism, all the way to the denial that objective truth is a coherent aim that inquiry may have. Indeed, according to postmodernism, the very development and use of the rhetoric of objectivity, far from embodying a serious metaphysics and epistemology of truth and evidence, represents a mere play for power, a way of silencing these "other ways of knowing." It follows, given this standpoint, that the struggle against the rhetoric of objectivity isn't primarily an intellectual matter but a political one: the rhetoric needs to be defeated rather than just refuted. Against this backdrop, it becomes very easy to explain the behavior of the editors of *Social Text*.

Although it may be hard to understand how anyone could actually hold views as extreme as these, their ubiquity these days is a distressingly familiar fact. A front-page article in the *New York Times* of October 22, 1996, provided an illustration.[5] The article concerned the conflict between two views of where Native American populations originated—the scientific archaeological account and the account offered by some Native American creation myths. According to the former, extensively confirmed view, humans first entered the Americas from Asia, crossing the Bering Strait more than 10,000 years ago. By contrast, some Native American creation accounts hold that native peoples have lived in the Americas ever since their ancestors first emerged onto the surface of the earth from a subterranean world of spirits. The *Times* noted that many archaeologists, torn between their commitment to scientific method and their appreciation for native culture, "have been driven close to a postmodern relativism in which science is just one more belief system." Roger Anyon, a British archaeologist who has worked for the Zuni people, was quoted as saying: "Science is just one of many ways of knowing the world. . . . [The Zunis' worldview is] just as valid as the archeological viewpoint of what prehistory is about."

How are we to make sense of this? (Sokal himself mentioned this example at a recent public forum in New York and was taken to task by Andrew Ross for putting Native Americans "on trial." But this issue isn't about Native American views; it's about postmodernism.) The claim that the Zuni myth can be "just as valid" as the archaeological theory can be read in one of three different ways among which postmodern theorists tend not to distinguish sufficiently: as a claim about truth, as a claim about justification, or as a claim about purpose. As we shall see, however, none of these claims is even remotely plausible.

Interpreted as a claim about truth, the suggestion would be that the Zuni and archaeological views are equally true. On the face of it, though, this is impossible, since they contradict each other. One says, or implies, that the first humans in the Americas came from Asia; the other says, or implies, that they did not, that they came from somewhere else, a subterranean world of spirits. How could both a claim and its denial be true? If I say that the earth is flat and you say that it's round, how could we both be right?

Postmodernists like to respond to this sort of point by saying that both claims can be true because both are true relative to some perspective or other and there can be no question of truth outside perspectives. Thus, according to the Zuni perspective, the first humans in the Americas came from a subterranean world, and according to the Western scientific perspective, the first humans came from Asia. Since both are true according to some perspective or other, both are true.

But to say that some claim is true according to some perspective sounds simply like a fancy way of saying that someone, or some group, believes it. The crucial question concerns what we are to say when what I believe—what's true according to my perspective—conflicts with what you believe—with what's true according to your perspective. The one thing not to say, it seems to me, on pain of utter unintelligibility, is that both claims are true.

This should be obvious but can also be seen by applying the view to itself. Consider: If a claim and its opposite can be equally true, provided there is some perspective relative to which each is true, then since there is a perspective—realism—relative to which it's true that a claim and its opposite both cannot be true, postmodernism would have to admit that it itself is just as true as its opposite, realism. But postmodernism cannot afford to admit that: presumably, its whole point is that realism is false. Thus, we see that the very statement of postmodernism, construed as a view about truth, undermines itself: facts about truth independent of particular perspectives are presupposed by the view itself.

How does it fare when considered as a claim about evidence or justification? So construed, the suggestion comes to the claim that the Zuni story and the archaeological theory are equally justified, given the available evidence. Now, in contrast with the case of truth, it is not incoherent for a claim and its negation to be equally justified, for instance, in cases in which there is very little evidence for either side. But prima facie, anyway, this isn't the sort of case that's at issue, for according to the available evidence, the archaeological theory is far better confirmed than the Zuni myth.

To get the desired relativistic result, a postmodernist would have to claim that the two views are equally justified, given their respective rules of evidence, and add that there is no objective fact of the matter which set of rules is to be preferred. Given this relativization of justification to the rules of evidence characteristic of a given perspective, the archaeological theory would be justified relative to the rules of evidence of Western science, and the Zuni story would be justified relative to the rules of evidence employed by the relevant tradition of mythmaking. Furthermore, since there are no perspective-independent rules of evidence that could adjudicate between these two sets of rules, both claims would be equally justified, and there could be no choosing between them.

Once again, however, there is a problem not merely with plausibility but with self-refutation. For suppose we grant that every rule of evidence is as good as any other. Then any claim could be made to count as justified simply by formulating an appropriate rule of evidence relative to which it is justified. Indeed, it would follow that we could justify the claim that not every rule of evidence is as good as any other, thereby forcing the postmodernist to concede that his views about truth and justification are just as justified as his opponent's. Presumably, however, the postmodernist needs to hold that his views are better than his opponent's; otherwise, what's to recommend them? By contrast, if some rules of evidence can be said to be better than others, then there must be perspective-independent facts about what makes them better and a thoroughgoing relativism about justification is false.

It is sometimes suggested that the intended sense in which the Zuni myth is "just as valid" has nothing to do with truth or justification but, rather, with the different purposes that the myth subserves, in contrast with those of science. According to this line of thought, science aims to give a descriptively accurate account of reality, whereas the Zuni myth belongs to the realm of religious practice and the constitution of cultural identity. It is to be regarded as having symbolic, emotional, and ritual purposes other than the mere description of reality. And as such, it may serve those purposes very well—better, perhaps, than the archaeologist's account.

The trouble with this as a reading of "just as valid" is not so much that it's false but that it's irrelevant to the issue at hand: even if it were granted, it couldn't help advance the cause of postmodernism. For if the Zuni myth isn't taken to compete with the archaeological theory as a descriptively accurate account of prehistory, its existence has no prospect of casting any doubt on the objectivity of the account delivered by science. If I say that the earth is flat and you make no assertion at all but instead tell me an interesting story, it has no potential for raising deep issues about the objectivity of what either of us said or did.

Is there perhaps a weaker thesis that, while being more defensible than these simpleminded relativisms, would nevertheless yield an antiobjectivist result? It's hard to see what such a thesis would be. Stanley Fish, for example, in seeking to discredit Sokal's characterization of postmodernism, offers the following (op–ed piece, in the *New York Times*):

> What sociologists of science say is that of course the world is real and independent of our observations but that accounts of the world are produced by observers and are therefore relative to their capacities, education and training, etc. It is not the world or its properties but the vocabularies in whose terms we know them that are socially constructed.[6]

The rest of Fish's discussion leaves it thoroughly unclear exactly what he thinks this observation shows, but claims similar to his are often presented by others as constituting yet another basis for arguing against the possibility of objective knowledge. The resultant arguments are unconvincing.

It goes without saying that the vocabularies with which we seek to know the world are socially constructed and that they therefore reflect various contingent aspects of our capacities, limitations, and interests. But it doesn't follow that those vocabularies are therefore incapable of meeting the standards of adequacy relevant to the expression and discovery of objective truths.

We may illustrate why by using Fish's own example. There is no doubt that the game of baseball as we have it, with its particular conceptions of what counts as a "strike" and what counts as a "ball," reflects various contingent facts about us as physical and social creatures. "Strike" and "ball" are socially constructed concepts, if anything is. However, once these concepts have been defined—once the strike zone has been specified—there are then perfectly objective facts about what counts as a strike and what counts as a ball. (The fact that the umpire is the court of last appeal doesn't mean that he can't make mistakes.)

Similarly, our choice of one conceptual scheme rather than another, for the purposes of doing science, probably reflects various contingent facts about our capacities and limitations, so that a thinker with different capacities and limitations, a Martian, for example, might find it natural to employ a different scheme. This does nothing to show that our conceptual scheme is incapable of expressing objective truths. Realism is not committed to there being only one vocabulary in which objective truths might be expressed; all it's committed to is the weaker claim that once a vocabulary is specified, it will then be an objective matter whether or not assertions couched in that vocabulary are true or false.

We are left with two puzzles. Given what the basic tenets of postmodernism are, how did they ever come to be identified with a progressive political outlook? And given how transparently refutable they are, how did they ever come to gain such widespread acceptance?

In the United States, postmodernism is closely linked to the movement known as multiculturalism, broadly conceived as the project of giving proper credit to the contributions of cultures and communities whose achievements have been historically neglected or undervalued. In this connection, it has come to appeal to certain progressive sensibilities because it supplies the philosophical resources with which to prevent anyone from accusing oppressed cultures of holding false or unjustified views.

Even on purely political grounds, however, it is difficult to understand how this could have come to seem a good way to conceive of multiculturalism. For if the powerful can't criticize the oppressed, because the central epistemological categories are inexorably tied to particular perspectives, it also follows that the oppressed can't criticize the powerful. The only remedy, so far as I can see, for what threatens to be a strongly conservative upshot, is to accept an overt double standard: to allow a questionable idea to be criticized if it is held by those in a position of power, Christian creationism, for example, but not if it is held by those whom the powerful oppress, Zuni creationism, for example. Familiar as this stratagem has recently become, how can it possibly appeal to anyone with the slightest degree of intellectual integrity, and how can it fail to seem anything other than deeply offensive to the progressive sensibilities whose cause it is supposed to further?

As for the second question, regarding widespread acceptance, the short answer is that questions about truth, meaning, and objectivity are among the most difficult and thorny questions that philosophy confronts and so are very easily mishandled. A longer answer would involve explaining why analytic philosophy, the dominant tradition of philosophy in the English-speaking world, wasn't able to exert a more effective corrective influence. After all, analytic philosophy is primarily known for its detailed and subtle discussion of concepts in the philosophy of language and the theory of knowledge, the very concepts that postmodernism so badly misunderstands. Isn't it reasonable to expect it to have had a greater impact on the philosophical explorations of its intellectual neighbors? And if it hasn't, can that be because its reputation for insularity is at least partly deserved? Because philosophy concerns the most general categories of knowledge, categories that apply to any compartment of inquiry, it is inevitable that other disciplines will reflect on philosophical problems and develop philosophical positions. Analytic philosophy has a special responsibility to ensure that its insights into matters of broad intellectual interest are available widely, to more than a narrow class of insiders.

Whatever the correct explanation for the current malaise, Alan Sokal's hoax has served as a flash point for what has been a gathering storm of protest against the collapse in standards of scholarship and intellectual responsibility that vast sectors of the humanities and social sciences are currently afflicted with. Significantly, some of the most biting commentary has come from distinguished voices on the left, showing that when it comes to transgressions as basic as these, political alliances afford no protection. Anyone still inclined to doubt the seriousness of the problem has only to read Sokal's parody.

Notes

1. *Social Text* 46–47 (Spring/Summer 1996): 217–252.

2. "A Physicist Experiments with Cultural Studies," *Lingua Franca* (May/June 1996): 62–64.

3. "Mystery Science Theater," *Lingua Franca* (May/June 1996): 54–61.

4. "Introduction," in Linda Nicholson (ed.), *Feminism and Postmodernism* (New York: Routledge, 1990).

5. "Indian Tribes' Creationists Thwart Archeologists," *New York Times*, October 22, 1996.

6. "Professor Sokal's Bad Joke," Op-Ed, *New York Times*, May 21, 1996.

3

A Plea for Science Studies

Philip Kitcher

Something has gone badly wrong in contemporary science studies.[1] Some of us have
spent large portions of our academic careers arguing for the importance of the critical study of science. Yet practicing scientists have not always responded favorably to
those arguments. Richard Feynman's famous (perhaps apocryphal) judgment that philosophy of science is about as useful to scientists as ornithology is to birds has been
quoted and echoed by Steven Weinberg, who entitles an entire chapter of *Dreams of
a Final Theory* "Against Philosophy."[2] More recently, humanists and social scientists
studying science have been viewed less as irrelevant dilettantes than as subversives
dedicated to undermining scientific authority. The recent books by Paul Gross and
Norman Levitt (*Higher Superstition*) and by Lewis Wolpert (*The Unnatural Nature of
Science*) make it plain that distinguished scientists find large portions of the work done
in the name of science studies ignorant, confused, and damaging. Alan Sokal's celebrated hoax reveals that the editors of one journal leaped at the chance to publish
pretentious nonsense because it resonated so well with what they wanted to claim
about science. When this episode is juxtaposed with the other scientific criticisms,
there's an obvious temptation to generalize and dismiss the entire field as a mess. So,
from my perspective, something has gone badly wrong. My aim is to try to work out
just what the trouble is and how science studies might do better.

Philosophers, historians, and sociologists might turn to areas of science because
they hope to illuminate issues that arise within their home disciplines. This is especially obvious in the case of philosophy, where traditional problems may be recast by
drawing on concepts and results from contemporary science: a knowledge of physics
may provide insights into determinism, and findings from neuroscience may shed light
on topics in the philosophy of mind.[3] Scientists should hardly view this kind of research as threatening (or even misconceived), although some may entertain the fantasy that they could do it better if only they had a spare Sunday afternoon or two.
More likely is the charge of "vulgar scientism" from historians, philosophers, and
sociologists who resent the notion that the intellectual purity of their disciplines should
be sullied by borrowing from the natural sciences. Much more controversial among

scientists is the thought that the arrow of illumination can run from history, philosophy, or sociology to science, that studies of science by outsiders might identify questions and answers that the protagonists miss. I want to begin by suggesting that this contribution from science studies is not just a theoretical possibility but something that has been achieved in a significant number of recent books and articles.

Historical, philosophical, and sociological perspectives can offer (1) valuable analyses of how contemporary scientific understanding has emerged, (2) conceptual and methodological clarification, especially in areas of theoretical dispute, (3) increased awareness of the social pressures that affect certain kinds of scientific research, and (4) investigations of the impact of scientific findings on individuals and on society, which can serve as the foundations of a more rational science policy. Among the examples of contributions in all these areas, I would cite (1) historical accounts of the development of Darwinism, eugenics, molecular biology, and the character of experimental research in high-energy physics;[4] (2) philosophical work on the sociobiology debate, the IQ controversy, the units of selection controversy, the implications of Bell's theorem, and causal methodology in the social sciences;[5] (3) socio-historical research on the ways in which excluding certain kinds of people from scientific research has affected the character of the science that is done;[6] and (4) studies of the social implications of contemporary molecular genetics.[7] The work done in these and similar areas seems to me to be an important part of scientific activity and often continuous with science itself. This continuity is expressed in the fact that historians, philosophers, and sociologists frequently collaborate with scientific specialists in the pertinent fields and sometimes publish in the most respected scientific journals.

Thus, the charge that science studies is populated by people who are ignorant in the areas of science about which they pronounce—a charge frequently voiced in the wake of Sokal's hoax—is absurd.[8] To set it firmly to rest, it may help to discuss in a little more detail one exemplary study. In 1986, a historian of earth science, Martin Rudwick, wrote a long and important book about a dispute that raged in geology in the 1830s, taking his title from the name given to the dispute by those involved in it—*The Great Devonian Controversy*.[9] Using an extraordinary wealth of sources, particularly journals, letters, and field notebooks, Rudwick was able to trace in great detail the ways in which a scientific debate was resolved. His theme was that not only the particular encounters with pieces of rock in a variety of places but also the social structures of British and European science affected the process of resolution. This was a work that could have been written only by someone with a rare combination of talents, for Rudwick is not simply a historian steeped in the culture of Victorian England and nineteenth-century Europe; he used to be, in addition, a paleontologist, whose purely scientific works are still used and cited.[10] Any scientists who believe that the field is populated by ignoramuses should read Rudwick, for he is an expert on the questions whose history he discusses, an expert by any measure that critics care to propose.

The obvious response to the citation of an individual case like this is the suggestion that it is exceptional, but, in fact, although cases of double Ph.D.s are relatively rare, many people who practice the history of science, the philosophy of science, or the sociology of science have substantial training in one or more area of science. Their scientific educations may differ from those of research scientists, may be less narrowly focused and not have the depth of knowledge in any area that researchers typically have. Indeed, historians, philosophers, and sociologists of science often have a peculiar mix of scientific knowledge, comparable to that of undergraduates in some respects, to graduate students in other respects, and akin in some ways to that of research professionals in still more respects. Thus a philosopher working on the measurement problem in quantum mechanics may know as much about recent mathematical results in this area as any professional physicist but be ignorant of the details of experimental procedures that any talented undergraduate physics major can carry out with ease. Some historians of contemporary biology have a broader knowledge of this science than most professional biologists do, although they might fail undergraduate exams designed to test students' ability to recognize organismal structures or cell types. Plainly, what is important is that scholars in science studies have the information that is pertinent to their projects, and it would be folly to chide them for being unable to perform tasks that are irrelevant to the questions they are attempting to answer.[11]

At this point, critics of science studies might concede that the field they are attacking is a mixture, containing some lines of research that are genuinely informed and valuable, but that these are largely outweighed by work that is more prominent because it makes flamboyant claims on the basis of ignorant and casual reflections on some aspect of the sciences. I believe that if this were publicly acknowledged, it would be an important step in the right direction, lessening the current intensity of the "science wars," for it would start to introduce distinctions in a literature that often seems to suppose a united band of humanist thugs. My aim in the next sections is to go further along this line by tracing some of the reasons that the shape of contemporary science studies is a matter for debate and to identify some of the flaws that concerned scientists have found in the parts of science studies they most dislike.

Some Points That Ought to Be Uncontroversial

Science studies ought to respond to two clusters of phenomena. Its systematic danger is to emphasize the themes in one cluster and to slight those in the other—even though both should be uncontroversial. A helpful first step in trying to understand disagreements about the role and status of science studies is to remind ourselves of these themes.

The Realist–Rationalist Cluster

1. In the most prominent areas of science, the research is progressive, and this progressive character is manifested in increased powers of prediction and intervention.
2. Those increased powers of prediction and intervention give us the right to claim that the kinds of entities described in scientific research exist independently of our theorizing about them and that many of our descriptions are approximately correct.
3. Nonetheless, our claims are vulnerable to future refutation. We have the right to claim that our representations of nature are roughly correct while acknowledging that we may have to revise them tomorrow.
4. Typically our views in the most prominent areas of science rest upon evidence, and disputes are settled by appeal to canons of reason and evidence.

5. Those canons of reason and evidence also progress with time as we discover not only more about the world but also more about how to learn about the world.

In declaring that these five theses ought to be uncontroversial, I am, of course, waving a red flag at those who think some of them are false.[12] Nonetheless, items 1 through 5 are, at least superficially, accurate descriptions of aspects of science that would strike those who reflect on most areas of science and their histories, so that scholars who wish to reject them have to take on the burden of explaining why appearances are deceptive.

We don't have to probe very deeply to find out why the realist–rationalist cluster is advanced. There are striking differences between the historical development of the arts and literature and the historical development of the sciences: older scientific claims live on in textbooks; the education of scientists frequently recapitulates, to some extent, the history of the disciplines in which they are trained; and older tools and techniques, both conceptual and physical, are still used to solve research problems, often with an explicit understanding of their limitations. In some areas of science, the visual representations produced show an impressive accumulation of detail—think of models of chemical molecules, genetic maps, and delineations of the sequence of geological strata and the fossils characteristic of them. Similarly, doubting the existence of the kinds of entities discussed by scientists—remote from sensory observation though they may be—often seems as strained as querying the existence of medium-size dry goods. We think, for example, that our current abilities to manipulate organisms and to produce yeast, flies, and mice (to name three much-transformed kinds of living things) with peculiar combinations of characteristics depend on the detailed genetic maps that molecular biologists have assembled and that the pattern of successful interventions would be impossible unless there were genes and, indeed, unless our genetic maps were approximately correct.[13] (Just as we believe that it would be miraculous for millions of tourists to navigate their ways around metropolitan subway systems unless the maps posted for their instruction were approximately accurate.)[14]

Nevertheless, even though we naturally take ourselves to have the right to believe the central claims most implicated in our successful interventions in nature, it is only proper to acknowledge our own fallibility. Our predecessors often thought, quite justifiably, that there were things in heaven or earth that turned out to be beyond the credence of later natural philosophy, and so our own judgments about what there is and how it is may prove faulty in some respects. Just as we see earlier inquirers as having parts of a correct picture of the phenomena they explored, so we can anticipate that our successors will make finer discriminations than we do, that we will take our place in the historical progression of scientific views, with our own insights and our own mistakes. We cannot tell, of course, which bits of what we believe they will throw away as misguided—for if we could, we would presumably make the changes ourselves—but we should suspect that there will be such bits. For the time being, we can only express rational confidence in the whole, perhaps committing ourselves most to those parts of our current science that seem most bound up with our predictive and manipulative successes.[15]

The sorting out of what is correct, worthy of being taught and built on, and what is not appears to depend on the advancement of evidence. Scientists do experiments, make observations, and review collections of specimens; they report what they find in ways that seem governed by agreed-upon rules; and they perform mathematical analyses and develop lines of reasoning that their colleagues scrutinize. At least according to the scientific self-image, the acceptance or rejection of scientific claims— including claims about the validity of instruments, experimental techniques, and competent performance, is the result of a process subject to canons of reason and evidence. (Historically, a central task for the philosophy of science has been to identify these canons.) With the growth of science, decisions about how to assess parts of science can be improved. We have a much greater awareness of statistical inference and statistical methodology than was available a few decades ago (let alone in the nineteenth century); part of Darwin's achievement consisted in his recognizing more clearly than had his predecessors that a theory might be supported by being able to systematize a wide body of observations, even if it did not issue in concrete predictions about the future; and experimental practices in biomedical science have benefited from greater understanding of the benefits of double-blind trials and the problems of placebo effects.

Much more could be said about each of the theses in the realist–rationalist cluster, but I hope that these brief remarks will indicate some features that those who would challenge them must explain away. Let us turn to another collection of themes, equally well supported by the historical and contemporary practice of science.

The Socio-Historical Cluster

1. Science is done by human beings, that is, by cognitively limited beings who live in social groups with complicated structures and long histories.
2. No scientist ever comes to the laboratory or the field without categories and preconceptions that have been shaped by the prior history of the group to which he or she belongs.
3. The social structures present within science affect the ways in which research is transmitted and received, and this can have an impact on intratheoretical debates.
4. The social structures in which science is embedded affect the kinds of questions that are taken to be most significant and, sometimes, the answers that are proposed and accepted.[16]

Again, I shall be relatively brief in defending these themes.

Although some idealized treatments of science proceed as if inquiry were carried out by subjects who were disembodied, logically omniscient, and alone, everybody knows better. Actual investigators live significant portions of their lives outside laboratories, having social relations not only with fellow scientists but also with those who support their research or are affected by it; they have positions in a wider society; and finally, their abilities to perform logical inferences and mathematical calculations are limited and fallible. Those who want to slight the first thesis in the socio-historical cluster surely do not contest these points but, rather, deny that they have any impact on the practice of science. Each of the three following theses identifies a way in which individual and group histories and/or social roles make a difference to scientific work.

Every time a scientist makes an observation, does an experiment, or proposes a line of reasoning, he or she draws on the categories and appeals to the standards current in a particular group, usually a relatively small group of specialists interested in a technical problem. Much of what the scientist takes for granted has not been independently checked but was absorbed in a period of training, so that the work may go forward rather than recapitulating, slowly and tediously, what has been done in the past. The dependence on concepts that were introduced long ago or on established standards that scientists have not questioned for generations is most obvious when there is broad revision, old concepts are discarded, or standards are modified.[17] So thesis 2 in the socio-historical cluster ought to be accepted.

Behind thesis 3 lies the obvious thought that scientists stand in complex relations of affiliation and opposition; they cooperate with some of their fellows and compete with others. There's little doubt that alliances have played important roles in the historical development of various sciences: the debates of the late seventeenth and early eighteenth centuries between Cartesians and Newtonians showed clearly how antecedent loyalties can incline the mind to respond to some considerations and ignore others; and Darwin's own cultivation of leading figures in the British scientific establishment was surely important to his securing an audience for evolutionary ideas. Perhaps it may be thought that this type of social impact is an unfortunate distortion of science, and when scientists are behaving "properly," they are indifferent to an argument put forward by a friend, a rival, or a detractor. Yet for reasons that ultimately stem from John Stuart Mill, we might believe that the possibility of debate among contending factions, each bound together by ties of solidarity, might contribute to the eventual articulation of superior positions, that a social system for science can take advantage of the facts of human competition and cooperation to work efficiently for the uncovery of truth.[18]

Finally, and perhaps most obviously, the kinds of problems singled out as important depend in part on the history of the field and on the wider interests of members of society. Contemporary studies of heredity suppose that some problems are especially significant—mapping and sequencing various genomes, identifying the structures and roles of particular molecules—partly because of the history of research on the large question "How are traits inherited?" which has defined the field from the beginning, partly because of what it is now possible to do, and partly because of the practical consequences of certain forms of inquiry when applied to the problems of certain kinds of societies (specifically the hope that the maps and sequences will help us address medical problems).[19] Less obviously, the practical demands and the history of research standards also help determine what will count as acceptable solutions, specifying, for example, the precision that an answer must achieve if it is to be applicable. The perennial worry voiced by some scientists about the distortion of a research agenda by practical concerns reinforces this thesis about the effects of society on science.

The challenge for science studies is to do justice to both clusters. The history of science studies and Science Studies (the capitals refer to the current and controversial work in the field) shows an initial period (up to the 1960s), during which the first cluster dominated—scientists were conceived as asocial, logically omniscient beings whose work was shaped only by what occurred in the lab. Since the 1970s, Science Studies has sometimes ignored the first cluster entirely—scientists have been conceived as brain-dead from the moment they enter the laboratory to the moment at which they leave. Curious stories are then told about the ways in which class or gender, toilet training or religious education, political disputes in the wider society, and large cultural styles determine the character of a researcher's work. Often these treatments are described with so broad a brush, connecting with the details of the scientific work at so high a level of generality—or even misunderstanding—that the research professional is easily moved to righteous indignation and, hence, some of the legitimate complaints about scientific ignorance raised by Gross, Levitt, Sokal, Wolpert, and others.[20]

Yet the realist–rationalist cluster is not always dismissed, even in works devoted to showing the subtle ways in which the themes in the socio-historical cluster play out. A study that fails to slight either cluster is Rudwick's *Great Devonian Controversy*, which I have already praised. Hence the task I have identified as central to science studies is sometimes undertaken, although I should concede that such ventures are much rarer than they ought to be.

It's precisely the overemphasis on the second cluster that provokes the critics. There's no denying that there are loony ventures styling themselves as contributions to Science Studies, that introduce fanciful pieces of terminology, play verbal games, and show an astonishing degree of incomprehension about aspects of science that high school students usually understand (the blunders are often accompanied by fervent denunciations of the evils of science).[21] In response to this is a tendency to link sophisticated scholars with interesting things to say, scholars like Helen Longino and Steven Shapin, to much less intelligent and informed authors. At this stage of my argument, however, it's important to note that the critics are broadly right to recognize a persistent danger of overemphasizing the second cluster and ignoring the first. My next aim is to understand the reason for this lack of balance.

The Source of the Trouble

The root of the problem is some bad philosophy that has been strikingly influential in contemporary history and sociology of science (and occasionally in some contemporary philosophy of science). Several ideas have been dramatically overinterpreted, to such an extent that they give rise (as we shall see later) to the Four Dogmas of Science Studies.

The Theory Ladenness of Observation

It's been a philosophical commonplace since the early 1950s that our observations of the world presuppose concepts and categories in terms of which we make sense of the flux of experience.[22] The temptation is to claim that we thus find in nature only what we bring to it, that the world—or, at least, the only world we can meaningfully talk about—is "shaped" or "constructed" by us so that it will conform to our prior categories.

Stripping down the argument in this way makes its absurdity evident. As Thomas Kuhn (one of the early defenders of the theory ladenness of observation) clearly saw, the fact that concepts and categories are involved in observation doesn't mean that the content of experience is determined by them or that we cannot be led by experience to reconceptualize the phenomena.[23] Nor does it imply that we are somehow "cut off" from the world or that the only world we can talk about must be "constructed."

It is easy to be seduced into accepting a false picture: we imagine ourselves sitting in a cave, or behind a screen, onto which images are projected and suppose that some of the features of the images are dependent on properties of the surface. How, then, can we ever discover what the "real objects" that are the sources of these images are like? For significant periods in the history of philosophy, thinkers have been tempted by this picture, but as many critics have pointed out, it has a serious flaw.[24] In perception, we are in causal contact with physical objects, and although this contact is mediated by our *having* certain kinds of psychological states ("perceptions," "representations"), we do not perceive by *perceiving* those states. There are interesting questions for perceptual psychology about the extent to which our prior beliefs, concepts, and training influence the character of our perceptual states, and we can look to physics, physiology, and psychology to illuminate them.[25]

So it would be more accurate to say not that the world is shaped by our categories but that our representations of the world are so shaped and that the shaping is open to empirical investigation. But at this point, the champions of social constructivism will surely object that science is being "privileged," that the defense is circular, that questions are being begged, and so forth. They are quite right to recognize that the approach I have outlined could not possibly succeed in answering a certain kind of skeptical question. If the invitation is to throw away all our beliefs, start from scratch, and justify the claim that the objects about which we form perceptual beliefs are as we represent them, then we could not offer our contemporary blend of physics, physiology, and psychology to advance the kind of picture of perception I have sketched. But neither can champions of Science Studies offer any rival picture, even one that uses screens, veils, or cave walls. Descartes launched philosophy on a quest for fundamental justification, and despite the many insights uncovered by him and his brilliant successors, we now know that the problem he posed is insoluble—just as we know that the problem of trisecting an angle with ruler and compass is insoluble and that the task of proving the consistency of arithmetic within arithmetic cannot be completed.[26] If the constructivist reminds us that we haven't shown on the basis of a set of principles that precede the deliverance of empirical science that our scientific opinions are reliable, the right response is to confess that we haven't. There is no such set of principles that will do that job, but by the same token, no set of principles will establish a constructivist picture. The only way to separate out the contributions of our histories of learning to our observations is to call on some parts of science in the way I have proposed.

Once this point is recognized, it's easy to see that the overinterpretation of the theory ladenness of observation leads to a kind of global skepticism that makes it impossible to say anything at all. If it's offered as a prelude to one of the usual claims about the role of society or social interests in the shaping of science, the enterprise will be vulnerable to the same kind of relentless request to justify categories. "You want to talk about air pumps, societies of gentlemen, vats of ferment, Renaissance courts, inscriptions. With what right do you employ these notions? Why do you tell commonsense psychological stories about the ways in which human motivation leads to action or think that any of the macroscopic objects—including people—are as our commonsense contemporary views take them to be? No privileging!" Consistency requires constructivists to take such criticism seriously, leading them to a point at which they can say nothing.

There are interesting problems about global skepticism and more refined debates about scientific realism, and philosophical inquiry can go much further with this dialectic.[27] Yet to appreciate the muddles of one prominent line in contemporary science studies, we need go no further than this. Convinced by the idea that they can never talk about things "as they are," some practitioners effectively demand a response to the global skeptical challenge for entities they don't like (the ontologies of the sciences) and then proceed to talk quite casually and commonsensically about things they do like (people, societies, human motives). There is a name for this kind of inconsistency; it is *privileging*.

The Underdetermination of Theory by Evidence

Every scientist knows that individual experiments can be ambiguous and that, if something goes wrong, it's possible to identify alternative hypotheses as blameworthy. Pierre Duhem formulated the point at the turn of the century, insisting that hypotheses are tested in bundles, and Quine, in a much more abstract idiom, proposed that "total science" is underdetermined by all possible experience.[28] Duhem thought that the scientist's "good sense" (*bon sens*) enabled him or her to sort things out. Although Quine typically makes vague references to "an ideal organon of scientific method," his principal point seems to be a logical one: For any inconsistent set of sentences containing a self-consistent statement S, there is a consistent subset of the original set containing S, and typically there are many alternative consistent subsets of the original set.[29]

This idea has been dramatically overblown by some historians and sociologists who have contended that it shows that the world can have no bearing on what scientists accept. To see how bizarre this is, we should note that the point also seems to show that society can have no bearing on what scientists accept. But once we take seriously the notion that there's more to methodology than being consistent, it's easy to recognize that the gyrations that social constructivists envisage as available responses to experience involve epistemic costs. By analyzing major protracted scientific debates, we can see that the impact of experience is complex and subtle and that rational scientists are eventually forced out of untenable positions.

Duhem started a line of thought that enabled us to see that there is no instant rationality in science, but it's wrong to conclude from this that there are not context-independent standards of good reasoning that, when applied to increasingly comprehensive experiences, resolve scientific debates. In the early phases of the chemical revolution, phlogistonians could offer alternative analyses of the chemical reactions that Lavoisier viewed as showing the absorption or release of oxygen. As the number of findings increased, it became more and more difficult—and ultimately impossible—to find any consistent and unified way of treating all the reactions. Hypotheses to the effect that one substance was a complex compound containing phlogiston, designed to work for one reaction, broke down for others, whereas Lavoisier's proposals about constitution were largely successful.[30] There is an obvious sense in which defenders of the phlogiston theory could have gone on: they could have proposed that the composition of substances varied with the presence of some external factor or that little green people came down and added or subtracted amounts of phlogiston to make the equations balance. Nobody should doubt the logical possibility of holding on to a pet hypothesis, come what may, but what Duhem saw—with his *bon sens*—was that this circumstance does not show that these possibilities are rational.

Some workers in science studies maintain, however, that it's legitimate, even correct, to approach an episode in the history of science (or in current science) without probing the details of the experiments and the reasoning from them, precisely because we know in advance that the world can make no impact on a scientist's beliefs.[31]

The appeal to underdetermination is, once again, the reformulation of a form of skepticism—tantamount to the freshman reader of Descartes who demands to be shown that it is inconsistent to suppose that one is alone in the universe with the sensations and thoughts of the moment. Just as Dr. Johnson replied to Berkeley by kicking a stone, so the critics respond to the overinterpretation of underdetermination by citing the successes of contemporary science. This is part of a correct answer, but it needs to be supplemented with a diagnosis of the philosophical errors that have induced serious scholars to forget all about scientists' research, experiments, and reasoning and to glory in the richness of their personal and social lives—in short, to lose sight of one cluster of themes in their fascination with the other.

The Variety of Belief

There is another line of argument that sometimes leads practitioners of Science Studies to the same point. Suppose we begin from the evident fact that people, including scientists, sometimes differ in their beliefs. How can we account for this fact? Not, it is suggested, by appealing to the world, for the nature that the believers confront is the same in both instances. So the explanation of variety in belief must lie elsewhere, in the different societies that the believers inhabit.

This argument needs only a clear statement to self-destruct (or should it be "self-deconstruct"?). People with different beliefs may confront the same nature, but their relations to nature can be strikingly different. Travel, it is supposed, broadens the mind, and in the history of science, those who travel often encounter things at odds with their own beliefs and with the beliefs of those they left at home. It's not hard to explain the differences in belief between those who have ranged widely and those who have stayed at home by recognizing the variation in experience of nature. In many instances of scientific controversy,[32] something like this is occurring: one group of scientists has a wider range of experiences of nature than the other, and sometimes the ranges just are different. So the argument goes astray near the beginning in supposing that explanations that appeal to nature have to take a particular form: Scientist *X* believes that thus-and-so because thus-and-so. Once we abandon this unpromising way of explaining belief, the leap to social explanation is revealed as the extraordinary leap that it is.

We can identify a genuine insight in the posing of a problem about the variety of belief, however, if we recall the philosophical practice against which early advocates of the Strong Programme in the sociology of knowledge were reacting. From the 1930s through the 1960s, philosophers of science were fascinated with the (perfectly legitimate) problem of understanding the justification of scientific beliefs, and they focused on true beliefs. Prevailing pictures of justification tended to identify relatively simple forms of inference that made it puzzling how any rational person could ever have opposed the great achievements in the history of science. The salutary point made by the rebellious sociologists emphasized the natural rationality of members of our species and made it particularly hard to conceive of the intelligent participants in protracted scientific debates as bigoted, prejudiced, or irrational. What was missing in this entire opposition was a clear conception of how intricate and difficult reasoning in complex scientific contexts often is. In the debate between Lavoisier and his opponents, there are no simple rules of instant rationality, and a careful philosophical reconstruction can explain how reasonable people can disagree for a very long time and yet, ultimately find themselves compelled by the evidence to reach consensus.[33]

Once this is recognized, we can identify the motivation for and the overdevelopment of one of the great shibboleths of much work in Science Studies, the principle of symmetry. In the early 1970s, David Bloor famously proposed that explanations of true and false belief should be "symmetrical"; that is, they should appeal to the same kinds of causes. There is an important insight here: human beings have broadly similar capacities, live in broadly similar ways, and the large-scale physiological, physical, and psychological determinants of their beliefs are the same. We don't usually explain a scientist's belief by attributing to him or her some special faculty that that person alone possesses—although we should note that on some occasions we do appeal to the fact that someone has an ordinary capacity developed to a high degree in some particular direction.[34] Such an appeal is quite compatible with the recognition that there are serious and important differences in the processes by which people form their beliefs: Terrie the traveler differs from stay-at-home Sam because Terrie has seen things that Sam hasn't. Even though their different beliefs have much in common (perception plays an important role for both), the details are different (they've had different opportunities for perception). Sometimes, we're rightly prepared to make judgments about the quality of the processes through which beliefs have been formed. If three students are supposed to use the data to compute the chance that a patient has a particular disease, we commend the first for an impeccable Bayesian analysis, correct the errors of a second who neglects the base rate, and are aghast at the performance of a third who simply mixes guesswork with an appeal to the gambler's fallacy. Of course, all three students' beliefs are generated by "the same types" of causes (all engage in computational processes), but the three are importantly different and differ in their degrees of reliability.

Neglect of these simple points leads to some curiosities of Science Studies discussions. Rudwick's study of the resolution of the "Great Devonian Controversy" was criticized for treating some of the actors "asymmetrically."[35] According to Rudwick's narrative, the community of geologists eventually came to agree on a view of the ordering of strata, although two figures continued to hold out for different conclusions. These two figures were an interesting mixture of the cases considered in the previous paragraph: both had far more limited experience than the large majority who achieved consensus, and both defended their beliefs by processes that were far less reliable. To chide Rudwick for failing to treat the "outliers" symmetrically is the same plea for phony equality that one might make in querying the judgment about a race. Down by the track are the official judges, with eyesight regularly tested, the best auxiliary equipment, and ample experience; all of them agree about the result. Up in the stands are two spectators. One of them has a partially blocked view, and the other has mislaid his spectacles; neither has ever judged a race before. Each issues a verdict at odds with the judges' consensus—and with the other, but we must be symmetrical, we must not privilege. In the delightful epigram of a fine logician, we must be so open-minded that our brains fall out.[36]

"Actors' Categories" and the Writing of History

One last muddle dominates much of the work in contemporary Science Studies. Just as sociologists of science abase themselves before the shibboleth of symmetry, so historians insist that narratives must be constructed in terms of "actor's categories": in telling the story of a scientific development, we must not employ concepts that were not available to the people involved.

The emphasis on actors' categories has a serious point. If a historian is able to make vivid the ways in which a group of past scientists represented the world around them, then it is possible to appreciate the course of their inquiries as they experienced them, and this serves an important explanatory purpose. Rudwick's account of the "Great Devonian Controversy" provides us with the participants' perspectives so that as we follow their investigations, we feel their surprises and see the lure of approaches that, some pages later, turn out to be fruitless. Yet it would be wrong to think that this is the only explanatory role that history should serve or that appeals to what we now accept are always out of place. Historians of mathematics have often found it illuminating to cite Frobenius's proof that there exist exactly three associative division algebras over the reals in explaining why Hamilton's inquiries into higher-dimensional analogues of the complex numbers broke down where they did.[37] Any such account will not help us see the inquiry as Hamilton saw it, but it will enable us to understand just why he faced the problems he did at various stages. Later knowledge can be employed in history to fulfill an explanatory function, different from that of immersing us in the world of the protagonists.[38]

Purists may worry about using any findings from modern science in understanding the past. As in other instances, purism leads quickly to absurdity. Should the military historian studying trench warfare between 1914 and 1918 abstain from drawing on a technical understanding of the effects of shell impacts on the landscape, of the spread of infectious disease, of the psychological consequences of life in the trenches—an understanding that may have been produced by reflecting on the events chronicled? No matter how resolute we may be in seeking actors' categories, any account of past people will involve assumptions about motivation and action, the character of the public world and human responses to it, and we rightly make those assumptions using the best information we have.[39] Once we recognize that trying to suspend some current beliefs can be valuable in giving us insight into the situations as they appeared to the participants and that not suspending those beliefs can be important in leading us to recognize ("from the outside") their problems and successes, we can give the historians' totem its precise due.

To hammer home the point, let me offer one last example. We know very well that Europe suffered from outbreaks of bubonic plague during the late Middle Ages, and modern science gives us an account of how the bubonic plague was spread. It's easy to recognize the legitimacy of two quite different styles of history of the plague years. One offers us the perspectives of the actors, uses their categories, and presents us with the options and difficulties as they saw them. The other draws on contemporary epidemiology to explain why the plague broke out where it did, why various strategies against it were ineffective, how some people who survived were enabled to do so, and so forth. Histories of both types can be genuinely illuminating, and the second should not be ruled out of court by a priori prejudices.

Many critics of Science Studies recognize the relativism that often runs rampant. In identifying four routes that begin from sensible starting points, I hope to have shown that the road to relativism is paved with the best of intentions and the worst of arguments. So practitioners come to inscribe on their hearts the Four Dogmas: (1) There is no truth save social acceptance; (2) no system of belief is constrained by reason or reality, and no system of beliefs is privileged; (3) there shall be no asymmetries in explanation of truth or falsehood, society or nature; and (4) honor must always be given to the "actors' categories."

It would be wrong, however, to leave the diagnosis of the malaise of contemporary Science Studies at just this point. When the Four Dogmas have been thoroughly absorbed, so that younger scholars start from their conclusions as if they were gospel, then enterprises of real peculiarity can be launched.[40] How can Science Studies be liberated from the asymmetrical treatment of society and nature achieved in the early phase of the sociology of scientific knowledge?[41] How can the lessons of Science Studies be applied to Science Studies itself? Wait! There are new fashions announced in Gallic haute couture.[42] Let us mix in some Lacan, some Lyotard, a dash of Deleuze. Let us play with Derrida.[43] Let us have actor networks, mangles of practice, emergent dialectical surfaces, multivocalized polygendered postphallogocentric transcategorially sensitive discourses. . . . Let us have solutions to problems that nobody has ever thought of posing about science before; indeed, let us forget about science entirely in our de-privileging of canonical texts and our elevations of context. Like Lear on the heath, "We shall do such things—What they are yet I know not—but they shall be / The terror of the earth."[44]

I exaggerate, of course, but only a bit. The thoughtful reader, taking up a book such as Latour's *We Have Never Been Modern* or Pickering's *The Mangle of Practice*, can only wonder at the height to which the seas of Science Studies have risen. Science seems no longer to be the principal subject (pride of place now being given to Science Studies itself); instead, we have entered a discourse as closed off from the phenomena that were once central to the field as some philosophical investigations of the 1950s with their exclusive obsessions with the blackness of ravens. Dimly, one sees that rival perspectives are being pitted against one another, but the exact character of the positions and the standards to which they are to be accountable are completely obscure. In the end, one can only ask, "If these are the answers, what, please, are the questions?"[45]

This is, of course, the point at which critics of Science Studies, both inside and outside science studies, should cry "Enough!" Just as the protagonists think that there is a seamless line of reasoning that leads them to their conclusions, their opponents buy into the same assumption and suppose that the entire enterprise was rotten from the beginning.[46] I share their impatience with the later stages of the project—the automatic assumption that the Four Dogmas are sound and that one must therefore undertake the projects I have parodied—but by trying to expose the exact points at which insight gives way to overinterpretation, I hope to prepare the way for a more sympathetic view of science studies, one that will not only offer a different picture of the sciences but also show how some pieces of scholarship that are icons in contemporary Science Studies might be put to better use.

The Real Challenges

The most general challenge today is to do justice to both clusters of themes. This task is not impossible, and in recent years, several books have appeared that, in different and occasionally incompatible ways, attempt to mix historical, philosophical, and sociological insights about science. Like Rudwick's study of the history of geology, Peter Galison's *How Experiments End* is a thorough investigation of historical episodes (this time in the context of twentieth-century physics), revealing the multiple constraints that operate in everyday experimental practice. Ronald Giere's *Explaining Science* links philosophical accounts of scientific reasoning to models of human cognition and takes some steps toward embedding human knowledge in a social matrix. The details of social exchange within a scientific group (the systematists who embrace "pattern cladism") are probed in David Hull's *Science as a Process*, and Hull shows clearly how a relentless concern for prestige can give rise to progressive conceptual evolution. From a more abstract perspective, in *Science as Social Knowledge*, Helen Longino explores the conditions for a well-ordered scientific community and argues that societal values play important roles in scientific decisions. John Dupre's *The Disorder of Things* sounds a similar theme about the relation between science and broader values as well as arguing for important differences and disconnections among the various sciences. Finally, in *The Advancement of Science*, I try to show the intricacy of the reasoning processes that figure in major scientific debates and to construct a formal framework for understanding how various kinds of social institutions, social relationships, and personal aspirations can play a positive role in the genesis of new knowledge.[47] Perhaps immodestly, I would like to see these works as grabbing hold of different pieces of the same (important!) elephant.

The books I have mentioned address two major groups of issues, in incomplete and inadequate ways. The first concerns the relation between the practice of science and the values of the broader society; the second focuses on the ways in which social relations and structures of various types figure in the doing of science. What kinds of value judgments enter into scientific decision making, and exactly where do they enter? Just in the funding agency? Just at the stage when research is being designed? At the point when conclusions are being reached? When those conclusions are disseminated? Or at all these points and more? Is there a tension between epistemic and other values, and if so, how should we think about this tension and its resolution?[48] How do such phenomena as reputation, lines of affiliation, competition for resources, and need for cooperation on large-scale projects affect the ways in which scientific questions are pursued and the answers that are accepted? What are the contemporary social institutions that shape scientific research, and are they well designed for the advancement of knowledge? Plainly, the two clusters of questions are intertwined, and it is hard to conceive of answering them independently of each other.

It should also be plain that these questions are important. Reflective scientists want to understand the ways in which existing arrangements foreclose certain kinds of opportunities. (Why should social systems that have evolved from the seventeenth century be expected to be particularly good at fostering contemporary scientific research?) Reflective people (whether scientists or not) want to know whether research in various areas is skewed by the values of particular groups and, at the broadest level, how science bears on human flourishing. A large part of the motivation for many scholars who enter science studies is to try to articulate ways in which science can be used for human good. Virtually all traditional philosophy of science ignores that motivation. A sad irony of contemporary Science Studies is that even though it may seem more responsive to broader concerns, its espousal of the Four Dogmas undercuts them.

Suppose that you are worried about the impact of scientific discoveries on human well-being. An immediate corollary is that no general picture that endorses a global skepticism about scientific achievement can be satisfactory.[49] For if we are led into blanket constructivism, rejection of notions of reason, evidence, and truth, then there is a terrible irony. The last thing that political liberals want to say about the excesses of pop sociobiology or *The Bell Curve* is that these ventures are just like the social constructions of Darwin and Einstein[50] or that because talk of reason is passé, there's no less reason to believe claims about the genetic determination of criminal behavior than to endorse the double-helical model of DNA. We need the categories of reason, truth, and progress if we are to sort out valuable science from insidious imitations.

It has been obvious for about half a century that research yielding epistemic benefits may have damaging consequences for either individuals or even the entire species. Philosophical stories about science have been narrowly focused on the epistemic. Faced with lines of research that have the capacity to alter the environment in radical ways, to transform our self-understanding, and to interact with a variety of social institutions and social prejudices to affect human lives, there is a much larger problem of understanding just how the sciences bear on human flourishing. There seems to be a strand in contemporary Science Studies that responds to this problem by trotting out every argument (however bad) that can be interpreted as debunking the sciences—as if its proponents were frightened of a monster and had resolved to cure their terror by insisting on its unreality.[51] Any such strategy is not only inaccurate but also politically jejune. Only by careful analysis of science and its relations to a wide range of human concerns—indeed, only by analysis that comes to terms with the themes in the two clusters—can we hope to start a public dialogue that can be expected to produce a "science for human use."[52]

If Gross and Levitt are correct to think that one of the motivations behind Science Studies is to make the world safe for humane concerns, then the Four Dogmas are a terrible bar to insight into serious issues. Not surprisingly, the contribution of Science Studies to exploring one set of questions that philosophers have neglected—the questions about values—has thus been limited. Yet given the pronounced emphasis on the social, one might think that recent work in Science Studies would have at least supplied tools for addressing the second, the issues concerning the ways in which social structures shape research. Any such hope is doomed to disappointment.

It is time for confession. In constructing a general approach[53] to the question "Do the structures of science interact with individual motivations to promote the reliability of collective learning?" I had to make up (guess) a lot of my own sociology. This was not negligence. There was no theoretical source to which I could turn for guidance about the character of the causal processes that affect research. To be sure, there are "case studies," investigations that deploy "folk" categories, like the ones I employ, but there is no systematic body of theory that would identify major causal factors—such as one might obtain from a sociologist of criminal behavior if one were interested in the social causes of crime. I suspect that David Hull, investigating his warring systematists, also had to go it alone, doing his illuminating "natural history of a scientific community" without benefit of guidance from theoretical sociology.

Sociologists of science sometimes offer interesting studies of historical or contemporary groups that deploy commonsense ideas about social interactions and individual interests: Shapin and Schaffer's study of the Hobbes–Boyle controversy is a case in point. I shall not reiterate the criticisms offered by others (or by myself on other occasions) but recognize—as indeed Gross and Levitt seem to do—the fine detail about the political disputes in which Boyle and Hobbes were embroiled. Yet like Hull and me, Shapin and Schaffer do not draw on any antecedent general view of social causation. Their views about what is important to people involved in political debates are entirely sensible—and entirely atheoretical.

An earlier generation of sociologists of science conceived their subject differently. Robert Merton and his successors (now typically—and unfairly—scorned by Science Studies) wanted to try to understand the causal processes in scientific communities; they hoped to do for science what other sociologists (then and since) have done for other areas of human life. Contemporary sociology has well-developed subdisciplines that study religious affiliation and organization, crime and socially deviant behavior, and so forth. Moreover, we expect that a sociologist in one of these areas will be able to advance our understanding of important phenomena, shedding light, for example, on how crime rates may be expected to increase or decrease with age distributions or economic trends. We anticipate that the sociologist will offer a causal model, identifying some factors as relevant and taking account of their interactions, and because the phenomena are complex, we may be prepared to tolerate only limited accuracy. Mertonian sociology envisaged advancing this kind of analysis for social phenomena in general and also for the parallel problems that arise with respect to science.

So, I suggest, contemporary science studies faces two large and important problems. Because of its adherence to the Four Dogmas (and its repudiation of connections to other parts of sociology), Science Studies fails to answer those problems. Nevertheless, I now want to suggest that certain contributions to Science Studies, including some that have been vigorously criticized, could prove genuinely useful in responding to the real challenges that confront us.

Beyond the "Science Wars"

So far, my discussion has been largely critical. Although I have defended science studies, my main aim has been to identify where Science Studies has gone wrong and how it leaves the most important issues unaddressed. But as I have noted in passing, I believe that there are valuable insights in works that have become icons of the field, even though those insights are compromised by argumentative overextensions. It is time to justify these remarks.

Plainly, I believe that critics of Science Studies such as Gross, Levitt, Sokal, and Wolpert have identified shortcomings in some contemporary discussions of science. Reactions to their criticisms, especially to *Higher Superstition*, have been intense: many workers in Science Studies (including those whose works have been attacked and those whose works have not) find the book ignorant and opinionated (to put their responses in relatively mild language). When pressed to elaborate, they typically complain that Gross and Levitt and other critics do not draw distinctions, that they treat peripheral people and central workers in the field as if they were minor variants of one another, tarring the latter with the sins of the former. A closer look at the criticisms reveals that this complaint is not, strictly speaking, accurate. *Higher Superstition* takes some pains to recognize the differences in the quality of individual work: Shapin and Schaffer, for example, are praised for producing a book that is "exhaustively and meticulously researched" (Gross and Levitt, 1994, p. 68) and for raising "serious and genuine" questions (65); Helen Longino and Evelyn Fox Keller are described as "anything but inept" (136) and are explicitly contrasted with cruder feminist writers (notably Sandra Harding). Yet I think that, at a deeper level, the defenders of Science Studies are correct in believing that the critics press good points too far. Just as Science Studies has overextended genuine insights to fashion the Four Dogmas, so do critics like Gross and Levitt want to tell a simple story, one that will attribute the same deep motivation to all those whose works they address.

The critics approach Science Studies as if it were driven by a common ideology that aims to "demystify science, to undermine its epistemic authority, and to valorize "ways of knowing" incompatible with it" (Gross and Levitt, 1994, p. 11). There is no doubt that some of the targets of the criticism do subscribe to this ideology and that even some of those who are (rightly) viewed as central to Science Studies do so: Sandra Harding announces that we need something different from the sciences as traditionally practiced, that we need instead "sciences and technologies that are *for* women and that are for women in *every class, race, and culture*."[54] Harding is a perfectly good "type specimen" of the views that Gross and Levitt want to oppose. The trouble with their treatment of other workers in Science Studies—other feminists like Longino and Keller and nonfeminists like Shapin and Schaffer—is that they are seen as variants of this same general type. Longino and Keller, we are told, also want to "defend ideology in the academy" (Gross and Levitt, 1994, p. 136); Shapin and Schaffer stick up for the "voiceless and excluded masses" against "snobbish, purse-proud, rank-conscious plutocrats" (69). All discussions in Science Studies are thus heard against the accompaniment of the most strident voices, and it is thus hardly surprising that critics can find no value in them.

This strikes me as terribly wrong. Thinking of Keller,[55] Longino, and Shapin and Schaffer as belonging to the same intellectual species as Harding is a bit like thinking of gibbons, chimpanzees, seals, and dolphins as being conspecific with opossums (they all are mammals, of course, but there the similarities end). Shapin and Schaffer want to understand an important episode in the birth of modern science, and even though they may have gone awry because of their fascination with Dogma 2, it would be uncharitable not to see that their work corrects some of the overrationalistic tendencies of earlier accounts and that it may point to a future story that does justice to both clusters of ideas. Longino's *Science as Social Knowledge* is notable for two main themes: first, that objectivity is a social notion and that to claim that a belief is objective is to maintain that it has emerged from a process of critical discussion in a society with particular features (especially a tradition of scrutiny from alternative perspectives, to which all members of the society have access); and second, that the values of particular subgroups in society have affected scientific research at a number of different levels, including the choice of what Longino calls "global assumptions."[56] Longino's two themes combine in her call for detailed scrutiny of the ways in which the scientific research actually carried out is partial, reflecting only the values and concerns of certain groups within the broader society, and in her vision of a relationship between science and society that is more democratic and open. This is hardly an attempt to enthrone ideology in the academy.

A principal problem with the assumption that all contributors to Science Studies are really variations on Sandra Harding is that it forecloses the possibility that some of them might offer valuable insights into pursuing the projects outlined in the previous section. If we are to understand the complexities of the relationship between science and social institutions, we will need rich descriptions of particular instances, and some parts of the sociology of science (as currently pursued) as well as the style of history that Shapin and Schaffer exemplify may aid our attempts to paint a more general picture. Similarly, Longino's thoughtful discussions of the ways in which values may surface in scientific research can help us formulate questions that have been too long neglected in studies of science. Once we see the importance of accommodating two sets of themes, the realist–rationalist cluster and the socio-historical cluster, it's clear that even works trumpeting the hegemony of the social can serve as parts of an eventual synthesis. Here, perhaps, is one place where there's a good argument for symmetry: just as philosophers of science would not want to dismiss traditional studies as devoid of insight (even though they were oblivious to the themes of the socio-historical cluster), so too we can hope to free the more penetrating achievements of Science Studies from the unfortunate influence of the Four Dogmas. Impassioned critiques of Science Studies, viewed as a monolithic ideology, endanger this important possibility. The critics seem to yearn to turn the clock back, to revive a world in which only the "friendly" themes of the rationalist–realist cluster are bruited and in which outsiders sing only happy songs around the scientists' campfires.

A View from the Marginalized Middle

I doubt that this essay will please anyone, for it attempts to occupy middle ground, and the heat of many of the exchanges of recent years make it plain that the middle is an uncomfortable place to be. Some of my scientist friends echo Cato, convinced that the destruction of science studies is the only remedy and inviting me to join them in a dance on the charred remains. Colleagues in Science Studies view as an act of betrayal any suggestion that the discussions in the field have identifiable shortcomings. So the middle is thoroughly marginalized, and those of us who occupy it have been moved, again and again, to repeat Mercutio's most famous expostulation.[57]

I have written this essay in the hope that, within science studies at least, we can transcend the culture wars and use the debate to fashion more productive approaches to important issues. It is hard to be optimistic. The trenches have been dug deeply, and the fire shows little signs of stopping. Even the title of this book reveals an important lack of mutual understanding. Whatever its faults, Science Studies is not "*a house built on sand.*" It is better conceived as a colony strung out on a difficult, but strategically important, seashore. Some of the buildings—gross and gaudy in self-advertisement—stand on pathetically slender foundations; they hardly need a tsunami to wash them away, the merest ripple will do. Others are a curious mixture of craftsmanlike work and jerry-building, often with a folly or a vast, unscoured stable attached. A few buildings, more modest, sneered at or ignored by the most ambitious architects, are constructed to last. Perhaps if this image is accepted, we can begin to see that we should neither announce utopia nor call for the bulldozers. What is needed is slum clearance and urban renewal, a project in which historians, philosophers, sociologists, and scientists all should all be invited to join.

Notes

I am grateful to John Dupre, Arthur Fine, David Hull, Norman Levitt, Martin Rudwick, Alan Sokal, and Gabriel Stolzenberg, all of whom offered me valuable advice, corrections, and comments (from many different perspectives). Special thanks are due to Noretta Koertge, whose careful critique of an earlier draft led to substantial improvements.

1. Here and throughout, I use "science studies" to refer to the field of the study of science by nonscientists, paradigmatically historians, philosophers, and sociologists; and "Science Studies" to refer to particular views about that field, specifically the grab bag of doctrines that have drawn the wrath of scientific critics. Capitalization of the title sentence is left as an exercise for the reader.

2. For a penetrating response to Weinberg, see Wesley Salmon's lucid essay "Dreams of a Famous Physicist," forthcoming in a collection of his essays to be published by Oxford University Press. In fairness to Weinberg, I should note that he does acknowledge that his chapter title is an overstatement.

3. These are only two among many obvious examples. For paradigms of this kind of philosophical work, see John Earman, *A Primer on Determinism* (Dordrecht: Kluwer, 1986); and Patricia Smith Churchland, *Neurophilosophy* (Cambridge, Mass.: MIT Press, 1985).

4. See, for example, Ernst Mayr and William Provine, eds., *The Evolutionary Synthesis* (Cambridge, Mass.: Harvard University Press, 1980), which is typical of much work on evolution after Darwin in its collaboration between historians of science and leading evolutionary biologists. For eugenics, see Daniel Kevles, *In the Name of Eugenics* (London: Penguin, 1987); for the first decades of molecular biology, see Horace Freeland Judson's magisterial *The Eighth Day of Creation* (New York: Simon & Schuster, 1979); and for experiments in twentieth-century physics, see Peter Galison, *How Experiments End* (Chicago: University of Chicago Press, 1987).

5. For the IQ controversy, see Ned Block and Gerald Dworkin's anthology *The IQ Controversy* (New York: Pantheon, 1974), especially the long essay by the editors. Block also wrote, in *Cognition* (1995), the single best diagnosis of the flaws of Richard Herrnstein and Charles Murray's *The Bell Curve*. Pioneering work on the units of selection controversy has been done by David Hull, William Wimsatt, and Elliott Sober; see, in particular, Sober's *The Nature of Selection* (Chicago: University of Chicago Press, 1992). The implications of Bell's theorem have been explored by numerous contemporary philosophers of science, including Bas van Fraassen, Abner Shimony, Arthur Fine, Jon Jarrett, and Geoffrey Hellman. Many of the most important essays on quantum mechanics by philosophers have appeared in *Physics Review Letters*; and for two recent studies that add new dimensions to the discussion of issues in the foundations of quantum mechanics, see David Albert's *Quantum Mechanics and Experience* (Cambridge, Mass.: Harvard University Press, 1992); and Tim Maudlin's *Quantum Non-Locality and Relativity* (Oxford: Blackwell, 1994). Groundbreaking work on causal modeling in the social sciences has been done by Clark Glymour, Peter Spirtes, Richard Scheines, and Kevin Kelly. See Glymour et al.'s *Discovering Causal Structure*. On the sociobiology debate, see Michael Ruse, *Sociobiology: Sense or Nonsense* (Dordrecht: Reidel, 1979); and my own *Vaulting Ambition: Sociobiology and the Quest for Human Nature* (Cambridge, Mass.: MIT Press, 1985).

6. See Londa Schiebinger, *The Mind Has No Sex?* (Cambridge, Mass.: Harvard University Press, 1989); and Kenneth Manning, *Black Apollo of Science* (New York: Oxford University Press, 1983). Paul R. Gross and Norman Levitt offer a somewhat condescending evaluation of Manning's work in their *Higher Superstition* (Baltimore: Johns Hopkins University Press, 1994), 285, n. 62),without offering any detailed criticism.

7. For example, see Troy Duster, *Backdoor to Eugenics* (New York: Routledge, 1990); Dorothy Nelkin and Laurence Tancredi, *Dangerous Diagnostics*, 2d ed. (Chicago: University of Chicago Press, 1994); Daniel Kevles and Leroy Hood, eds., *The Code of Codes* (Cambridge, Mass.: Harvard University Press, 1992); and my own *The Lives to Come: The Genetic Revolution and Human Possibilities* (New York: Simon & Schuster, 1996).

8. Although Gross and Levitt are also often read as making this charge, this seems to be more a matter of the tone of *Higher Superstition* than of what they say by way of criticizing prominent figures in Science Studies. True, they rightly take Bruno Latour and Sandra Harding to task for their ignorance on various technical matters (matters that are directly relevant to the topics they discuss), but Gross and Levitt explicitly note technical competence in other cases. For example, they praise Harmke Kamminga's exposition of chaos theory (95), single out Scott Gilbert's "distinguished" exposition of developmental biology (117, 121), and remark on Evelyn Fox Keller's extensive training in science (140).

Here and throughout this chapter, parenthetical page references to Gross and Levitt are to *Higher Superstition*. I also doubt that Sokal would accuse all practitioners of Science Studies of scientific illiteracy. Yet the myth is now widespread among scientists, and it needs to be debunked.

9. Published by the University of Chicago Press in 1986. When four societies in science studies (the History of Science Society, the Society for the Social Study of Science, the Society for the History of Technology, and the Philosophy of Science Association) met together that year (the only occasion on which all four have ever met together), Rudwick's book was the focus of a unique multidisciplinary symposium in which scholars from all four societies commented on it.

10. Stephen Jay Gould pays tribute to Rudwick's paleontological work in his review of *The Great Devonian Controversy* (see *An Urchin in the Storm* [New York: Norton, 1987], 78), but perhaps the real compliment is the fact that Rudwick's work on brachiopods is cited as an important illustration in one of the most widely read essays in recent evolutionary theory, Gould and Lewontin's "The Spandrels of San Marco and the Panglossian Paradigm: A Cri-

tique of the Adaptationist Programme," *Proceedings of the Royal Society of London* B 205 (1979): 581–98.

11. Of course, scholars in science studies sometimes fail to recognize that particular pieces of scientific information would be useful to them, and professional researchers can play a valuable critical role in pointing this out. This is quite a different objection, however, from the charge that science studies is filled with ignorant dilettantes.

12. Here I should acknowledge that some of these theses are matters of sophisticated philosophical debate and that some philosophers have offered serious challenges to straightforward readings of them. In particular, Hilary Putnam, Arthur Fine, Nelson Goodman, and Richard Rorty all would object to the most obvious interpretation of 2. It seems to me important to separate philosophical worries about the understanding of the relation of thought and language to reality from the much cruder suggestions put forward in most of contemporary Science Studies, and also to recognize that for Putnam and Fine at least, there is a sense in which 2 can be construed to be true. (This may also hold for Goodman and Rorty, although here I am less confident.) The major objections to 2 that condemn strong versions of realism is that philosophers have added metaphysical encumbrances to the ordinary practices of describing the accomplishments of the sciences, not that the practice of identifying these accomplishments ought to be radically revised. This is clearest in Fine's commendation of "the natural ontological attitude"; see *The Shaky Game* (Chicago: University of Chicago Press, 1986).

13. Of course, some philosophers would demur. See Bas van Fraassen, *The Scientific Image* (Oxford: Oxford University Press, 1980). Van Fraassen's views have been subject to extensive discussion and criticism. See, for example, P. Churchland and C. Hooker, eds., *Images of Science* (Chicago: University of Chicago Press, 1984).

14. Here it's important to recognize that in both instances, the appropriate notion of accuracy depends on conventions of map reading. One should not infer from standard genetic maps that chromosomes are beautifully straight, any more than from the familiar map of the London Underground that the directional relationships among various stations are precise.

15. *Seem* is the right word here, for our knowledge of what parts of our beliefs genuinely do work for us is itself partial.

16. It's worth noting that Gross and Levitt seem to acknowledge this point (139), although they seem mostly unwilling to accept the idea that historical, philosophical, or sociological investigations might reveal in detail how it applies to particular scientific fields (one exception is their praise for some of Stephen Jay Gould's historical studies).

17. The *locus classicus* for this theme is obviously Thomas Kuhn's *The Structure of Scientific Revolutions* (Chicago: University of Chicago Press, 1962; expanded 2d ed., 1970). The theme essentially develops Kuhn's account of "normal science," although as I have suggested, the characteristics of normal science may best be appreciated by looking at those convulsive changes that Kuhn calls "revolutions." It is important to note, however, that everything I have said can be accepted without endorsing Kuhn's account of scientific revolutions. There is no need to suppose that the large changes involve "conversion experiences." For my own attempt to develop the sociohistorical themes without the nonrationalist elements that sometimes surface in Kuhn's writings, see my *The Advancement of Science* (New York: Oxford University Press, 1993).

18. See J. S. Mill, *On Liberty*, chap. 2. Paul Feyerabend has urged that science should always find a place for the voicing of heretical views (see his *Against Method* [London: Verso, 1975]), and Elisabeth Lloyd argues that his defenses of heterodoxies are designed to exemplify the strategy of playing the devil's advocate, whose participation Mill saw as so important. From a quite different direction, I have argued that there is no reason to think that social institutions that take advantage of our rivalries and loyalties are necessarily opposed to the

advancement of knowledge. See my "The Division of Cognitive Labor," *Journal of Philosophy* 87 (1990): 5–22; and chap. 8 of *The Advancement of Science*.

19. Of course, critics claim that the optimism is ill based. See Richard Lewontin, "The Dream of the Human Genome," in his *Biology as Ideology* (New York: Harper, 1992). I try to arrive at a realistic assessment in chaps. 4 and 5 of *The Lives to Come*.

20. I should note that many of the most egregious examples cited by Gross and Levitt are by people whom practitioners of Science Studies would not view as central to the field. As one who has attended numerous fora in Science Studies during the past decade, I have never heard of Steven Best, Katherine Hayles, Maryanne Campbell, or Morris Berman. I don't think that I have heard presentations that cite Stanley Aronowitz or Jeremy Rifkin and only a couple that allude to Carolyn Merchant. By contrast, the following figures are omnipresent in the presentations and discussions: Bruno Latour, Donna Haraway, Steven Shapin, Simon Schaffer, Helen Longino, Evelyn Fox Keller, and Sandra Harding (all of whom are discussed by Gross and Levitt) as well as Harry Collins, Peter Galison, Lorraine Daston, Paul Forman, Norton Wise, Trevor Pinch, Michael Lynch, Andrew Pickering, and Ian Hacking (none of whom receives a mention).

21. Many, though not all, of the most bizarre ventures are by people whom central practitioners in Science Studies would see as both confused and peripheral to the enterprise.

22. A seminal work is Wilfrid Sellars's essay "Empiricism and the Philosophy of Mind," originally published in 1956 and reprinted in Sellars, *Science, Perception, and Reality* (London: Routledge & Kegan Paul, 1963). Philosophers of science often encountered arguments akin to Sellars's through the presentations of Norwood Russell Hanson, *Patterns of Discovery* (Cambridge: Cambridge University Press, 1958); and Thomas Kuhn, *The Structure of Scientific Revolutions*.

23. See Kuhn, *The Structure of Scientific Revolutions*, chap. 6.

24. Among the best treatments is that by J. L. Austin, *Sense and Sensibilia* (Oxford: Oxford University Press, 1962); and Jonathan Bennett's discussion of the "veil of perception" doctrine in his *Locke, Berkeley, Hume: Central Themes* (Oxford: Oxford University Press, 1971), esp. chap. 3.

25. See Jerry Fodor, "Observation Reconsidered," *Philosophy of Science* 51 (1984): 23–43; also Paul Churchland's exchanges with Fodor in *Philosophy of Science* 55 (1988): 167–87, 188–98.

26. I apologize here for a slightly unrigorous formulation of both problems, but mathematicians and logicians will easily see how to add the appropriate qualifications.

27. See, for example, Barry Stroud, *The Significance of Philosophical Skepticism* (Oxford: Oxford University Press, 1984); Hilary Putnam, *Reason, Truth and History* (Cambridge: Cambridge University Press, 1981); Arthur Fine, *The Shaky Game*; and Richard Rorty, *Objectivism, Relativism, and Truth* (Cambridge: Cambridge University Press, 1991).

28. Pierre Duhem, *The Aim and Structure of Physical Theory*, trans. and reprinted (Princeton, N.J.: Princeton University Press, 1954); and W. V. Quine, "Two Dogmas of Empiricism," in his *From a Logical Point of View* (New York: Harper, 1953).

29. Formulating Quine's thesis of the underdetermination of theories is much harder than it initially appears. If one believes that there is a privileged class of observational (evidential) statements, then it's possible to propose that there are many alternative theories equally well supported by all true observational statements. Arguing for this is not easy, however, unless one thinks that mere compatibility suffices for maximal support and is prepared to resist objections that some theories are simply trivial semantic variants of one another. Quine's own views about meaning and synonymy make it hard for him to dissect the latter issues, and his suggestions about confirmation are not very detailed. But the principal worry about the underdetermination thesis is that Quine's own formulation appears to have as one of its main

results (indeed, insights) the problematic character of the distinction between theoretical and observational statements. If we simply abandon this distinction, we arrive at the relatively banal thesis offered in the text.

30. For my reconstruction of parts of this example, see *The Advancement of Science*, 272–90. The intricate details of many of Lavoisier's experiments and his reasoning from them are provided by F. L. Holmes in *Lavoisier and the Chemistry of Life* (Madison: University of Wisconsin Press, 1985).

31. Thus Steven Shapin and Simon Schaffer reconstruct the debate between Boyle and Hobbes without ever going through the details of Boyle's numerous experiments with the air pump or investigating the ways in which an opponent of vacua might have tried to account for Boyle's findings. See their *Leviathan and the Air Pump: Hobbes, Boyle and the Experimental Life* (Princeton, N.J.: Princeton University Press, 1985). Why do they proceed in this way? The meticulous historical scholarship shows that they are not lazy, and their other writings demonstrate that they are eminently capable of dealing with the technicalities of science. The answer is that they "know" from the start that any experiment can always be interpreted in many different ways: "Hobbes noted that all experiments carry with them a set of theoretical assumptions embedded in the actual construction and functioning of the apparatus and that, both in principle and in practice, those assumptions could always be challenged" (112), accompanied by a footnote announcing the "resonance" with the Duhem–Quine thesis. There is a limited (Duhemian) sense in which the point is correct, but for reasons I give in the text, that limited version won't support Shapin and Schaffer's neglect of the experimental details. They have been guilty of overextending the argument from underdetermination in just the way I have described. Does this vitiate their entire study? It leaves many of their major conclusions unargued (and incorrect), but as I shall indicate later, parts of the study remain valuable contributions to the study of science.

32. But not in all. As Duhem saw very clearly, there are some cases in which scientists share the same experiences of nature and draw different conclusions. I've already argued that we shouldn't leap from conclusions about transient underdetermination to the view that these differences can't be resolved by further experience.

33. Again, see *The Advancement of Science*, 272–90.

34. For example, among the Morgan group, Bridges was notable for his ability to spot mutant fruit flies, and his exceptional skill might account for some differences in belief between him and others.

35. See Trevor Pinch, "Strata Various," *Social Studies of Science* 16 (1986): 705–13; and Harry Collins, "Pumps, Rock and Reality," *Sociological Review* 21 (1986): 819–28. It is worth pointing out that these reviews make many insightful points about Rudwick's work, despite the overinsistence on "symmetry."

36. The remark is originally due to Alan Ross Anderson, who used it in days long before the advent of symmetry fetishism.

37. I choose this example because it is a case in which first-rate traditional work in the history of science confronts recent trends in Science Studies. See Andrew Pickering, *The Mangle of Practice* (Chicago: University of Chicago Press, 1995), 141, n. 26. In emphasizing the importance of following scientists through their inquiries, Pickering seems quite blind to the insights that come from adopting an external perspective, and he dismisses what he calls "scientist's accounts" (3). As I suggest in the text, this is a serious blunder.

38. Because the point is so easily misunderstood, it is worth noting explicitly that my claim involves no "Whiggism" or "teleology." There's no suggestion that "the truth must out" or that the actors are somehow "drawn" toward it. Instead, I am simply making the obvious point that current knowledge enables one to see why historical actors do or not face problems (or encounter "resistance," in Pickering's phrase) in the pertinent phases of their inquiry. We

see why Hamilton's earlier efforts embroiled him in inconsistencies and also why later he was able to develop an apparently consistent theory. We understand why early bubble chambers didn't work, why geologists failed to find unconformities in Devon, why Mendel's studies showed apparent independence of assortment—in terms of the properties of condensation, the character of the Devon strata, and the chromosomes of pea plants, respectively.

39. "But in using contemporary science, we might be wrong!" Indeed. But all this shows is that the history we write today is fallible and that future developments in science may provide cause for rewriting. This should not be surprising. After all, it would be very odd for historians to plead that the knowledge they amass is especially invulnerable to revision, that it can survive changes in belief and context!

40. Anyone who has tried to talk to people who have recently been trained in Science Studies will know that the conclusions of the four arguments I have criticized are treated as axiomatic. There is just no questioning them, and one's raising of questions reveals that one must be a strange relic of the unenlightened past.

41. The old practice of explaining scientific developments in terms of the actors' interests privileges society over nature. This is asymmetrical and thus must give way to something better. We need a simultaneous construction of both nature and society from something more basic, so we go down the road to Latour's actor-network theory or Pickering's mangle of practice. There is no need to venture toward such murky destinations if we can avoid making the mistakes I have criticized.

42. It is worth repeating an insightful footnote of Larry Laudan: "Foucault has benefited from that curious Anglo-American view that if a Frenchman talks nonsense it must rest on a profundity which is too deep for a speaker of English to comprehend." *Progress and Its Problems* (Berkeley and Los Angeles: University of California Press, 1977), 241, n. 12. I think that Laudan overstates Foucault (whose treatments of madness, the clinic, and punishment seem to me insightful) but that his diagnosis of a peculiar tendency among Anglo-American academics is correct.

43. Alan Sokal was surely moved to perpetrate his hoax by the grandiose silliness of much postmodern discourse. That hoax may have made dialogue in Science Studies and between students of science and scientists far more difficult (as Arthur Fine pointed out, Sokal's action made it far harder for a philosopher of physics to interact with already suspicious physicists— although Sokal has gone on record praising the work of several philosophers of physics), but it has probably served a valuable purpose in departments of literature. Reaction to the hoax has made it clear that some emperors are naked, so Sokal may have given scholars who focus on Dante or Jane Austen the courage to deny that the study of urban graffiti has quite the same depth or interest.

44. Russell originally used this quotation to characterize Nietzsche's philosophy (*History of Western Philosophy* [London: Allen and Unwin, 1946], 734). That seems to me unfair, but to apply much more aptly to contemporary work in Science Studies. Perhaps some future scholar will view my judgment as unjust. I think it more likely that both the writings I criticize and my criticisms will vanish into dust.

45. To be fair, Pickering's book sometimes emerges from its preoccupation with finding a proper idiom for Science Studies to offer some descriptions of parts of scientific practice. In my judgment, these descriptions would be far more illuminating if they were stripped of the web of metaphors that seems to be the book's primary purpose to promulgate.

46. As David Hull pointed out to me, my tracing of the route to the Four Dogmas and beyond offers an intellectual history of Science Studies, but it would be interesting to accompany this with a social history. He indicates the outlines along which the history might go (using the kind of analysis deployed in his study of the wars among competing cladists (*Science as a Social Process* [Chicago: University of Chicago Press, 1988]). In the beginning, a group of young Turks

say some provocative things, flaunting orthodox views about the sciences, hinting that scientific knowledge isn't what it's cracked up to be. They become influential and, in a decade or so, point out that their statements have been misinterpreted, that their critiques are more nuanced than has been supposed. Younger Turks view this as a cop-out and decide to make a niche and a name for themselves by going beyond their insufficiently radical predecessors. And so it goes. Writing a history of this kind—revealing the social interests at work in the development of Science Studies—would be an interesting project, whether or not Hull's intriguing sketch is accurate.

47. Other recent studies that point toward more adequate interdisciplinary work in science studies include important essays by Arthur Fine, "Science Made Up," in *The Disunity of Science*, ed. Peter Galison and David Stump (Stanford, Calif.: Stanford University Press, 1996), 231–54; and Nancy Cartwright, "Fundamentalism and the Patchwork of Laws," *Proceedings of the Aristotelian Society*, 1994.

48. This question was raised forcefully by Isaac Levi in a penetrating discussion of *The Advancement of Science*. See his contribution to the symposium on that book in *Philosophy and Phenomenological Research*, September 1995, 619–27; and my response, 671–73. In *The Lives to Come*, I address some of Levi's concerns in a concrete case (the ethical and social issues around the Human Genome Project). I offer a more general discussion in my essay "An Argument about Free Inquiry" (*Nous* 31, 1997, 279–306), but this is only a first start at addressing a complex of neglected issues.

49. Gross and Levitt appreciate this point (45), as does Evelyn Fox Keller; see her *Secrets of Life, Secrets of Death* (New York: Routledge, 1992), 3–5. Keller once remarked to me that many of the most serious concerns about the ethical and social implications of the Human Genome Project result from recognizing that people do have genotypes (really) and that their genotypes can be discovered.

50. Although I independently made this point on several occasions, I owe this elegant formulation of it to some witty remarks by Richard Boyd.

51. I suspect that some of the authors who provoked Gross and Levitt's critique were moved by this strategy.

52. I owe this phrase to Jonas Salk. For decades, Salk was interested in promoting the understanding of the interaction between the sciences (particularly the biological sciences) and human concerns. Shortly before his death, Patricia Churchland and I had several conversations with him about the ways in which such understanding might be advanced, and in one of these Salk used the phrase to characterize our projected joint enterprise.

53. In chap. 8 of *The Advancement of Science*.

54. *Whose Science? Whose Knowledge?* (Ithaca, N.Y.: Cornell University Press, 1991), 5. Italics in original. Harding thinks that such sciences will also benefit "female men." Both this book and her previous book are admirably clear about what she is claiming, although the clarity of the claims does tend to show the poverty of the argument.

55. There are important differences between Keller and Longino and also among positions that Keller has defended at different stages of her career. For reasons of space, I do not deal with Keller's complex views here, but I urge those who know her only through Gross and Levitt's critique to read her most recent book, *Secrets of Life, Secrets of Death*.

56. See Hull, *Science as a Social Process*, 66–81, 86–98. I note, for the record, that I do not agree with Longino's account of objectivity. The differences between our positions can be gleaned from our respective contributions to F. Schmitt, ed., *Social Epistemology* (Lanham, Md.: Rowan and Allanheld, 1994).

57. "A plague o' both your houses!" *Romeo and Juliet*, 3.1.89. Note that despite his position in the middle, Mercutio is closer to the Montagues than to the Capulets.

Part II

Myths, Metaphors, and Readings

Scientific prose (with the notable exception of computer manuals) is characterized by clarity, precision, and economy. The overriding goal is to describe observations, experiments, hypotheses, theories, and inferences in plain, unambiguous language. Metaphors and analogies can be found in science writing, of course, and careful philosophical or historical analyses of their heuristic role in discovery and communication often enrich our understanding of the scientific process. (See, for example, Hesse's classic book on models and analogies as well as Lovejoy's account of the enduring metaphoric power of the idea of the Great Chain of Being.)

But sometimes a scientist's metaphors can be misleading: although Harvey often compares the heart with a pump, current scholarship suggests that his research was actually dominated by vitalist conceptions more akin to the Aristotelian tradition than to the mechanistic approach that became popular later in the century. The lesson is clear—one must not read the metaphors in scientific texts in isolation from the literal matrix in which they are embedded. They can be understood only as part of the scientific context.

Postmodernists, however, suspicious of the very concept of literal meaning, neither read others nor expect to be read literally. Approaching scientific texts as if they were a species of myth, poetry, or a Freudian dream, they search for hidden allusions, hints of unpleasant ideological convictions, and symbols of unconscious agendas. The resulting "interpre-

tations" are occasionally ingenious but more often ludicrous in their ignorance. In this section, we will encounter examples of some especially bizarre readings of scientific episodes.

Paul Gross traces the construction of a foundational myth of the cultural studies critique of science—the "case" of the passive egg and active sperm. First developed by the Biology and Gender Study Group, this myth has now been publicized in Newsweek as a paradigmatic case of how gender ideology influences the content of science. Drawing on his own research, which investigated the role of maternal RNA, plus a review of the history of theories of fertilization, Gross shows that the active role of the egg in reproduction was a commonplace throughout this century, especially following the elegant microscopy studies published by E. E. Just in 1919. That is, the myth of the bashful egg and macho sperm is a social construction of the science studies community and bears no relationship to what actually happened in biology.

Philip Sullivan takes on two more popular cases in which ideology is supposed to have had a profound influence on science: Hayles's account of the role of gender in the development of fluid mechanics and an episode from the history of statistics as presented by a member of the Edinburgh school. Sullivan shows that much of Hayles's interpretation rests on a conflation of the technical and ordinary language senses of terms such as *linearity, turbulence,* and *continuity.* MacKenzie's case study goes astray in part because he fails to analyze the mathematical aspects of the debate between Pearson and Yule and instead posits implausible linkages to their divergent political views.

In his second essay, Paul Gross deals briefly with more imputations of gender influences on biology, but his primary aim here is to expose a variety of rhetorical strategies that figure prominently in the writings of "science wars" correspondents. Gross objects not to their polemical tone but to dubious practices such as cobbling together misleading quotations and gratuitously impugning motives. He calls for a return to the scholarly ideals associated with the humanities.

Michael Ruse brings us back to the topic of gender and biology with an analysis of the possible political components of evolutionary theory. He points out that at times the theory of natural selection has been invoked by feminists as well as by misogynists but admits that some passages in Darwin and elsewhere reflect the sexist and racist ideology of the times. He insists, however, that the language used in popular science writings must be distinguished from that in scientific treatises. The popularizations often tell us more about what the audience would find interesting than they do about the way scientists conceptualize research problems.

4

Bashful Eggs, Macho Sperm, and Tonypandy

Paul R. Gross

Pure Tonypandy. A dramatic story with not a word of truth in it. If
you can bear to listen to a few sentences of the sainted [Sir Thomas]
More, I'll give you another sample of how history is made.
"Inspector Grant," in Josephine Tey's *The Daughter of Time*

The Story and the Facts

Tey's (Elizabeth MacKintosh's) classic detective story was first published in 1951. In
it, Grant, with the help of friends whom he dispatches to musty London archives for
such facts (not stories!) as can still be dug up, discovers that the foul murder of the
little princes—for which Thomas More, Shakespeare, and every schoolchild in England since then has condemned Richard III—could not possibly have been committed by him or his agents. The case against Richard, it seems, was a political fabrication of the scheming Henry VII. Grant finds a metaphor for such stories: *Tonypandy.*

"Tonypandy," Grant said . . . "is a place in the South of Wales."
"If you go to South Wales you will hear that, in 1910, the Government used troops
to shoot down Welsh miners who were striking for their rights. You'll probably hear
that Winston Churchill, who was Home Secretary at the time was responsible." . . .
"[In fact] all contact with the rioters was made by unarmed London police. The only
bloodshed in the whole affair was a bloody nose or two. The Home Secretary was severely criticized in the House of Commons incidentally for his 'unprecedented intervention.' That was Tonypandy. That is the shooting down by troops that Wales will
never forget." (Tey 1977, 102)

This excerpt brings me to my subject: Tonypandy, *class* "science studies," *order*
feminist history and epistemology, and the consequences of some *species* thereof. A
characteristic consequence is the article in the April 21, 1997, issue of the mass-
circulation *Newsweek.* Its title is "The Science Wars," and the banner reads: "How
much is research influenced by political and social fashions? An important debate is
making scientists re-examine their assumptions of objectivity." This, you might agree,
is not trivial. A vast audience is told that scientists have good reason to doubt the
objectivity of science. What consequence, for intellectual and cultural life, could be
more important?

The reporter, Sharon Begley, must be presumed innocent.[1] If she did what reporters usually do, she obtained this alarming information mostly by listening to

opinions. In the article, she reports, among other things, on the proceedings of a conference held at the University of Kansas, "Science and Its Critics." There, she asserts, "working scientists presented compelling examples of how science got wrong answers when social and political values influenced the work" (1997). This revelation must have been music, if it really happened, to the ears of social analysts of science present at the meeting. A current tenet of their discipline is indeed that "social and political values" are decisive for the content of science, that is, for knowledge.[2] To support her assertion, Begley quotes Cornell physicist Kurt Gottfried, author of a paper in *Nature* on the science wars (Gottfried and Wilson 1997): "Cultural and other extraneous factors are more important in the creation of science than most people realize."

Yes, probably. But his opinion, for which no "compelling" evidence is provided or needed, is not the burden of Gottfried's article (cowritten with K. G. Wilson).[3] Instead, the article is devoted to the analysis of a *locus classicus* of science studies: Andrew Pickering's *Constructing Quarks*, which Gottfried and Wilson say "remains the most instructive, ambitious, and outrageous Edinburgh study of modern physics" (1997). Much of my essay, then, specifies this "outrage": confusion, endemic to social studies of science, of "'science-as-practice' as compared to 'science as knowledge,' to use Pickering's terms." Gottfried and Wilson consider Pickering's conclusions "preposterous."

A second canonical work of science studies, *The Golem*, by Collins and Pinch, receives equally rough treatment in the article, which then ends with the thought: "Forcing scientific knowledge into the Strong Programme's doctrinal straitjacket has often produced sophistry, and compromised studies of practice" (Gottfried and Wilson 1997). This doesn't look to me like Professor Gottfried handing out bouquets for questioning scientific objectivity. It looks more like a *beau geste*, distancing himself from "extremists" and softening his criticism of a particular brand of science studies.

What, then, *is* Begley reporting as a failure of scientific objectivity, whose discovery is supposed to have forced scientists to reassess their assumptions? Four cases are featured in the *Newsweek* piece, each nicely summed up in an illustrated sidebar, contrasting "old think" and "new think." The four have to do with (1) conception, (2) primatology, (3) heritability of gaud in the peacock's tail, and (4) the value of the Hubble constant.

Here I examine one of those: old think and new think on "conception." I leave to experts on the other subjects an appropriate response to each (although I doubt that many experts will respond). The "conception" case, though, is one for which I feel a certain urgency. Among other reasons, it has been my field of research for forty-three years. As presented in the *Newsweek* account—and now in hundreds of college classrooms across the country—the story is pure Tonypandy. I thought that we had dealt with it in sufficient detail in *Higher Superstition* (1994, 117–26), but not many in science studies, reporters, or professors of women's studies, even among commentators on that book, seem to have actually read it. This "case" of a failure of scientific objectivity, however, is already one of those, like Francis Bacon's penchant for rape (discussed elsewhere in this volume), that Wales will never forget.

The Claim

The claim in *Newsweek*'s own words (the quotation marks are in the original):

> In the 1960s biologists studying conception described the "whiplashlike motion and strong lurches" of sperm "delivering" genes required to "activate the developmental program of the egg," which "drifted" along passively. The model portrayed sperm as macho adventurers, eggs as coy damsels.

The following is the sidebar, illustrated with an electron micrograph of a spermatozoon penetrating the egg surface:

> *Old think*: Macho sperm swim powerfully and purposefully toward the drifting egg. At conception, genes in sperm activate the developmental program of the passive egg. *New think*: Sperm are ineffectual swimmers. The egg actively grabs the sperm, and genetic material in the egg alone guides development in the first few hours after fertilization.

Clear enough. The article asserts that new think is a result of reassessing "objectivity," old think having been badly subjective in its social and political—that is, masculinist—bias. Readers of *Newsweek* now have it straight on two counts: scientific knowledge can be as biased as any other kind (note the implicit redefinition of knowledge), and social analysis has either ferreted out this particularly odious bit of male bias in science or actually inspired the new think.

Now, you may say, as do some of my Struthian colleagues in the life sciences, "That's nonsense; nobody pays attention to this male bias." But "nobody," it seems, refers to themselves. Important people do pay attention to it. An example is an important person attending not to Begley but to the same Tonypandy as it circulated long before Begley reported on a conference in Kansas:

> For our entertainment she recounts it, objecting however that it makes her and her eggs into passive little patsies—Sleeping Beauties waiting for Prince Charming's wake-up call—whereas in biological/historical fact both ova and ovulator are assertive, even aggressive actors in our life's story. (Barth 1997, 167)

This passage is from *The Tidewater Tales* by the distinguished John Barth, a splendid novel despite its postmodern superstructure, first published in 1987 and now handsomely reprinted as a paperback. There is more of the same in the book, in fact an entire closet drama, in which actively swimming eggs[4] excoriate the spermatic hordes down at the "confluence" as "macho whipcrackers" and "cantino macho pigs." Barth knows the new think on these matters, or at least his delightful (and pregnant) heroine does.

More important, though, even than the proficient Barth or the reportorial Begley, is the source with which I am most familiar: college and university faculty. I hear versions of this passive eggs / warrior sperm story (some of them much nastier than those just quoted) from their students. In 1995, the last time I taught the 400-level developmental biology course at my university, the issue was raised quite aggressively, early in the semester: Half a dozen students wanted to discuss the masculinist bias of their textbook and of the instructor in regard to the "passivity" of eggs. Their source was, as far as I could determine, an anthropology course and seminars or discussions elsewhere, for example, at the Women's Center.

Here is an irony: the author of their textbook, Professor Scott Gilbert of Swarthmore College, is one of those responsible for this bit of Tonypandy (as I will explain later). There is certainly no masculinist bias in his textbook. In any event, bashful eggs and macho sperm are a story about science, offered as sociopolitical analysis, that has consequences for what large numbers of people, some of them future political and intellectual leaders, believe. My colleagues are thus wrong: lots of people pay attention, and many of them believe such stuff.

Sources

Let us therefore retire Sharon Begley, John Barth, and the students: they are blameless. They have gotten this story from sources supposed to be knowledgeable about the relevant scientific literature, about how reproductive biologists and embryologists actually think, investigate, and write on these subjects. What or who are they? It turns out that there are many sources, all of them feminist publications and teachings, endlessly citing one another. Space allows me to offer only a sample, but it will be representative. Bear with me, please; we will get to the facts of the case after we've sampled what has been said—by those who should know better—about it.

A recent entry: in her book *Im/Partial Science*, Bonnie B. Spanier observes the following:

> Both the Biology and Gender Study Group at Swarthmore and Emily Martin have traced the effects of gender ideology in descriptions of the interactions of egg and sperm cells in fertilization. . . . Despite recent evidence of the egg's activity in binding and drawing in the sperm as well as blocking out extra sperm and the evidence for the very weak propulsion of sperm tails . . . the stereotype of the active individual sperm persists. (Spanier 1995, 24)[5]

So Spanier is not really a source, just a propagator, but she lists the Biology and Gender Study Group (BGSG) as one of her sources. Note that she says that evidence of the egg's activity is recent and complains that culture-laden (masculinist) stereotypes persist even in the face of these startling new findings.

Even more recently, Evelyn Fox Keller and Helen E. Longino edited a book, *Feminism and Science*, in which this claim is given prominence. In their introduction, they argue that "the female gamete, the egg, is repeatedly treated as a passive object towards which the energetic sperm propels itself and on which it acts" (Keller and Longino 1996, 6).[6] They then refer to their source, an included essay by Emily Martin. Martin's chapter is concerned with metaphors, with the claim that classic texts express "intense enthusiasm" for the male side of reproduction but show none for the female side (Martin 1996).[7] "How is it that positive images are denied to the bodies of women?" she demands (105). It is here that we see actually quoted the "whiplash motion and strong lurches" that can "burrow through the egg coat," citing for those words the textbook *Molecular Biology of the Cell*, by Bruce Alberts and others.[8] Martin concludes with "One clear feminist challenge is to wake up sleeping metaphors in science, particularly those involved in descriptions of the egg and the sperm" (114).

For this source, too, there are precedents. In *The Less Noble Sex* (1993), Nancy Tuana complains, "Female passivity is also seen in theories of conception. Theorists continue to perpetuate a view of the active sperm and the passive ovum awaiting penetration," (170) citing as her sources Donna Haraway and Evelyn Fox Keller. But the most influential source, by any measure, is that widely read and reprinted manifesto of the BGSG led by Scott Gilbert, which appeared, among several other places, in *Feminism and Science* (1989), edited by Nancy Tuana. This is the article to which almost all who propagate the story of that most egregious failure of scientific objectivity—the passive egg—refer, and justifiably so: it is a magisterial mining of metaphors (and metonymy and synecdoche), but its own, however, not those of its subjects.

Gilbert's piece is famous for the masculinist metaphors of fertilization it exhibits. It is here that readers can find "Sperm Goes A'Courtin'," "heroic sperm struggling against the hostile uterus," "fertilization as a kind of martial gang-rape," "the egg is a whore, attracting the soldiers like a magnet." I spend little time on it here, since it is discussed with full citations in detail in *Higher Superstition*. Suffice it to say that these quotation-marked metaphors are inventions of the BGSG. Never, in more than four decades of work in the field, have I heard any colleague speak in that way or encountered such writing in journals of the discipline. The real quotations offered by the BGSG, as detailed in *Higher Superstition*, are from ancient or secondary sources; have little to do with the state of professional understanding since the nineteenth century; and are, in any case, an order of magnitude less colorful than the metaphors of the BGSG itself. They do, however, cite an apparently legitimate scientific source of the "new" knowledge to which they refer: an article by G. and H. Schatten.

This article, "The Energetic Egg" (1983), is a semipopular report of the state of knowledge of fertilization, written by Gerald and Heide Schatten (Emily Martin misnames her) for *The Sciences*, the public magazine of the New York Academy of Sciences. The Schatten piece is a good effort of its kind, although it gives much more attention to the work of the authors and their colleagues of the Berkeley and Stanford groups than to findings on fertilization of others working at that time, or before that time. Nevertheless, it *was* a useful summary of the cell biology of fertilization at that time. The part of it later exploded into an indictment of masculinist bias is a summary of electron microscope studies by G. Schatten when he worked with the late Daniel Mazia at Berkeley. The important paper—in short, the first real scientific paper relevant to the claims and the argument—is a 1976 Schatten–Mazia paper in *Experimental Cell Research*.

Here are its method and findings. In 1976, the technology of scanning electron microscopy (SEM) had advanced sufficiently for direct observation in three dimensions, and at very high resolution, of the details of sperm/egg interaction. Schatten and Mazia added to the technology certain tricks of preparation that maximized the chances of catching the process efficiently and at its earliest stages. They glued eggs to polylysine-coated glass, added sperm, then quickly fixed the preparations while the interaction was still in progress, for SEM. They could then visualize sperm attachment, membrane fusion, and entry of the sperm into the egg interior. These were competent studies that yielded several widely reproduced electron micrographs. They showed the following (I quote from the paper):

> The first activity of fertilization is the discharge of the acrosome of the spermatozoon. . . . This filament . . . brings the sperm head into intimate contact with the egg surface. . . . At the first moments of fusion, the surface of the egg is uplifted in a small cone. . . . When the membranes fuse, the internal mass of the sperm head is seen enclosed in a column which proximally is egg membrane and distally sperm membrane. (331)

And so on. They saw what others had seen earlier, was already in the textbooks, and was known decades earlier to happen,[9] but they saw it more clearly. The acrosome filament (an autopolymerizing mechanism in the sperm head that creates a sort of probe) initiates the contact; the egg surface responds by elevating a "fertilization cone"; surface membranes of the two cells fuse; and the sperm head (in some species the tail, too) enters the egg cytoplasm.

Let me now summarize the astonishing claims, built on this (and similar) work, including studies on the biophysics of sperm-flagellar motion, about what Begley calls "conception." Please bear in mind that the following claims are, in turn, grounds for the superclaim that the objectivity of science is often illusory.

Claim 1. Science ignored, until recently (or until prodded, perhaps, by feminist—epistemological criticism), the possibility that the egg may play an "active" role in fertilization.

Claim 2. As late as the 1960s, science ignored the possibility that the egg might play an "active" role in "guiding" development; masculinist bias prevented the consideration of such a possibility.

Both claims are false. Their offering by anyone who pretends to know something about developmental or reproductive biology is either shoddy scholarship or not scholarship at all, just politics. Here are the facts.

The Egg "Grabs" the Sperm: 1878, 1919

EXAMPLE: In 1919, E. E. Just published his research on fertilization in the sand dollar, *Echinarachnius parma*. At this stage of his career, Just was a young but influential observer of fertilization and early development in marine animals (which have been the material of choice for everyone else, including Schatten and Mazia). Just's conclusions about the first moments of the process are like the common textbook account:

> Immediately on insemination, sperm pierce the jelly hull [not the egg itself], reaching the vitellus [the egg cell] with rapid spiral movements. The moment the tip of the sperm reaches the cortex of the vitellus all movements cease, the head and the tail in a straight line at right angles to a tangent of the egg surface. Penetration follows as an activity of the egg; the spermatozoon does not bore its way in—the egg pulls it in. (Just 1919, 5)

Just, a scholar, was not content merely to report what he had seen (usually with greater care than did many of his contemporaries).[10] For his emphatic conclusion, he cited, therefore, available precedent: "Kupfer and Beneche in 1878 made similar observations on the lamprey egg and reached the conclusion that the sperm is engulfed" (5). *Engulfed*: note that, please. No monograph or serious textbook on fertilization or embryology that I know failed, after the 1920s, to figure or at least to mention the fact of engulfment or the "fertilization cone," the egg structure that does the engulf-ing. Recognition of an "active" role of the egg in fertilization therefore dates back almost to the time of discovery of fertilization itself. All that happened later was that with the improvement of optical microscopy, and even later with the invention of the electron microscope, the process was imaged in ever-finer detail. The fundamen-tals have not changed. After the late nineteenth century, the egg was not seen, in any scientifically meaningful way, as "passive." What the Schattens are referring to in their joint paper, "The Energetic Egg," is G. Schatten's earlier work, whose technique allowed detailed observation of the fertilization cone. Schatten and Mazia found that the cone is a morphologic rearrangement of egg-surface microvilli and that this rear-rangement, in turn, is a profound reorganization of cytoskeletal macromolecules in the egg cortex. There was no doubt, of which I am aware, from the 1920s onward, about such structural remodelings of the cortex or of the reality of the fertilization cone and its part in bringing the sperm nucleus into the egg cytoplasm.[11]

To reinforce that last statement, I quote the relevant words from a favorite devel-opment textbook of the 1960s and 1970s, that of Berrill (finally, Berrill and Karp): "The first visible response to sperm contact is the engulfment of the sperm by a protrusion of cytoplasm, the *fertilization cone*" Berrill and Karp 1976, 118). "New think," there-fore, cannot be that the egg participates in the incorporation of the sperm head (or, if you love aggressive metaphors, "grabs" it). That is very old think, and it is also true. New think can only be that this is a recent discovery or that it was suppressed by patriarchal false consciousness. And that is false.

Did the literature of developmental and reproductive biology wax enthusiastic over male contributions to fertilization and development while discounting those of the female? No. If you care, try this experiment for yourself: count the papers in any sci-entific journal devoted to embryology from, say, 1880 to the present, on the egg and also those on the spermatozoon. If you are inclined to Tonypandy, you will be shocked to find that the studies on eggs overwhelmingly outweigh those on sperm, and with good reason, for the egg is vastly more important to development.[12]

"Those acolytes of scientific objectivity," says Begley (1997), "were spectacularly wrong." I can't imagine where she got that: perhaps from the musical-chairs citations in feminist science critique or from speeches in Kansas, at which there were either no embryologists present or some present who thought better of challenging Tonypandy. But if by "acolytes of scientific objectivity" Begley means embryologists and developmentalists working on fertilization, the claim is false. In fact, they got it spec-tacularly right, long before the technology could examine the ultrastructure of the sperm/egg interaction at very high resolution.

The Egg Can Go It Alone

It would be understating the argument against this "case" to close the book on Claim 1 without mentioning parthenogenesis. In 1913, Jacques Loeb, then a member of the Rockefeller Institute for Medical Research, published a retrospective volume on his studies of parthenogenesis and fertilization dating from 1902 (Loeb 1913).[13] There are three points to be made about this famous book. First, it is a useful sampler of Loeb's uncompromising and premature "isotropism" (as I called it in a review of the real arguments in embryology at that time: Gross 1985)—his insistence not only that development is reducible to physics and chemistry but also that one part of the egg is as good as another. Second, it includes a scholarly survey of the literature on natural parthenogenesis, or normal development, that is, from the egg, without benefit of sperm. Needless to say, this isn't what you would expect from a scientist and a discipline fixated on spermatozoa as heroes and ova as "passive." And finally, Loeb supplies summaries here, in full—often tedious—detail, of his experiments on artificial parthenogenesis—forcing eggs that normally develop only after fertilization by sperm to do the whole thing on their own. Did science ignore it? Not on your life. This work—rather than Loeb's later, more fundamental research—made him an international scientific celebrity. By the time Loeb wrote his parthenogenetic swan song, biologists the world over knew that among many species, the egg is self-activated and self-developing and that it could probably be made so in most other species if one tried hard and cleverly enough.

What the Egg Knows and Does

So much for coy and passive eggs versus shock-troop sperm. But what about the second claim—that (in the 1960s) science was indifferent to the egg's role in the control of development? Or as in Begley's (1997) summary of what the social analysts of science must have told her, that "new think" shows that "genetic material in the egg alone guides development in the first few hours after fertilization?"

This last statement is true and important, too! It underlies the elegant work for which a Nobel Prize was recently awarded to Christiane Nüsslein-Volhard. But it is not "recent." The first unabashed and mechanistically specific statement of that truth (along with hard evidence in favor of so judging it) was published by me and my then graduate student, the late Gilles Cousineau—in 1964. Independent but confirming results were being obtained and published by A. Spirin in Moscow, Albert Tyler and his students in Pasadena, and Jean Brachet in Brussels and Naples. Nor was our paper obscure: it was quickly reprinted in several collections on development. The following is from its concluding summary:

> It is suggested, in connection with the results described, that the unfertilized egg contains a store of masked template material (messenger RNA), whose information concerns in part the proteins which must be made for cell division. (Gross and Cousineau 1964, 393)

That was the beginning, not the end of it. The role of the "template material" (which I later named "maternal mRNA") in guiding early development became a preoccupation of the new molecular biology of development.[14] A full review of it was published as early as 1967 (Gross 1967). Twelve years later, the October 19, 1979, issue of *Science* carried a long summary by Gina Kolata of the state of the field, entitled "Developmental Biology: Where Is It Going?" including the now fully mature "maternal mRNA" story. This was a report of a symposium held at the Marine Biological Laboratory, Woods Hole, September of that year, at which achievements of the prior decade were discussed. Kolata's essay summarizes the role of the unfertilized egg's information, especially that embodied in the maternal mRNA:

> Now, investigators think they know at least some of the molecules that act as morphogenetic determinants in early development. They have been able to show that cell determination begins even before an egg is fertilized and is mediated through these substances . . . the substances are inactive and are inhomogeneously distributed. When the egg is fertilized and divides, the substances become active and, presumably, alter gene expression. Since the substances were unevenly distributed in the egg . . . daughter cells are different from the start. (315)

It is indeed a fact that the egg—eggs of many species, at least—contain morphogenetic information, some but not all of it as a sort of secondary genome, the maternal messages, indispensable for the first stages of development. But it is obviously false to say that this was not recognized "in the 1960s," false to suggest that science was blind to the possibility. In fact, what convinced us that it would be worth the tedious biochemical work necessary to test such a possibility was the existing embryological literature, including the old literature. Any serious student of it recognized that the facts of parthenogenesis and of the development of hybrids could not be explained on the basis of contributions from the sperm. Sperm were understood to be dispensable, at least in principle, but the egg is indispensable. What the spermatozoon does is make possible the variability (the key substrate of natural selection) that results from syngamy, which is the combination of paternal and maternal chromosomes. Over evolutionary time, the male gamete probably acquired a share in the chain of molecular processes called (properly) "activation" of the egg, but what the egg does and knows is internal to it. One way or another, this last notion motivated research in "experimental embryology" from its beginning in the nineteenth century. Today a few mammals have been cloned from ova. In the 1950s and 1960s, frogs were routinely cloned from frog eggs. As far as I know, nobody, however brainwashed in masculinity, has tried to clone animals from spermatozoa. Given the supercompacted state of DNA in the sperm head and its vanishingly small amount of cytoplasm, that would be asking for failure.

The Sociocultural Spin on Fertilization

So there is nothing to it. It is Tonypandy. Whatever metaphors people may have used to speak or write about fertilization and early development, the extraordinary progress of this discipline has had little or nothing to do with them. Some metaphors (not the metaphors created by the BGSG, which are sui generis) may have caused individual investigators to think wrongly—or rightly—about the subject; they might have determined, for individuals, what experiments to do or not to do. But the body of knowledge we call developmental biology depends on those metaphors not at all. The fertilization cone is an object, not a metaphor. The restoration of diploidy consequent to fusion of sperm and egg pronuclei happens; it is not a metaphor. DNA is not a metaphor, except possibly in English departments, nor is messenger RNA. These are substances that can be purified and stored in the refrigerator and that do remarkable things as chemical reactants in vitro and in vivo. The differences between "mosaic" and "regulative" development are referred to in words that have metaphoric entailments, of course! There are no other kinds of words. Those differences are, moreover, not absolute, but every scholar who knows the subject knows what they mean, in a dozen languages, and knows the same meanings. "Mosaic" and "regulative" development are phenomena dependent on chemical reactions and molecular structures.

Nothing about the "objectivity" of these things and processes is being "reassessed"—by serious developmental biologists—as a result of social or political insights. Oh: Josephine Tey's "The Daughter of Time" reflects, she notes, an old proverb: "Truth is the daughter of time." Yes, but it can be an awfully long time, especially when politics demands a particular story line and Wales needs something never to forget.

Notes

1. Begley's piece is tagged, at the end, as having been written with Adam Rogers.

2. The advanced-sociological position is, of course, that there is no such thing as value-free knowledge, that is, no "objectivity" according to the standard definition. Therefore the goal should not be to eliminate sociopolitical values from science, which is impossible, but to have enough scientists with the right sociopolitical values. The right values are, of course, identified with those of the social analyst.

3. Given that most citizens know little or nothing about the content or the methods of science, it would not be surprising if they had little idea of what percentage of either depends on social forces.

4. I know of no eggs of relevant interest—vertebrate or invertebrate—that "swim." Many invertebrate embryos do, however, and of course so do spermatozoa.

5. There are shelves full of evidence on "active" and "inactive" sperm in biological and medical libraries, especially the latter, in such fields as human fertility and sterility. "Active" sperm do the trick. If conception is at stake, mostly inactive ones, when discovered in the clinic, call for a donor who produces the "active" kind.

6. Leaving the matter of "passivity" aside for the moment, there is no question about the spermatozoon's self-propulsion, energetic or otherwise, or about its acting on the ovum. It does both; Antony Van Leuwenhoek saw the propulsion, and although Buffon didn't, everybody who has looked since then has seen it. When the sociopolitical analysts of the science get around to citing actual research (which they do rarely), the findings cited show that the spermatozoon "acts on" the egg, and vice versa. Read on.

7. This essay is remarkable, even among its stablemates, for angry tendentiousness. It is concerned entirely with rhetoric and narrative strategies, and even there it imputes qualities to cited works that they do not display. The only scientific information discussed, and very superficially, has to do with some measurements of the force of sperm propulsion and additions to the already well-known facts of sperm/egg adhesion through the agency of binding molecules and receptors on each. The data do nothing to alter the standard picture—many decades old—that the interaction is mutual and that the spermatozoa swim and the eggs don't. Least of all do they suggest any disparagement of ova and special enthusiasm for sperm—to, that is, any ordinary and reasonable reader.

8. Leaving aside, this time, "whiplash motion and strong lurches," which seem to me fair descriptions of what one sees under the microscope when observing a fresh suspension of sperm, there is no doubt about their "burrowing through the egg coat." Not, that is, if by

"coat" is meant the egg's external investments (such as the jelly layer surrounding most marine invertebrate eggs). That's exactly what spermatozoa do. For years, while there were sea urchins available for quantities of gametes, my students in embryology lab watched it happen.

9. See also the 1968 textbook by C. R. Austin, *Ultrastructure of Fertilization*, for a sense of how much of the general position was already explicit in the literature, especially of transmission electron microscopy (TEM).

10. Because he was black, E. E. Just had a hard time advancing his career in American science, and he was deeply and properly resentful about it. Eventually he left the country to end his days in Europe. There is evidence that he was not particularly hostile to Nazis. But as a young investigator in Woods Hole, he was known as a gifted observer and as the most reliable source of methods for handling sperm, eggs, and embryos. I should mention that his personal history does not support the idea that he was a feminist—just in case that idea should arise.

11. "Sol-gel transformations" of that kind, in the cortex and interior of the "activated" egg, were the stock-in-trade of my teacher (and Daniel Mazia's), Lewis V. Heilbrunn, who studied and wrote about them decades before World War II.

12. My own output in this game: some 200 papers in refereed journals, having to do with fertilization or early development, and as I count at this moment from my perhaps incomplete bibliography, three at most on spermatozoa, all the rest on the egg. Those on spermatozoa are concerned, moreover, with the structure of sperm DNA, not with the sperm's role in fertilization.

13. I have before me Loeb's own copy, which he later signed and gave to the MBL Library in Woods Hole.

14. Emily Martin is apparently aware of the timing of these discoveries, but she is quoted by Begley as saying that they "just sat there. . . . No one knew what to do with it." I, and two dozen of my Ph.D. students knew what to do with it, and they are today, in consequence, full professors at Harvard and MIT and Vanderbilt and Indiana, and the like.

References

Austin, C. R. 1968. *Ultrastructure of Fertilization*. New York: Holt, Rinehart & Winston.

Barth, J. 1997. *The Tidewater Tales*. Baltimore: Johns Hopkins University Press.

Begley, S. 1997. The Science Wars. *Newsweek*, April 21, 54–57.

Berrill, N. J., and G. Karp. 1976. *Development*. New York: McGraw-Hill.

Gilbert, Scott, Beldecos, A., et al. The Biology and Gender Study Group. 1989. The Importance of Feminist Critique for Contemporary Cell Biology. In *Feminism and Science*, ed. N. Tuana, Bloomington: Indiana University Press, pp. 172–87.

Gottfried, K., and K. G. Wilson. 1997. Science as a Cultural Construct. *Nature*, 386 (10 April): 645–647.

Gross, P. R. 1967. The Control of Protein Synthesis in Development and Differentiation. In *Current Topics in Developmental Biology*. Vol. 2, ed. A. A. Moscona and A. Monroy. New York: Academic Press, pp. 1–46.

———. 1985. Laying the Ghost: Embryonic Development, in Plain Words. *Biological Bulletin*, 168 (Suppl.): 62–79.

Gross, P. R., and G. H. Cousineau. 1964. Macromolecule Synthesis and the Influence of Actinomycin on Early Development. *Experimental Cell Research* 33: 368–395.

Gross, P. R., and N. Levitt. 1994. *Higher Superstition: The Academic Left and Its Quarrels with Science*. Baltimore: Johns Hopkins University Press.

Just, E. E. 1919. The Fertilization Reaction in *Echinarachnius Parma*. *Biological Bulletin* 36: 1–10.

Keller, E. F., and H. E. Longino. 1996. *Feminism and Science*. Oxford: Oxford University Press.

Kolata, G. B. 1979. Developmental Biology: Where Is It Going? *Science* 206 (19 October): 315–316.

Loeb, J. 1913. *Artificial Parthenogenesis and Fertilization*. Chicago: University of Chicago Press.

Martin, E. 1996. The Egg and the Sperm: How Science Has Constructed a Romance Based on Stereotypical Male–Female Roles. In *Feminism and Science*, ed. E. F. Keller and H. E. Longino. Oxford: Oxford University Press, pp. 103–117.

Pickering, Andrew. 1984. *Constructing Quarks*. Chicago: University of Chicago Press.

Schatten, G., and D. Mazia. 1976. The Penetration of the Spermatozoon through the Sea Urchin Egg Surface at Fertilization. *Experimental Cell Research* 98: 325–337.

Schatten, G., and H. Schatten. 1983. The Energetic Egg. *The Sciences* (Sept./Oct.): 28–34.

Spanier, B. B. 1995. *Im/Partial Science: Gender Ideology in Molecular Biology*. Bloomington: Indiana University Press.

Tey, J. 1977. *The Daughter of Time*. New York: Pocket Books.

Tuana, N. 1989. *Feminism and Science*. Bloomington: Indiana University Press.

———. 1993. *The Less Noble Sex: Scientific, Religious, and Philosophical Conceptions of Woman's Nature*. Bloomington: Indiana University Press.

5

An Engineer Dissects Two Case Studies

Hayles on Fluid Mechanics and MacKenzie on Statistics

Philip A. Sullivan

O ccasionally scientists and engineers have an opportunity to compare predictions with outcomes for an event never before observed, and they usually regard such incidents as vindicating their work. For me, one such opportunity began on April 16, 1970. I was absorbed in the minutiae of a departmental meeting when a secretary interrupted, informing us that a colleague had been called to help with the rescue of *Apollo* 13. The meeting broke up, and my colleague assembled a team of advisers, of which I was a member.

The *Apollo* craft comprised three modules: a service module providing both life support and rocket thrust for most of the voyage, a lunar excursion module (LEM) to land on the moon, and a module for both the voyage and terrestrial reentry. But when an explosion completely disabled the service module, the LEM became a lifeboat, with its life support and rocket thrust—intended only for lunar landing and return to lunar orbit—becoming essential to the rescue. Normally the LEM would have been jettisoned just after completing its mission, by severing the tube connecting it to the reentry module. This tube, which also served as the LEM access tunnel, was to be cut by a ring of explosive located just 4 inches from the reentry module's hatch. To ensure that shock waves from the explosion did not damage the hatch, before detonation the 5 psi oxygen atmosphere in the tunnel would have been evacuated. The service module's rockets would then have been used to back away from the LEM. But because these rockets were inoperative, NASA's engineers proposed using the oxygen pressure as a spring to jettison the LEM just before reentry. A previous incident suggested that retaining the full 5 psi in the tunnel could cause shock damage to the hatch. It was on this point that an engineer at the LEM manufacturer called my colleague for advice.

With a telephone line held open to allow us immediate access to data on spacecraft geometry, masses, and other quantities, we worked in two groups. One used Newton's laws of mechanics to estimate LEM separation speeds attainable with various tunnel pressures. The second group estimated the strength of the pressure pulse generated by the explosive charge. They adapted formulas verified, in the first instance, by comparisons with photographs of the first atomic explosion at Alamogordo, New Mexico.[1] We concluded that a tunnel pressure of 2 psi would provide sufficient separation speed while minimizing the risk of damage to the reentry module. We assumed that other groups were consulted, but we subsequently learned that our advice was the main basis for a decision to lower the tunnel pressure and thus to complete a successful rescue.[2]

Such incidents convince us that science provides an efficient and objective way of obtaining, organizing, and using the knowledge in the disciplines we practice. Typically, the history of space exploration provides numerous examples in which Newton's laws of mechanics and gravitation have been used to accurately predict both the trajectories and timing of missions. In a word, science works! Consequently, when we learn that certain philosophers, sociologists, and other humanists claim that scientific knowledge is not objective or value free or that scientific laws are inventions and not discoveries, our instinctive reaction is to dismiss such views as based on ignorance or even envy.[3] But some of us, increasingly concerned that these views are adversely influencing public attitudes toward science, have decided to scrutinize our critics' work. This essay is therefore one engineer's reaction to two samples of the literature. I conclude that the initial instinctive reaction of scientists is well founded, that the problems these philosophers, sociologists, and humanists have with scientific knowledge reflect problems in those disciplines and not problems in science.

In this chapter, I state my sense of both the scientific enterprise and the major claims made by our critics. Then as a prelude to reviewing two papers, I discuss the process of scientific discovery. In the first paper, Katherine Hayles, a professor of English, attempts a hermeneutic interpretation of my discipline, fluid mechanics, claiming that it is a highly gendered subject in which progress has been hampered by characteristically male prejudices. But Hayles misconstrues every one of the mathematical and physical issues she addresses, making her associated discussion meaningless. In the second paper, British sociologist Donald MacKenzie investigates an early twentieth-century dispute between two statisticians, and he attempts to show that cultural perspectives have entered statistical theory. But seemingly determined to find a social explanation, MacKenzie misrepresents both the mathematical issues in the dispute and subsequent developments. I then suggest a major source of our critics' confusion over science and conclude by commenting on the role of cultural studies of science and on broader implications for scholarship.

On the Nature of Science

Scientists usually work in a framework of unstated assumptions. According to English astronomer John Barrow, these include the following (1988, 24–26):

There is an external world separable from our perception.
The world is rational: "A" and "not A" cannot be simultaneously true.
The world can be analyzed locally; that is, we can examine a process without having to take into account all the events occurring elsewhere.
We can separate events from our perception of them.
There are predictable regularities in nature.

The world can be described by mathematics.

These presuppositions hold everywhere and at all times.

In discussing these assumptions, English biologist Lewis Wolpert observes that they "may not be philosophically acceptable, but they are experimentally testable and they are consistent with the ability of science to describe and explain a very large number of phenomena" (Wolpert 1993, 107). In other words, notwithstanding the ongoing debates among philosophers and theoretical scientists about the implications of certain features of quantum mechanics, the day-to-day practice of scientists is best characterized by the doctrine that philosophers call *scientific realism* (Sparkes 1991, 187).

In contrast, a characteristic feature of twentieth-century Western thought is the growing influence of the doctrine of *relativism*, which asserts that there are no absolutes in morals, values, or knowledge. Experimental confirmation just after World War I of Einstein's theory of relativity in physics played a major role in spreading acceptance of the doctrine among intellectuals (Johnson 1984, esp. 1–11), but it has no philosophical relationship to Einstein's theory. Australian philosopher A. W. Sparkes suggested that those who believe so allow their thinking to be dominated by a "mere pun" (Sparkes 1991, 206).

The application of this doctrine to scholarship has been, at best, problematical and is increasingly controversial. More of a moral position than a coherent theory of knowledge, the main consequence has been its advocates' denial of the possibility of objectivity, even as an ideal. Classicist Mary Lefkowitz described its effect on history, noting that taken to its logical conclusion, relativism reduces history to fiction. Motive becomes more important than evidence, and the credibility of views of an individual on controversial matters is judged by that individual's class, race, or gender (1996, 49–52). That is, a form of ad hominem argument takes precedence over the evaluation of the intrinsic merits of evidence (Sparkes 1991, 105–6).

Relativism seems to be taken as axiomatic by current cultural critics of science. For example, as described by Australian philosopher Alan Chalmers, sociologists distinguish between "weak" and "strong" programs. In weak programs, social factors affect both the choice of topic to be investigated and the pace of investigation, but in strong programs, "social influences can affect the content of good science" (Chalmers 1990, 96). Although nobody having any sense of the history of science denies the weak version, most sociological accounts accept the strong or relativistic version. Philosopher Susan Haack listed the claims made by relativists:

Social values are inseparable from scientific inquiry.

The purpose of science is the achievement of social goals.

Knowledge is nothing but the product of negotiation among the members of the scientific community.

Knowledge, facts and reality are nothing more than social constructions.

Science should be more democratic.

The physical sciences are subordinate to (i.e., are a subdiscipline of) social science.[4]

The critiques I present here illustrate two points. First, advocates of relativism have not provided a single convincing instance of cultural factors entering the theoretical propositions of what Chalmers calls "good science." Second, as claimed by biologist Paul Gross and mathematician Norman Levitt, because most cultural studies of science do not understand the content of the subject they critique, they are severely flawed.[5] Thus they divert attention from serious issues in the practice of science worthy of detached analysis.

The Process of Scientific Discovery

Comprehension of arguments concerning the nature of scientific knowledge requires a clear understanding of the way in which scientists make their discoveries. Consequently, to illustrate the process, I describe a case history: the development of a theory of sound leading to a prediction of its speed of propagation. This story is especially apt because it begins with notoriously self-serving behavior by the greatest of all scientists, Isaac Newton (1642–1727), and because it became a controversial issue for eighteenth- and nineteenth-century physics.

In the first edition of his *Principia* (1687), Newton suggested that sound consists of minute oscillations of air particles along the direction of propagation and that these oscillations were accompanied by waves of compression and expansion. He used physical arguments to relate this behavior to his solution of the pendulum problem, thus in effect investigating a pure tone. In modern notation, with P_a and ρ_a being, respectively, atmospheric pressure and density, he concluded that the speed of propagation $c = \sqrt{(P_a/\rho_a)}$. As implied by the formula, his theory predicted that c would be independent of frequency or pitch and of intensity or loudness. Newton was initially satisfied with the theory because it was consistent with measurements by others—obtained by observing the elapsed time between the flash and report of a cannon—and his own—obtained by synchronizing the return of a wall echo with the beats of a pendulum. But subsequent measurements made his prediction about 20% too low.

Consequently, in the second edition of the *Principia* (1713), Newton described modifications of his theory that brought his prediction into exact agreement with an average of the measurements he considered most accurate. However, these modifications were both logically inconsistent with his original theory and made arbitrary assumptions having no empirical basis whatever.[6]

Biographer Richard Westfall observes that Newton did this because he "had no intention of exposing a 20 percent deficiency to the barbs of continental critics" (1980, 735). But no one was fooled, and in 1727—when Newton was still alive—the prolific Swiss mathematician Leonhard Euler (1707–83) wrote, "Recent [investigators] found that sound consists in a trembling of the air, but about that trembling, their notions were . . . confused. The most acute Newton tried . . . but with little happier event."[7] Westfall comments that "the passage [describing the modifications] is one of the most embarrassing in the whole *Principia*" (735). Furthermore, others were aware of the severe limitations in the physical arguments of the first edition. For instance, in 1759, the French mathematician Joseph Louis Lagrange (1736–1813) wrote that Newton's theory "may have satisfied the physicists . . . but it is not the same with the geometers, who in studying the demonstrations on which it rests have failed to find there the same degree of solidity and conviction which characterizes the rest of [Newton's] works."[8]

In 1759 both Lagrange and Euler published mathematical formulations removing the limitations inherent in Newton's physical arguments, thereby showing that his formula applied to all sounds and not just pure tones.[9] But this mathematical advance did not resolve the discrepancy. It remained one of the "big questions" of science until the 1860s, attracting the best scientific minds of the period, with many theories being proposed and many experiments being performed. As late as 1864, John Le Conte, a professor of natural philosophy at South Carolina College, quoted the following observation in an 1858 fluid mechanics text as evidence that the speed of sound is independent of pitch or frequency: "On a fine and still evening of June 1858, the *Messiah* was performed in a *tent*, and the Hallelujah Chorus was distinctly heard, *without loss of harmony*, at a distance of *two English miles*" (italics in original) (1864, 10). But Le Conte also remarked, "Notwithstanding the acknowledged incompleteness and obscurity which characterizes his physical reasoning on this subject, it is now universally admitted that the true principles of the theory of the propagation of sound were first enunciated (in 1687) by the immortal author of the [*Principia*]" (1).

In 1802, the French mathematician Pierre Simon, Marquis de Laplace (1749–1827), was the first to suggest the source of the discrepancy. Because a theory of heat had not been available to Newton, he in effect assumed that the fluctuations in air density take place at a constant temperature, whereas we now know that fluid particles undergoing such fluctuations also undergo temperature fluctuations. In 1823 Laplace derived the formula now used; it assumes that the fluctuations occur so rapidly that adjacent fluid particles cannot transfer heat from one to another.[10] But notwithstanding Laplace's stature as a mathematician, many rejected his idea as extreme, and final acceptance came only with other theoretical and experimental developments. In 1851 the Anglo-Irish mathematician Sir George Stokes (1819–1903) published an analysis of the effects of heat radiation on sound propagation, leading him to conclude that Laplace's hypothesis was correct. Indeed, this explanation has withstood the test of time, although we now know that at frequencies far higher than audible, it must be modified.

This story illustrates all the characteristics of scientific discovery, and two points are relevant here. First, unless there is political interference, flawed or weak arguments of even the most respected scientists are rapidly exposed. Furthermore, eager for recognition and priority, individual scientists are quick to promote novel explanations. Thus, when scientists finally do agree on the solution to a problem, this agreement is not solely the result of negotiation; it is forged by the evidence.

The second point is that interpretation of this evidence is often complex, being in many ways akin to the assembly of an elaborate jigsaw puzzle. The result of any individual theoretical or experimental investigation is at best ambiguous, and its ultimate meaning is dependent on other investigations. In the initial assembly stages, the available puzzle pieces may suggest many interpretations and may provoke much controversy, so rhetoric and reputation can play major roles in persuasion. As additional investigations are undertaken, however, any consensus on interpretation that develops often has such compelling consistency or can allow such spectacular predictions that it becomes increasingly difficult to deny that it reflects objectivity.[11] Finally, as the sound–speed story shows, for questions at the frontier of science, assembly of the puzzle may take decades and longer.

It follows that to make a convincing case, advocates of relativism in science cannot simply point to the disputes and controversies surrounding a puzzle that is still being assembled. Rather, they must demonstrate that social factors have entered the content of Chalmers's "good science." For the purposes of this discussion, I understand "good science" to be those propositions that are part of an accepted scientific consensus and that have an established record of successful prediction. Both of the two case studies I examine here concern aspects of mature disciplines that, when used appropriately, are capable of making accurate predictions.[12]

Hayles's Analysis of Fluid Mechanics

Katherine Hayles begins her paper "Gender Encoding in Fluid Mechanics" (1992) on a promising note. The only other published commentary, a psychoanalytic interpretation by Luce Irigaray, alleges that a *"complicity of long standing between rationality and a mechanics of solids"* has led to a *"historical lag in elaborating a 'theory' of fluids"* (1985, 106–107; italics in original). Hayles criticizes Irigaray's analysis because it makes no reference to the fluid mechanics literature and contains only vague allusions to the physics and mathematics of the subject. She is sympathetic to scientists such as myself who dismiss Irigaray's analysis as nonsense.

Hayles is also sympathetic to the larger projects of Irigaray and other feminists such as Haraway and Keller who search for gender encodings within the subject matter and methods of various scientific disciplines. Hayles wants to argue that the subject matter of fluid mechanics is laden with unacknowledged masculinist metaphors and values that originate in classical Greek thought. As a consequence, she believes that "if [physical] laws represent particular emphases and not inevitable facts, then people living in different kinds of bodies and identifying with different gender constructions might well have arrived at different models for flow" (31). Furthermore, by conceptualizing science in general and fluid mechanics in particular as masculinist, as other feminists have done, Hayles finds it difficult to imagine women being part of these enterprises.

Unlike Irigaray, Hayles has consulted the literature on the subject and its history, so portions of her essay could easily appear authoritative to the lay reader. But every assertion she makes about the content of fluid mechanics is either wrong or grossly misleading. This basic fact vitiates her subtle and elaborate attempts to reveal gender encodings.

The basic strategy of Hayles's essay is as follows: First she looks for conceptual dichotomies in the science; second, by applying psychoanalysis to the history of these ideas, she attempts to establish that these dichotomies are gendered; and finally, she claims that this gender encoding has impeded the development of the subject. The dichotomies she claims to identify are (1) continuity versus rupture, (2) conservation versus dissipation, (3) linearity versus nonlinearity, and (4) steady flow versus turbulence. The first is supposedly male and the second, female. They are alleged to correspond to male stereotypes of manly rationality, simplicity, and straightforwardness, on the one hand, and feminine hysteria, complexity, and duplicity, on the other. Every step of her argument can be challenged, but I concentrate here on the dichotomies she purports to find within fluid mechanics. If they do not exist, then the question of their engenderings and later influence does not arise.

Hayles describes her first two dichotomies as "assumptions, inherited from the Greeks, that put an enduring spin on future developments. They can be stated as a pair of hierarchical dichotomies in which the first term is privileged at the expense of the second: continuity versus rupture, and conservation versus dissipation" (22).

To describe her errors, I shall summarize the basic principles of fluid mechanics. First, matter is described not in molecular terms but as it is perceived in everyday experience, as a *continuum*.[13] Second, although fluid mechanics addresses the flow of liquids and gases, these ideas are subsumed into a concept called a *fluid*. Third, three basic ideas of physics are used to formulate principles that are commonly called *conservation laws*: (1) conservation of mass, or *continuity*; (2) *conservation of momentum*, or Newton's laws of mechanics; and (3) conservation of energy. Although these ideas are simpler to comprehend than many others in physics, fluid mechanics is a challenge because even though insights into many fluid-flow phenomena can be obtained using simple mathematics, problems exist that require the most sophisticated mathematics known and that remain unsolved to this day. Consequently, a major emphasis has been to devise both physical and mathematical simplifications that make a problem soluble.

Hayles's first two dichotomies refer, respectively, to the first and third of the conservation laws. Consider first her notion of "continuity versus rupture" and conservation of mass. This principle asserts that matter is neither created nor destroyed in a fluid flow and that this simple fact constrains the motion. The idea can be expressed in a variety of ways. One is to recognize that as a given identifiable lump of fluid moves, its shape and volume change continuously, and part or all of it may switch back and forth between liquid and gas phases, but that at every instant its mass remains constant. Another way to describe this idea is identical in principle to the daily fluctuations of a bank account balance: Given a volume fixed in space, in any given interval of time, the increase in fluid mass inside the volume equals the difference between the mass that flows in and flows out of that volume.

Both ideas lead to the same mathematical equation. One physical simplification is called *conservation of volume*. In some circumstances, it is possible to assume that the volume of a fluid lump does not change as it moves; the flow is then said to be *incompressible*. Examples of incompressible flows are the propagation of surface waves on water and the displacement of air by a moving automobile. But to describe problems such as the propagation of sound in both liquids and gases, one must revert to conservation of mass, in which case the flow is said to be *compressible*.

Curiously, even though every introductory text describes conservation of mass as the basic principle, with conservation of volume as a useful simplification,[14] and even though the literature is replete with investigations showing that these ideas work, Hayles gets it the wrong way around: "Conservation of mass, or more accurately, volume" (1992, 26). This mistake causes her to make an error that invalidates her argument: "Important instances exist where continuity and conservation are not applicable, for example cavitation" (31). A liquid flow is said to *cavitate* if the dynamics causes the pressure at one or more positions in the flow to fall below the value at which the liquid can boil; vapor bubbles then form. This frequently occurs on high-performance marine propellers, and analysis of such flows is very difficult. One might think of the local boiling as "rupturing" the liquid flow, but this is not a useful idea.

In cavitating flows, the conservation of volume simplification cannot be used, and one must revert to conservation of mass. In any case, Hayles's alleged dichotomy of continuity versus rupture does not exist.

Consider now conservation of energy and Hayles's second misconceived dichotomy: "conservation versus dissipation." In fluid flows not involving chemical reactions, the energy in a lump of fluid can take three forms: *kinetic* energy of motion, *potential* energy arising from movement through a gravitational field, and *internal energy*, or heat. The conservation principle may be stated as follows: As an identifiable lump of fluid moves, the rate of change of the sum of the kinetic, potential, and internal energies for this lump is equal to the rate at which work is done at the boundaries of the lump by forces exerted by adjacent lumps of fluid and by any solid boundaries, plus the rate at which heat is conducted into the lump across the boundaries. The principle is formulated using *thermodynamics*, the science concerned with the production of work by means of heat. A fundamental principle of this subject, known as the *first law of thermodynamics*, is described by Hayles as "a conservation principle, stating that the energy of a closed system is constant" (30).

Contrast Hayles's version with that given by the German physicist Rudolf Clausius (1822–88), one of the founders of the discipline: "In all cases in which work is produced by the agency of heat, a quantity of heat is consumed which is proportional to the work done; and conversely, by the expenditure of an equal quantity of work an equal quantity of heat is produced" (Truesdell 1980, 188–89). In other words, the first law of thermodynamics makes a statement about the relationship between two types of energy; thus it says something much more specific than implied by Hayles's characterization. A modern statement of this law, given in an introductory text, is "The total energy change of a closed [constant mass] system is equal to the heat transferred *to* the system minus the work done *by* the system [on its surroundings]" (Abbott and Van Ness 1989, 9–10, italics in original).

In a fluid flow, energy is continually interchanged back and forth among the three different types, but it is always subject to the first law of thermodynamics. The term *dissipation* in Hayles's second dichotomy refers to a particular aspect of this interchange process. The forces acting at the boundaries of a fluid lump arise from both pressure and friction. Pressure forces convert work into heat in such a way that all that heat can be converted back to work; that is, the process is reversible. In contrast, frictional forces convert work into heat irreversibly: a fraction of this heat cannot subsequently be made to perform work. This is the idea behind the *second law of thermodynamics*. Thus dissipation occurs *within* the processes in which energy is conserved, and so Hayles's dichotomy does not exist.

Hayles's discussion of conservation of energy suggests that she has confused it with a useful physical simplification. In some flows, friction plays a secondary role; for example, oceanic waves can propagate thousands of miles from a storm center. By omitting friction, one can calculate many of the properties of such flows. If such *frictionless* flows are also incompressible, conservation of energy simplifies considerably: the internal energy of a lump of fluid remains constant and does not interact with the other forms of energy, and there is no dissipation. This simplification is called *conservation of mechanical energy*. Only in this sense might one speak of dissipation as "the opposite of conservation," as Hayles would have it. But as for continuity, such a distinction is not useful. Flows exist in which mechanical energy is not conserved but for which dissipation is negligible; examples are the rising atmospheric air currents created by solar heating and the flow about a moving automobile. The conditions under which these various simplifications can be used are well understood, and every competent text carefully explains them.

In her discussion of the conservation laws, Hayles repeatedly uses the word *privilege*, implying the possibility of choice; that is, males supposedly prefer one element of a dichotomy over the other. In 1897, the Indiana state legislature passed a bill making the legal value of the ratio of the length of a circle to its diameter (π) equal to 4 instead of the irrational number 3.141592653 . . . , its inconvenient traditional value (Casti 1989, 121). If one knows that π is determined by geometry and logic and that this determination agrees precisely with measurement, the bill seems ridiculous. In the same vein, anyone having a sense of both the history of science and its achievements finds the notion that fluid mechanicists might have any choice when setting up conservation laws equally ridiculous.

As an example, consider the discovery of the first law of thermodynamics as it was described by Clausius. This law is mainly the achievement of one man, the English scientist James Joule (1818–89). Well into the nineteenth century, scientists were split between two schools of thought regarding the nature of heat, one holding that it was a substance called *caloric* and the other that it was a form of internal motion in matter. The evidence was inconclusive. Joule, holding the latter view, set out to determine what he called the *mechanical equivalent of heat*.

Over a period of 40 years from 1838 onward, Joule performed a series of experiments that he claimed showed that the amount of heat generated by expending mechanical work was a universal constant independent of the nature of the conversion process. His first experiments used falling weights to drive an electrical generator; the electrical current was dissipated as heat in the rotating part or armature, which also functioned as a calorimeter. He exercised great ingenuity to compensate for such effects as atmospheric friction, obtaining a value of the mechanical equivalent, now denoted by the symbol J in his honor, of 838 foot-pounds of work for each British Thermal Unit of heat; this is within 8% of the modern value of 778.2. But when Joule's first experiments were not being ignored, owing to the very small temperature rises he measured, they were—justifiably—dismissed as subject to large errors. Nevertheless he persisted, devising increasingly elaborate experiments. His last determination used falling weights to drive paddles stirring water and obtained $J = 772.6$, which is within 0.8% of the modern value.[15]

By the time of his death, Joule's idea had become universally accepted, and many increasingly sophisticated determinations have since been performed, so that scientists now acknowledge J to be a universal constant of nature. The modern international system of units (SI) recognizes this by using the same unit for work and heat, and appropriately, this unit is called a *joule*. This is an excellent example of the role of evidence in forging a consensus and in showing that—as Chalmers puts it—"the natural world does not behave one way for capitalists and in another way for socialists, in one way for males and another way for females" (Chalmers 1990, 112).

Hayle's discussion of her third dichotomy—linearity versus nonlinearity—misrepresents elementary mathematical concepts. To solve fluid-flow problems, calculus is used to express the three conservation principles as a set of *differential equations*; these equations specify the relationships among the rates of change of various fluid properties at a given position, the fluid properties at that position, and the spatial gradients of these properties there. The properties typically include density, pressure, and temperature, together with the velocities in each of the three directions in space. In their most general form, these equations are extremely difficult—even impossible—to solve, so that progress has depended on using mathematical simplifications in addition to the physical simplifications of the type just cited. For example, the speed of sound in air can be accurately predicted by making the physical simplification that both friction and heat conduction can be ignored and then by making the mathematical simplification that sound propagation involves only very small fluctuations of atmospheric pressure and density. The simplified equations are then *linear*, and it is in relation to this concept that Hayles commits an egregious error.[16]

The mathematical concept of linearity, described in note 16, has nothing to do with the term *linear thinking* as it seems to be used in the cultural studies literature, following a single line of logic to the exclusion of others (see Gross and Levitt 1994, 104–5). Hayles nonetheless uses the term in this sense. First, she gives a misleadingly restrictive definition of the mathematical sense of nonlinearity: "An equation is said to be nonlinear when products of derivatives, not merely single derivatives, appear" (Hayles 1992, 20).

This definition, however, omits entire classes of nonlinear forms, causing Hayles to make a major error. The equations describing frictionless flow of a fluid are named after Euler, who first derived them in their modern form (Hughes and Brighton 1991, 48). According to Hayles, "Euler's equations [are] linear." Not so, they contain a type of nonlinearity omitted by her definition and discussed in note 16. She then makes the characteristic cultural studies confusion: "Nonlinearities . . . begin appearing when environments are interactive, with multiple components that continuously affect one another through feedback loops. . . . Linear analysis works well with straightforward causal sequences; it does not work well with the recursive feedback loops characteristic of complex flow" (Hayles 1992, 21). Every undergraduate engineering student knows this is wrong. The curriculum invariably includes an introductory course in control system theory dealing with several linear processes occurring simultaneously; such processes can involve elaborate interactions and feedback. The mathematics of this introductory course is based entirely on systems of linear differential equations (see, e.g., DiStefano, Stubberud, and Williams 1967).

Hayles's confusion over the meaning of nonlinearity gets her into trouble in her discussion of her fourth dichotomy, steady flow versus turbulence. The flow from a kitchen sink faucet illustrates the main features of turbulence. At low flow rates, the falling column of water appears steady and glassy and has a simple tapering shape. Furthermore, it strikes the bottom of the sink making little or no noise. But when the flow is increased, at a certain critical rate an abrupt change occurs: the surface of the water column becomes mottled and opaque, and the column itself wavers slightly around a mean shape and position. In addition, the impact with the sink is noisy, and the flow in the faucet emits a distinct hiss. The first kind of flow is called *laminar*. In such flows, the fluid particles move along paths that—if the overall flow is steady— do not move about in time. In contrast, in a *turbulent* flow, individual fluid particles move along extremely complicated paths that move erratically in time. Turbulence is ubiquitous in nature. Examples are wind gusts, sunspots, and jet aircraft noise. When thinking about such flows, it is important to note that certain overall features remain steady: in the kitchen sink experiment, even when turbulent, the flow rate out of the faucet remains steady. Consequently, it is useful to consider turbulent flows as two interacting parts: a steady *mean* flow on which are superimposed fluctuations or *turbulence*.

Discussion of fluid mechanicists' ideas about turbulence is important to Hayles because one eighteenth-century scientist compared the complexities of such flows to stereotypes of the female character, leading her to claim that turbulence is on the female side of a gender dichotomy and steady flow is on the male side (Hayles 1992, 27). Quoting from a loose description of turbulence in a current text, she asserts,

> The "main flow" has "superimposed" on it "random fluctuating components." These swerves from predictability are "haphazard movements" rather than dynamics intrinsic to the environment. What can be modelled is normal. What cannot is an aberration, a chance event, a superfluity. Complex flow, then, is a phenomenon which must but cannot be accounted for within an analytic tradition whose basic assumptions privilege constancy over change, discrete factors over dynamic interaction, linear sequence over nonlinear recursion. (22)

But this account of the prodigious attempts by fluid mechanicists to understand and predict turbulent flows is nothing short of a travesty. To illustrate why, I summarize the main developments.

First, analysis of turbulent flows is universally acknowledged to be one of the most difficult problems in all of science. To wit, an eminent fluid mechanicist is reputed to have declared, "When I meet my Maker I shall have two questions for Him; one on quantum mechanics, and one on turbulence. I shall expect a straight answer on the first." Nevertheless, progress has been made, especially since instruments for recording the complicated details of such flows came into widespread use in the 1950s. In broad-brush terms, the physics is understood. Under certain conditions, laminar flows become unstable, breaking down into eddies comparable in size to the flow itself. These eddies also are unstable, breaking into smaller eddies, which, in turn, are unstable. This process transports kinetic energy from the mean flow to the smallest eddies, which dissipate it through friction. A considerable body of evidence now shows that the smallest eddies are much larger than any characteristic molecular dimension, thus confirming that turbulent flows can be modeled as a continuum. Furthermore, this evidence supports the idea that the differential equations used to predict with great accuracy the details of many laminar flows also completely describe turbulent flow.

Until recently, the accepted approach to analyzing turbulent flows has been based on a procedure known as *Reynolds averaging*. Named after the English engineer Osborne Reynolds (1842–1912), who discovered a simple criterion for predicting the onset of turbulence, the approach capitalizes on the observation that mean flows are often steady. The equations of unsteady laminar flow are averaged over an interval of time that is much greater than the longest time of the turbulent fluctuations. This procedure introduces additional terms known as *Reynolds stresses* into the laminar equations, which account for the effect of the turbulence on the mean flow. Much effort has been expended by fluid mechanicists to describe the physics of the Reynolds stresses and to provide mathematical descriptions of them suitable for prediction. For example, one approach uses an averaging procedure to derive an expression for the kinetic energy of the turbulence in a lump of fluid. The resultant expression contains terms representing the creation of turbulence by the mean flow, its transport and diffusion by the mean flow, and its dissipation by friction. Simple physical arguments are then used to relate the Reynolds stresses to the kinetic energy of the turbulence. For some problems, such as turbulent flow in a pipe or the frictional processes in the airflow over the wings of an aircraft, this procedure is effective. More recently, the availability of powerful digital computers has promoted the development of alternatives to Reynolds averaging. Called *large eddy simulation*, these techniques follow in detail the unsteady motion of the larger eddies, averaging only over the smaller eddies. This is currently a "hot" topic and may ultimately become the approach of choice.

How do these developments square with Hayles's characterizations? When not meaningless, her characterizations are wrong. First, as anyone who has taken a course in fluid mechanics knows, her dichotomy of steady flow versus turbulence is false, as many flows are neither steady nor turbulent. Some unsteady flows, such as the propagation of a system of oceanic waves from a storm, are complex. Second, in no sense do the mathematical models of flows "privilege constancy over change," because by definition, flows are processes that involve changes in time and space. In these processes, conservation laws act like bank balances imposing relationships between the elements participating in the changes. Third, neither I nor any of my colleagues could attach any meaning to Hayles's claim that we privilege "discrete factors over dynamic interaction." In any case, as I just stated, turbulence modeling involves the consideration of interactions between the mean flow and the turbulence. Finally, the phrase "privileging linear sequence over nonlinear recursion" is both garbled and wrong. The term *recursion* refers to a type of mathematical formula unrelated to the notion of nonlinearity discussed here, and the Reynolds stress terms used to account for the effect of turbulence on the mean flow are a direct consequence of nonlinear terms of the type discussed in the notes, terms that Hayles characterizes as linear.

In addition to posing four false dichotomies, Hayles makes numerous other mistakes. Discussion of all these would occupy many pages, so I comment here on two central to her claim that the fluid mechanicists' culture is especially inhospitable to women: "Because hydraulics has never been able to theorize successfully phenomena central to its enquiries, the craft tradition of an apprenticeship to an experienced practitioner remains essential to the field" (37). Hydraulics is a subdiscipline of fluid mechanics dealing with the flow of water in pipes and channels, but the context of Hayles's remarks suggests that she is not aware of the distinction. The two mistakes she makes are misunderstanding of the concept of a particle of matter, or *mass point*, in Newtonian mechanics, and misconstruing the history of the calculus.

Concerning the first mistake, Hayles argues that

> Euler . . . wanted to develop a theory of hydraulics that would be based on the trajectories of particles as they moved through the flow. The simplest way to deal with the particles was to treat them as mathematical points. But, by definition a point has no extension and consequently no mass. When a point moves, nothing in the material world changes. . . . Euler sought to circumvent the difficulty by defining a "fluid particle,"

an infinitesimal body that could be treated mathematically as a point but that would have just enough "smearing" to enable it to possess such physical properties as volume, mass, density, etc. Hydraulicians are not generally as finicky as mathematicians in insisting on rigor; given their subject matter they cannot afford to be. (27)

Every undergraduate physics student knows that Hayles's statement about points is nonsense. Newton's three laws of motion show that the motion of a body under the action of external forces can be calculated exactly by concentrating all the body's mass at a single point in its interior known as the *center of mass* (Spiegel 1967, 167, theorem 7.2). Furthermore, her statement about the level of rigor achievable in formulating the equations of continuum fluid mechanics is wrong; it is comparable to other areas of physics.

Consider now Hayles's discussion of the calculus. This branch of mathematics is concerned with two problems: finding the slope of a curve and finding the area enclosed by a curve. Folklore has it that these problems were solved by Newton and the Prussian polymath Gottfried Leibniz (1646–1716). But calculus developed over a period of 2,500 years beginning with the Greeks. Historian Carl Boyer concludes that Newton's and Leibniz's contributions "differed from the corresponding methods of their [immediate] predecessors . . . more in attitude and generality than in substance and detail" (Boyer 1959, 299). Furthermore, their arguments were unclear, leading to a period of confusion that was not resolved until the nineteenth century. As Boyer put it, "The founders of the calculus had clearly stated the rules of operation which were to be observed. And the astonishing success of these when applied to mathematical and scientific problems . . . led men to overlook somewhat the highly unsatisfactory state of the logic and philosophy of the subject" (224). Basically, Newton's and Leibniz's arguments implied undefined mathematical operations such as the division of zero by zero, which nineteenth-century mathematicians replaced with the concept of the *limit of a sequence* (Courant 1937). This achieved *rigor*: clear and logical arguments that all could understand.

How does Hayles characterize this development? First she quotes Boyer: "Boyer, among others, argues that '[if] the assumptions of mathematics are quite independent of the world of the senses, and if its elements transcend all experience, the subject is at best reduced to bare formal logic and at worst to symbolic tautologies'" (Boyer 1959, 3; Hayles 1992, 25).

Hayles then repeatedly claims that "rigor is equated with the excision of the observer—a fundamental premise in the ideology of objectivity" (1992, 23). But her quotation is taken from Boyer's introduction, in which he describes various philosophical positions on the nature of mathematics. This is Boyer's conclusion: "Mathematics determines what conclusions will logically follow from given premises" or "hypothetical truth" (1959, 308). Moreover, mathematics arises from experience, but "in the final and rigorous formulation and elaboration of [concepts such as calculus] mathematics must necessarily be unprejudiced by any irrelevant elements in the experiences from which they have arisen. . . . The calculus is without doubt the greatest aid we have to the discovery and appreciation of physical truth" (301). Thus rigor in mathematics is associated with excision of "irrelevant elements." But what meaning can be attached to Hayles's assertion that rigor requires "excision of the observer": that mathematical arguments cannot be observed?

Having shown that Hayles's dichotomies do not exist, that she has misunderstood an elementary concept of physics, and that she has misconstrued the history of calculus, it is not necessary for me to discuss her psychoanalytically based attempts to "genderize" the content of fluid mechanics. But to give the reader a sense of her approach, I note one anecdote. Hayles discusses an incident in the career of an eminent twentieth-century hydraulicist, Hunter Rouse. He celebrated his successful defense of his Ph.D. dissertation at Karlsruhe, Germany, by having a swimming party just before his marriage. He ruptured his eardrum in a dive, leading Hayles to comment that this "suggest[s] deflowering by intercourse with water, the subject of his professional research" (1992, 39). What does this mean? That, soon to be married, his unconscious mind prompted him to celebrate his two achievements by a symbolic but physically destructive act?

Hayles's analysis of both the content and practice of fluid mechanics is consistent with the relentlessly relativist claims of other feminists. My overall assessment of such work is succinctly expressed by philosopher Margarita Levin:

> Despite the enormous scope of feminist claims,[17] the actual evidence offered for the androcentric distortion of science is extremely thin. . . . Of a piece with feminist unwillingness to confront the success of science is a glaring failure to give any concrete account of what a feminist science would be like. Attempts to address this question have so far been evasive. . . . [Feminist] promissory notes address only the more morally elevated attitudes that feminist scientists would, presumably, have and the applications of scientific results they would make. . . . One still wants to know whether feminists' airplanes would stay airborne for feminist engineers. (1988, 104–5)

MacKenzie's Case Study in Statistics

Donald MacKenzie's paper "Statistical Theory and Social Interests: A Case Study" investigates a dispute that developed at the end of the nineteenth century out of attempts to evaluate the effectiveness of vaccines in reducing mortality from diseases such as cholera and smallpox. The protagonists were two British statisticians: Karl Pearson (1857–1936) and his sometime student George Udny Yule (1871–1951). Yule proposed a simple procedure for interpreting vaccination data as they were then collected, whereas Pearson argued that this procedure had serious mathematical flaws. Both men were highly respected. Pearson was a prolific and creative mathematician who contributed to the strength of materials before turning his attention to statistics and devising techniques that are now in every text (Timoshenko 1983). In 1911, Yule wrote a popular introductory text that, by 1950, had gone through fourteen editions and was still being reprinted in 1965. Chalmers states that MacKenzie's study "is frequently cited as an exemplary one of its kind" but denies that MacKenzie offers evidence supporting the strong sociological programme.[18] Chalmers does not, however, discuss the mathematical issues, so I examine them here.

MacKenzie gives an account of these issues as a prelude to discussing the sociological aspects. Noting that the two men never completely resolved their differences and asserting that "logic and mathematical demonstration alone were insufficient to decide between the two positions" (1978, 52), MacKenzie bases his case on two claims.

First, those who worked with Yule favored his approach, whereas those who worked with Pearson favored his; second, statisticians have not subsequently developed a consensus. MacKenzie's examination of the social backgrounds of the persons involved leads him to suggest that Pearson's approach was an extension of attitudes that might be expected from an "upwardly mobile" professional class, whereas Yule's was representative of "downwardly mobile" descendants of landed gentry. But MacKenzie misrepresents Pearson's views on the central technical point and then incorrectly describes both subsequent developments in the literature and the approach taken in current statistics texts.

At first blush, the dispute does appear to support relativist claims, for two reasons. First, those people using statistical techniques frequently do not understand the difficult mathematics necessary to comprehend fully the meaning and limitations of these techniques. Consequently, as one recent text explains, certain procedures have been "consecrated . . . by popular use" (Press et al. 1989, 489). The second reason is the marked contrast in the protagonists' social outlook. Whereas Pearson was an activist for causes deemed progressive in the period between 1880 and 1920, Yule was a much more conventional character. Pearson was a feminist, founding in 1885 a club to discuss what was then called "the woman question," and he became a leader of a cause that was as politically correct then as it is anathema today: eugenics (Kevles 1985). He was the first holder of the eugenics chair endowed at the University College, London, by Francis Galton (1822–1911). Galton, a cousin of Charles Darwin, was the acknowledged founder of the eugenics movement. In contrast, Yule opposed both eugenics and giving women the vote (MacKenzie 1978, 58).

Statistical data are often presented in *contingency tables*. An example is a tabulation of the numbers of votes in an election by party and by electoral district. Vaccination data also are conveniently summarized in a 2 × 2 contingency table:

CATEGORIES	Recoveries	Deaths	TOTALS
Vaccinated	a	b	$N_V = a+b$
Unvaccinated	c	d	$N_U = c+d$
TOTALS	$N_R = a+c$	$N_D = b+d$	N

In this table, a is the total number of persons who were vaccinated and who recovered, and N_U is the number of persons who were not vaccinated; the meaning of the other symbols follow from their categories. At first glance, this table seems easy to interpret: if the proportion $p_V = a/N_V = a/(a+b)$ of vaccinated survivors is greater than the proportion $p_U = c/N_U = c/(c+d)$ of unvaccinated survivors, then one might infer that the vaccine is effective. Also, the difference $p_V - p_U$ is an indicator, or *measure*, of the vaccine's effectiveness.

Such an approach raises many questions, however. Essential to understanding both these questions and the dispute between Pearson and Yule is a distinction that statisticians make among nominal, ordinal, and continuous variables. A *nominal* variable has discrete attributes and no natural ranking; an example is the number of votes cast for each candidate in an election. An *ordinal* variable is discrete but has a natural ordering; examples are position in a race or grade in an examination. A *continuous* variable is a quantity having numerical values that are additive; examples are time, distance, and temperature (Press et al. 1989, 476).[19] Statisticians have devised techniques appropriate to each type of variable. Note that the classification is hierarchical: a continuous variable is also ordinal, and an ordinal variable is nominal. The dispute developed mainly because Yule regarded vaccination data as obviously nominal, whereas Pearson initially analyzed these data by adapting methods for continuous variables.

The currently accepted approach to continuous variables was developed by Galton in 1885 and by Pearson in 1886.[20] To illustrate the method, consider the heights h and weights w of a sample of N adult humans. In some average sense, taller persons tend to be heavier. We can use a 2×2 table to quantify this observation. Choose a value of h dividing persons into "tall" and "short" categories, and a value of w dividing them into "light" and "heavy"; putting the numbers into the four cells of the table shows any trend. But there is a more precise way; first plot weights against heights on a graph. With the persons numbered $i = 1, 2 \ldots N$, any one's weight w_i and height h_i appear as a point. Since at any given height, individual weights can vary considerably, the collection of N points appears as a cloud known as a *scatter diagram*.

We can express any discernible trend by fitting a curve through the points. Frequently the trend and scatter are such that using only the simplest mathematical relationship, the equation for a straight line $w = Ah + B$ is appropriate. We choose A and B using a technique known as *the method of least squares* to minimize the total scatter of the points from the trend line. The extent of the scatter from a straight line is measured by the *linear correlation coefficient, r*, which is determined in the process of choosing A and B. The minimum and maximum possible values of r are -1, and $+1$, respectively, with no discernible trend corresponding to $r = 0$. As r increases from 0, there is an increasing tendency for w to increase with h, with all the points falling precisely on the line $w = Ah + B$ when $r = 1$ and A is positive. Correspondingly, as r decreases from 0, there is an increasing tendency for w to decrease with increasing h, with $r = -1$ occurring when all the points fall on the line; in this case A is negative. To calculate r, Pearson developed the *product–moment* formula given in every textbook.[21]

In 1900 Yule addressed the problem of developing a criterion equivalent to r for nominal variables and 2×2 tables: he used the term *association* to distinguish the problem from the analysis of the continuous case. He proposed the measure

$$Q = \frac{[ad - bc]}{[ad + bc]}$$

because it has three properties:

1. If vaccination does not affect the proportion of survivors, then $Q = 0$.
2. If all those vaccinated survive ($b = 0$) or if all those unvaccinated die ($c = 0$), then $Q = 1$.
3. If all those vaccinated die ($a = 0$) or if all those unvaccinated survive ($c = 0$), then $Q = -1$.

In regard to property (1), if the number of survivors is unaffected by vaccination, then one expects $a/N_V = c/N_U$ or, equivalently, $a/b = c/d$; it follows that $ad = bc$ and $Q = 0$. Aware that continuous variables are also nominal, Yule discussed the relationship between Q and r, concluding that it required investigation because the subject "bristles with difficulties and possibilities of fallacy" (1900, 278). One such follows from the preceding expression: Q does not distinguish between survival caused solely by vaccination and survival caused partly by other factors. This follows because if survival depends solely on vaccination, one expects—in contrast to property (2)—both b and c to be 0, again giving $Q = 1$. An equivalent ambiguity occurs at $Q = -1$.

A few months after Yule proposed Q, Pearson showed that if data in a 2×2 table are derived from continuous variables, by making an assumption about the statistical nature of the data, one could estimate r from the four numbers a, b, c, and d without actually identifying the variables.[22] Pearson called this estimate of r the *tetrachoric correlation*, r_T. No explicit expression is available to calculate r_T, so it must be determined by solving a complicated equation. Pearson argued that this procedure could be used to rank the effectiveness of vaccines because there were underlying continuous variables. For example, these might be "strength to resist small-pox when incurred" and "degree of effectiveness of vaccination" such as might be indicated by elapsed time since vaccination. Furthermore, death or recovery, Pearson averred, is simply a dichotomous classification of the severity of a patient's illness. He also presented examples suggesting that Q was not a good estimate of r.

In the same paper, Pearson proposed a second estimate of the correlation of continuous data in a 2×2 table, and he denoted this by the symbol r_{hk} to distinguish it from r. Unlike r_T, it is given by an explicit formula:[23]

$$r_{hk} = \frac{[ad - bc]}{\sqrt{(a + b)(b + d)(d + c)(c + a)}}$$

Note that like Q, $r_{hk} = 0$ if $a/b = c/d$. But unlike Q, $r_{hk} = 1$ only if both $b = c = 0$, and $r_{hk} = -1$ only if $a = d = 0$. This removes the ambiguity in Q noted earlier. Moreover, owing to a theoretical development Pearson made between 1900 and 1904, r_{hk} has since become the method of choice for measuring association in 2×2 tables.

In the context of contingency tables, Pearson's theoretical development arose from two questions. First, given that survival can depend on many factors other than vaccination, what is the probability that the difference $p_V - p_U$ is caused by all these unidentified factors and not by vaccination? Second, given that such data are usually a sample of a much larger population, what is the probability that the difference is due to unrepresentative sampling? To answer both these questions, Pearson investigated the probability properties of a quantity known as *chi-square* (χ^2). For a 2×2 table, with $p_a = a/N$, $p_b = b/N$, $p_c = c/N$, and $p_d = d/N$ being the proportions in the four categories,[24]

$$\chi^2 = \frac{N(ad - bc)^2}{(a + b)(b + d)(d + c)(c + a)} = \frac{N(p_a p_d - p_b p_c)^2}{(p_a + p_b)(p_b + p_d)(p_d + p_c)(p_c + p_a)}$$

The first of these two expressions shows that $\chi^2 = 0$ if $ad = bc$, so as for Q and r_{hk}, $\chi^2 = 0$ if the number of survivors is unaffected by vaccination. Also, if all the vaccinated survive and all the unvaccinated die, then $b = c = 0$ and $\chi^2 = N$. The second expression shows that for fixed-category proportions p_a, p_b, p_c, and p_d, χ^2 is proportional to N, reflecting the greater probability that larger samples accurately represent the proportions in the population. It also shows that for fixed N, increasing χ^2 corresponds to increasing p_a and p_d and thus to the increasing probability that survival is caused by vaccination.

Thus, increasing χ^2 corresponds to the decreasing probability that the difference $p_V - p_U$ is the result of unidentified factors or sampling errors. If this probability is less than a certain value, say 1 in 20 or 1 in 100, then we can accept that $p_V - p_U$ is caused by vaccination. The results are then said to be *statistically significant*. Using certain assumptions, Pearson calculated χ^2 as a function of the probability of its occurrence. His approach has now become a standard test for statistical significance, and tables of his calculations are given in every text (see Spiegel 1988, 245–61, 489).[25]

In 1904 Pearson proposed a measure of association based on χ^2 and denoted by the symbol ϕ; for a 2 × 2 table:

$$\phi = \sqrt{\frac{\chi^2}{N}}$$

The first expression for χ^2 given earlier shows that $\phi = r_{hk}$. The theoretical development leading to ϕ is independent of variable classification; it can be nominal, ordinal, or continuous. Thus Pearson removed the restriction on its use implicit in his original derivation as r_{hk}. For this and a second reason, ϕ is now known as the *correlation of ranks or attributes* (see Spiegel 1988, 248).[26]

The second reason developed out of Yule's work. In the 1911 edition of his text, he proposed a measure of association for 2 × 2 tables he called the *product–sum correlation* r_{PS} (see Heron 1911, Yule 1912). His argument, which involved an adaptation of the product–moment formula for calculating r, was subsequently generalized by the British psychologist Charles Spearman to the correlation of ordinal variables. Spearman's generalization has now become the standard technique (Press et al. 1989, 489). The argument is essentially as follows: The coefficient r extracted from a scatter diagram is independent of both the origin and magnitude of the scales used to describe the variables; in other words, it depends only on the geometrical proportions of the positions of the points in the scatter diagram.[27] This suggests the following approach: Consider, to be specific, the numerical grades of a class of students for examinations in two different subjects. To ascertain the extent to which performance in one subject is a good predictor of performance in the other, the grades are used to numerically rank the students in each subject, and the product–moment formula is then applied to the rank numbers (see, e.g., Nie et al. 1975, 288; Press et al. 1989, 489). This is known as the *Spearman rank-order correlation coefficient* r_S; its meaning is that for the continuous variable case. Thus $r_S = 1$ if the ranking is identical for the two subjects and $r_S = 0$ if there is no discernible relationship. For a dichotomy as in a 2 × 2 table, one has just two ranks for the two variables, and r_S reduces to Yule's product–sum correlation r_{PS}.

But a few months after Yule published his text, David Heron, an associate of Pearson, pointed out that in describing r_{PS}, Yule had blundered. Apparently because he used a symbolic logic notation rather than the customary algebraic notation, he did not recognize that r_{PS} is equal to r_{hk} and thus to ϕ. This led Yule to describe r_{PS} as "the theoretical value of the correlation coefficient,"[28] thus confusing it with r; but Pearson had shown in 1900 that r_{PS} is not equal to r. Heron also used examples to show that for 2 × 2 tables derived from continuous data having a given fixed r, both Q and r_{PS}

varied strongly as the criteria dividing the data into dichotomies changed. In other words, they are unstable, making them a poor estimate of r.[29]

In a 1912 paper, Yule described a fourth property of Q: it is independent of the proportion vaccinated or the severity of the epidemic.[30] He regarded this property as being especially valuable because it allowed comparison of data from epidemics of different severity and from different jurisdictions in which the authorities had varying degrees of success in enforcing vaccination. He replied to Heron's observations on the instability of Q when applied to continuous data by showing that the tetrachoric estimate of r, r_T, was also unstable when applied to census data on the ages of husband and wife divided into various "old" and "young" categories.[31] Yule also reiterated in sarcastic terms his belief that vaccination data were nominal and that methods based on correlation concepts were inapplicable. Finally, he argued that Q, r_{PS}, and r_T are measures of association that have different properties and are designed for different purposes, so that they should not be directly compared.[32]

In 1913, motivated partly by Yule's sarcastic remarks, Pearson and Heron published a lengthy and strongly worded reply that describes two serious conceptual problems for Q. First, Yule had acknowledged that his choice of Q was "empirical" (Yule 1912, 585). Pearson and Heron show that, as a direct consequence, there is an infinite number of functions satisfying Yule's four properties; selective use of these gives conflicting estimates of association. Since Yule had not provided any rationale in terms of the laws of probability for the particular functional form he chose as Q, Pearson and Heron asserted that Q is "arbitrary" (1913, 177). In mathematical terms, Yule had not solved the problem he had posed. The second conceptual problem is a consequence of the fourth property of Q. Pearson and Heron show that by using this property to adjust the entries in a table, one can generate a "series of tables in which an ascending order of Q is accompanied by a descending order of ϕ" (175). That is, the two coefficients will flatly contradict each other in their ranking of a set of tables. They assert, "Q and ϕ cannot both be valid" (175).[33]

In 1915 Yule and a physician colleague published an assessment of the value to epidemiology of r_T, $\phi = r_{PS}$, Q, and a fourth measure of association, his *coefficient of colligation* ω.[34] Reviewing both the theoretical difficulties described by Pearson and Heron and biological difficulties brought to his attention by epidemiologists, Yule concluded that all four had serious problems of interpretation. The paper expresses the hope that since two of the coefficients were defined by Yule himself, its "condemnation of [all] these coefficients" (Greenwood and Yule 1915, 189) would be seen as impartial. This paper ended the public phase of the Pearson–Yule dispute.

I make four comments on MacKenzie's critique of these issues. First, apparently not recognizing that r_{PS} is equal to ϕ, MacKenzie claims that Pearson rejected r_{PS} (1978, 63). But this is not correct. Pearson and Heron asserted, "Is [r_{PS}], that is, ϕ, applicable to every [2 × 2 table]? If so, why does not Mr. Yule use it and drop his [Q]?" (1913, 167). MacKenzie also claims that Pearson had a "general cosmological bent towards continuity and variation rather than homogeneity and discrete entities" (1978, 63), hinting that he made elementary variable classification errors. On the contrary, Pearson and Heron made it abundantly clear that they were aware of the classifications I just cited: they discussed in precise terms the conditions under which vaccination data can be associated with continuous variables, including presenting scatter diagrams as supporting evidence. MacKenzie also suggests that Pearson's "affection" (1978, 42) for r_T led him to use it uncritically; contrast this assertion with Pearson and Heron's own remarks: "If the problem [is] . . . the evaluation of a new treatment, we should certainly not use [Q]. We should probably today not use a tetrachoric r_T, except as a control. We should most likely use ϕ" (1913, 176).

My second comment is that MacKenzie incorrectly portrays subsequent developments:

> The general approach of modern statisticians is . . . closer to that of Yule than that of Pearson. Yule's Q remains a popular coefficient, especially among sociologists. Pearson's tetrachoric coefficient . . . has almost disappeared from use except in psychometric work. It is interesting to speculate whether this situation can be explained in terms of, on the one hand, the sharing by most statisticians of Yule's lack of an overall, specific goal-orientation and, on the other, the continuing influence of hereditarianism on the other. (1978, 65)

But the two sources MacKenzie cites as the basis of these remarks do not support his assertions.

The first, written by James Davis in 1971, is a monograph on Q written for persons with little mathematical training; both its limited contents and its colloquial style make it unsuitable as a statistics text. Nevertheless Davis's assessment of Q is revealing: "For 40 years Q did not appear to have any intrinsic meaning, but was merely a useful gadget. However in 1954 Leo Goodman and William Kruskal published an important paper which not only made life meaningful for Q but also led to a considerable re-thinking about measures of association in general" (1971, 44–45). Goodman and Kruskal's analysis, which is MacKenzie's second source, notes that although measures of association based on χ^2 or r can be used to rank-order sets of data, the numerical value of these quantities has no clear meaning in terms of probabilities; this is a frequently voiced concern (Press et al. 1989, 479). They propose a measure for ordinal variables to rectify this deficiency. Given the ordering of the ranks of two cases on one variable, it is the probability of correctly guessing the ordering of the ranking on the other variable. When applied to 2×2 tables, this procedure gives Q, thus finally providing a theoretical basis for Yule's proposal. However, Goodman and Kruskal also observe, "In those cases in which [contingency tables] arise jointly from a multivariate normal distribution on an underlying continuum, one would naturally turn to measures of association based on correlation coefficients. These in turn might well be estimated from a sample by the tetrachoric correlation coefficient method or a generalization of it" (1954, 736). In the 1950 edition of his text, Yule supports this assessment.

My third comment is that MacKenzie's portrayal does not adequately reflect the picture in current texts. No one that I consulted recommended Yule's Q or Pearson's tetrachoric r_T as a measure association in 2×2 tables; they all recommended Pearson's ϕ. This includes an outline text notable for its eclectic choice of examples from disciplines as diverse as psychology, agriculture, and production engineering (Spiegel 1988) and a widely used text for the social sciences.[35] Thus, contrary to MacKenzie's assertion, a consensus on interpreting such tables has emerged. Moreover, the choice of measure of association is related to assumptions used to formulate the mathematics. These transcend cultural perspectives and have nothing to do with MacKenzie's goal orientations.

My fourth comment is my initial reaction to MacKenzie's paper. Given the obvious mathematical difficulties that Pearson and Heron identified for Yule's Q, why does MacKenzie try to make a sociological case out of it? In my judgment, the difficulties were of a type that, unless rebutted in detail, would be sufficient to warrant regarding Q with suspicion. In this respect, I noted earlier Yule's own comments, and according to MacKenzie's own source—Davis—this is what ensued.

Finally, the development of statistics provides one of the best demonstrations of the value-free nature of applied mathematics. Even as the school Pearson founded was developing techniques in pursuit of hereditarian goals, others were learning the school's techniques for totally unrelated purposes. For example, W. E. Deming, the engineer who pioneered the use of statistics to radically improve manufacturing quality, studied for a year in the mid-1930s under the tutelage of J. R. Fisher, Pearson's successor in the Galton eugenics chair (Gabor 1992). And no matter what their intended field of application, every current statistics text mentions Pearson when describing his contributions; his eugenics motivation is rarely, if ever, noted.

I conclude that MacKenzie's stance is a good example of a disturbing trend in the sociology of science noted by Canadian political scientist J. W. Grove: "The sociologists' assumption that, once the [knowledge] production process has been laid bare, there is nothing more to be understood is false. . . . The sociology of science . . . is only half the story, and totally misleading in the absence of the other half" (1989, 22–3).[36]

Discussion

Both case studies illustrate the importance of understanding the mathematical constraints on using scientific theories to solve problems and, in particular, the importance of simplifying assumptions and mathematical approximations in the detailed articulation and application of these theories. Hayles and MacKenzie both made the same complex of mistakes: They did not sufficiently analyze the details of the mathematical arguments that all scientists—regardless of their position on particular scientific questions—would agree are of central importance. Having failed to understand what was really at issue, Hayles and MacKenzie cast about for social explanations.

More generally, given the pattern of scientific discovery I described earlier, one can easily assess cultural critiques of the sciences. First, relativists have completely failed to provide any convincing evidence of social values entering the theoretical propositions of good science. Second, when relativist critiques of scientific knowledge are not incompetent or when they are not evading the issue,[37] their case histories serve only to illustrate the role of three extraneous factors:

1. Social pressures influence the choice of puzzles and the progress made in investigating them.
2. The hypothesis validation process is in its formative stages; that is, the puzzle is still being assembled.
3. Scientists or engineers are using good science to solve problems with perceived social value that are so complex that they must be simplified in some way to obtain a solution, thus allowing the reintroduction of social factors.

The frequent failure of cultural studies of science to make these distinctions, together with the failure to distinguish between science and technology, renders sterile accounts of scientific activity that are of intrinsic scholarly value and, in many cases, of direct public interest.

Conclusion

As both knowledge and practice, science is now intrinsic to our cultural ethos; indeed, our daily lives become increasingly dependent on it. Moreover, if recent predictions about the likelihood of an asteroid hitting the Earth in the not-too-distant future are correct, science may yet be called on to save humankind from catastrophe.[38] The cultural studies preoccupation with relativist accounts deflects attention from the elucidation of important issues arising from our dependence on science. These include the distinction between science and pseudoscience and between science and technology, the role of the state in promoting science and science education, and—possibly the most important—the misuse of science by scientists themselves and by special-interest groups.

Finally, I discern a disturbing implication of relativist accounts of the sciences. A generation ago, the British sociologist Stanislav Andreski depicted the social sciences as sorcery, as gibberish designed to placate special-interest groups (Andreski 1972).[39] More recently, Alan Bloom compared the humanities to the old Paris flea market: Among masses of rubbish, one can, by diligent searching, find the occasional undervalued intellectual nugget (Bloom 1987, 371). One's first reaction is to dismiss the authors of these remarks as crabbed reactionaries. But my encounter with cultural studies of science leads me to conclude that such views must be taken seriously. I agree with Gross and Levitt's conclusion that natural scientists can no longer afford to be indifferent to the scholarly practices in the social sciences and the humanities, especially when their product involves a misrepresentation of the nature of scientific knowledge.

Notes

The author expresses his gratitude for the insightful and constructive comments of many colleagues and in particular for those of the editor of this volume, Noretta Koertge, without whose suggestions this chapter would have been especially opaque to readers not familiar with the relevant technical issues.

1. In 1950, English fluid mechanicist G. I. Taylor published his wartime analysis of the propagation of the blast wave generated by an explosion involving a specified energy release. He also used a sequence of photographs published in a U.S. magazine to verify the propagation law he derived and to estimate the energy release of the Alamogordo explosion. This estimate was within 15% of the value determined by the U.S. Army using measurements on the ground surrounding the explosion. Although this value had been announced by President Harry Truman, the method of determination was then a classified secret, and so according to his biographer, George Batchelor, Taylor's publication "caused quite a stir" (1996, 202–7).

2. Private communication from R. A. Oman, Grumman Aerospace Corporation, Bethpage, New York, to J. B. French, University of Toronto, May 4, 1970.

3. According to Wolpert (1993, 122), "It is not . . . surprising that, as Howard Newby, chairman of [Britain's] Economic and Social Research Council, put it, because of their 'massive inferiority complex' social scientists have 'descended with glee on those who have successfully demystified the official credo of science and who have sought to demonstrate that science is but *one* means of creating knowledge.' For them it becomes quite unnecessary to . . . try to emulate traditional science."

4. See Rita Zurcher, "Farewell to Reason: A Tale of Two Conferences," *Academic Questions* 9(2) (1996): 52–60. Zurcher quotes Haack on p. 59. While acknowledging that science is a social activity, in that "co-operative and competitive engagement of many people, within and across generations, in the enterprise of scientific enquiry, contributes to its success," Haack denies every one of these claims.

5. Gross and Levitt's discussion of the ignorance issue is on pp. 88–92; apparently some writers actually boast of their ignorance.

6. Furthermore, in both editions, he expressed his result in a form that confused investigators for the next 150 years. With g being the acceleration due to gravity and $p = \rho g H$ being the pressure at the bottom of a column of fluid of height H and density ρ, he stated (in words) $c = \sqrt{(gH_a)}$, where H_a is the height of a hypothetical constant density atmosphere. This form gives a spurious dependence on g. See Le Conte (1864) for a history of investigations of this problem.

7. See Truesdell 1954, xxv. Truesdell remarks, "For Euler while yet a boy to dismiss in a single line the theory of the great Newton, then still living, was typical of Euler's fearless rectitude." Euler later dismissed one of Newton's adjustments as a "subterfuge."

8. Truesdell 1954, xxxv. The eighteenth-century term for a mathematician was "geometer."

9. They showed that if the fluctuations in the pressure p and the density ρ are related through an equation of the form $p = f(\rho)$, then $c = \sqrt{(df/d\rho)}$. Newton assumed Boyle's law, for which $p/\rho = P_a/\rho_a$. Thus $f = (\rho P_a)/\rho_a$, and $df/d\rho = P_a/\rho_a$, giving Newton's formula.

10. Truesdell 1980, 29–32. Laplace, invoking a now-discredited theory of heat, nevertheless derived the currently accepted formula. If the compression–expansion process involves no heat exchange between adjacent fluid particles, then $p/(\rho^\gamma) = P_a/(\rho_a{}^\gamma)$, where the exponent γ is the ratio of specific heats at constant pressure and constant volume. Then $c = \sqrt{(\gamma P_a/\rho_a)}$; $\gamma = 1.40$ for air. Relativists can take no comfort from Laplace's use of a discredited theory of heat, because Truesdell shows that the mathematical argument is independent of any speculations about the nature of heat.

11. One does not have to look to such bizarre predictions as simultaneous interaction at a distance in quantum mechanics or black holes in astrophysics to find examples defying the prevailing common sense that have subsequently been verified. Sound waves involve the propagation of very small fluctuations in density. The German mathematician Georg Riemann (1826–66) investigated the large fluctuation case, concluding that he could not obtain solutions without postulating the existence of surfaces in the flow across which density, pressure, and temperature jump discontinuously. Initially, these jumps were dismissed as artifacts of his mathematics, but we now know that they have counterparts in physics known as *shock waves*.

12. Although most wish otherwise, statistics has little to say about outcome of single events; its predictions can be accurate when large numbers of events are involved.

13. This is an excellent approximation for all features of fluid flow taking place on scales that are large compared with the molecular scale (which is roughly ten-billionths of a centimeter).

14. See, for example, Hughes and Brighton 1991. This text is a succinct summary of current ideas. Conservation of mass, with conservation of volume as a simplification, is described on pp. 35–36.

15. Joule's idea gained considerable credibility when his measurements were used to determine the ratio of specific heats γ, finally achieving precise agreement between prediction and measurements of the speed of sound. Previous determinations of γ were subject to large error. This a good example of the interlocking or jigsaw puzzle nature of scientific evidence, since it lent credibility to both the first law of thermodynamics and Laplace's explanation of the sound speed problem. See LeConte 1864.

16. The concept of a linear differential equation is accessible to all who have had a first course in calculus. Consider a problem in every first-year physics course: A mass m is supported by a spring having stiffness k. The mass is given an initial displacement and released; the subsequent motion is a decaying oscillation. With the history of this motion described by the displacement x as a function of time t, written symbolically as $x = x(t)$, Newton's laws of mechanics, together with assumptions about the nature of frictional forces, leads to the following differential equation:

$$m \frac{d^2x}{dt^2} + c \frac{dx}{dt} + kx = 0$$

Any function $x(t)$ that satisfies this expression is called a *solution* of this equation. Typically, with suitable choices of the constants A, B, α and ω, the function

$$x = e^{-\alpha t} [A\sin(\omega t) + B\cos(\omega t)]$$

is one such solution. Now suppose we find two such solutions, which we label $x_1(t)$ and $x_2(t)$, respectively. Define a new variable $y(t) = x_1(t) + x_2(t)$. Then the rules of calculus show that $y(t)$ is also a solution, that is,

$$m \frac{d^2y}{dt^2} + c \frac{dy}{dt} + ky = 0$$

Similarly, if we define $z(t) = Cx(t)$, where C is any constant, then $z(t)$ is also a solution. Finally, if D is another constant, these two rules show that the *linear* combination $w(t) = Cx_1(t) + Dx_2(t)$ is a also solution; this may be verified by direct substitution.

Now replace the frictional term $c(dx/dt)$ with $cx(dx/dt)$, giving the following equation:

$$m \frac{d^2x}{dt^2} + cx \frac{dx}{dt} + kx = 0$$

Supposing again that we find two solutions $x_1(t)$ and $x_2(t)$; substitution of $y(t)$ into this equation shows that

$$m \frac{d^2y}{dt^2} + cy \frac{dy}{dt} + ky = c \left(x_1 \frac{dx_2}{dt} + x_2 \frac{dx_1}{dt} \right),$$

that is, $y(t)$ is no longer a solution of the differential equation for x; the same is true of $z(t)$ and $w(t)$.

Differential equations for which a linear combination of two solutions is also a solution are said to be *linear*; all other differential equations are *nonlinear*. Thus the first differential equation is linear, and the second has a particular type of nonlinearity; it is of the type that occurs in Euler's equations of frictionless flow. But Hayles's definition excludes it.

17. For example, prominent feminist science critic Sandra Harding asserts, "The [feminist] critiques appear skeptical that we can locate anything morally and politically worth redeeming or reforming in the scientific worldview, its underlying epistemology, or the practises these legitimate." See Harding 1986, 29.

18. Chalmers (1990, 104) concludes that MacKenzie "fails to provide an adequate answer to [the] question of how the theoretical propositions themselves embody social factors."

19. Social scientists sometimes classify continuous variables into *interval* and *ratio* variables, but this does not have any statistical significance.

20. See pp. 160–66 of Pearson and Heron 1913 for a history before the dispute.

21. See, for example, Spiegel 1988, 294–323. To calculate r, first calculate the mean or average height h_{av} and the mean weight w_{av}. With a total of N pairs of measurements h_1 and w_1 for the first individual, h_2 and w_2 for the second, and so on, the averages are

$$h_{av} = \frac{1}{N}\{h_1 + h_2 + \ldots + h_N\} \; ; \; w_{av} = \frac{1}{N}\{w_1 + w_2 + \ldots + w_N\}$$

Using the symbol Σ to denote the sum of all such quantities for the N individuals, we can write

$$h_{av} = \frac{1}{N}\Sigma\, h_i \; ; \; w_{av}\, \frac{1}{N} = \Sigma\, w_i$$

Next express the heights and weights in terms of their deviations from the means

$$x_i = h_i - h_{av} \text{ and } y_i = w_i - w_{av}$$

for each of $i = 1$ to N, and then calculate the *standard deviations*, s_h for the h_i, s_w for the w_i, and the *covariance* s_{hw} given by

$$S_h = \sqrt{\frac{\Sigma x_i^2}{N}}; \; S_w = \sqrt{\frac{\Sigma y_i^2}{N}}; \; S_{hw} = \frac{\Sigma(x_i y_i)}{N}$$

Then Pearson's product–moment formula for r is

$$r = \frac{S_{hw}}{S_h S_w} = \frac{\Sigma\,(x_i y_i)}{\sqrt{(\Sigma x_i^2)(\Sigma y_i^2)}}$$

22. Pearson (1900) assumed that the underlying probability density function is bivariate Gaussian; that is, each variable obeys a "bell curve" distribution.

23. Pearson (1900) described r_{hk} as "the correlation between the errors in the position of the means."

24. To formulate the test, we first establish the consequence of the *null hypothesis*, which is the assumption that vaccination has no effect. The value of a that one would expect in this case is such that the fraction $a/N_V = N_R/N$, with similar fractions for b, c, and d. Hence, denoting these expected values with the subscript e, under the null hypothesis, we should have

$$a_e = \frac{N_R}{N}\,N_V; \; b_e = \frac{N_D}{N}\,N_V; \; c_e = \frac{N_R}{N}\,N_U; \; d_e = \frac{N_D}{N}\,N_U.$$

To measure the deviation from these values, Pearson proposed using a quantity denoted by the symbol $\chi 2$ which, for a 2×2 table is given by

$$\chi^2 = \frac{[a - a_e]^2}{a_e} + \frac{[b - b_e]^2}{b_e} + \frac{[c - c_e]^2}{c_e} + \frac{[d - d_e]^2}{d_e}$$

After suitable algebraic manipulation, we obtain the expression given in the text. See Spiegel 1988, 244, 254.

25. As often occurs in science, Pearson was not the first to propose such a test; his main contributions were to develop it and to overcome the objections of skeptics. See Stigler 1986.

26. Spiegel also (incorrectly) describes ϕ as a tetrachoric correlation.

27. The formula for r developed in n. 21 has two useful properties. First, it is independent of the units used to measure h and w and independent of the origin or reference point used to

make those measurements. Second, it is symmetrical with respect to the variables; that is, we obtain the same results if we switch the roles of w and h. It is the first property that suggested r_{PS} to Yule. We attach the value +1 to vaccination and recovery and –1 to death and lack of vaccination. There are, for example, b persons having $x_i = +1$ and $y_i = -1$. Proceeding in this way and applying the product–moment formula gives r_{PS}.

28. In spite of the dispute, Yule used this misleading description in the 1924 edition of his text. See Heron (1911, 113) and Yule (1924, 216).

29. In terms of the height–weight example, this corresponds to changing the values of h, classifying persons into "tall" and "short," and an equivalent change for w. In one example, Heron showed that for $r = 0.1$, corresponding to a very weak relationship, Q could be made to vary from 0.13 to 0.84. In the same range, r_{PS} varied from 0.063 to near 0.

30. If N_V is changed by a factor k to kN_V without altering the proportion of vaccinated survivors, then a multiplies to ka, and b to kb. Substitution of ka in place of a and kb in place of b in the formula for Q shows that the factor k cancels out. Changing N_D by a factor m without altering the proportion vaccinated shows that the factor m also cancels out.

31. For these data, r was available; the variation in r_T occurred because the data did not satisfy Pearson's "bell curve" assumptions.

32. This is common in statistics. Consider, as an example, the weights of a set of N objects. The *mean* weight is the value that, when multiplied by N, gives the total weight of the set. The *median* weight is that value below which exactly one-half the set has a lower weight. The two, called *measures of central tendency*, are related; for example, as one increases, so does the other. Also for most populations, as N increases, the two approach each other.

33. Notwithstanding Yule's valid assertion that Q and ϕ are different measures of association and are thus not directly comparable, the starkly contrary behavior described by Pearson and Heron is such that most mathematicians would seek an explanation.

34. See Greenwood and Yule 1915. In 1912 Yule defined

$$\omega = \frac{\sqrt{ad} - \sqrt{bc}}{\sqrt{ad} + \sqrt{bc}}$$

He attached theoretical significance to ω because for "symmetrical tables" in which $a = d$ and $b = c$, $\omega = \phi$.

35. Nie et al. 1975 recommend ϕ for 2×2 tables (p. 224) and note that any dichotomy can be treated as a continuous variable (p. 5), thus in effect recommending ϕ for all three types of variable. In their discussion of larger tables, they recommend Goodman and Kruskal's procedure as one of several possibilities for ordinal variables, noting that it reduces to Q for 2×2 tables (p. 226).

36. Wolpert (1993, 122) is equally scathing: "By ignoring the achievements of science, by ignoring whether a theory is right or wrong, by denying progress, the sociologists have missed the core of the scientific enterprise."

37. As Levin remarks, feminist critic Sandra Harding's response to this crucial question is an evasion: "Why should we continue to regard physics as the paradigm of scientific knowledge-seeking? . . . I will argue that a critical and self-reflective social science should be the model for all science" (Harding 1986, 41-44). Apparently she is not troubled by the fact that natural science clearly works, whereas much of social science does not. Consider, for example, Alan Bloom's assessments: "Social science is...an imitation, not a part [of natural science.]," and "practically every social scientist would like to be able to make reliable predictions, although practically none have" (Bloom 1987, 358, 360).

38. How many appreciate the irony implied in the proposed use of rocket-launched nuclear bombs to deflect these objects? See Tom Gehrels, "Collisions with Comets and Asteroids," *Scientific American*, March 1996, 54-59.

39. Among other disturbing characteristics, on pp. 129-33, Andreski describes the use, by a respected anthropologist, of pseudomathematical expressions such as "fire = (water)$^{-1}$."

References

Abbott, M. M., and H. C. Van Ness. 1989. *Schaum's Outline of Theory and Problems of Thermodynamics*. 2d ed. New York: McGraw-Hill.

Andreski, S. 1972. *Social Sciences as Sorcery*. London: Andre Deutsch.

Barrow, J. D. 1988. *The World within the World*. Oxford: Clarendon Press.

Batchelor, G. K. 1996. *The Life and Legacy of G. I. Taylor*. Cambridge: Cambridge University Press.

Bloom, A. 1987. *The Closing of the American Mind*. New York: Simon & Schuster.

Boyer, C. B. 1959. *The History of the Calculus and Its Conceptual Development*. New York: Dover.

Casti, J. 1989. *Paradigms Lost: Images of Man in the Mirror of Science*. New York: Morrow.

Chalmers, A. 1990. *Science and Its Fabrication*. Minneapolis: University of Minnesota Press.

Courant, R. 1937. *Differential and Integral Calculus*. Vol. 1, 2d ed., trans. E. J. McShane. London: Blackie and Son.

Davis, J. 1971. *Elementary Survey Analysis*. Englewood Cliffs, N.J.: Prentice-Hall.

DiStefano, J. J., A. R. Stubberud, and I. J. Williams. 1967. *Schaum's Outlines of Theory and Problems of Feedback and Control Systems*. New York: McGraw-Hill.

Gabor, A. 1992. *The Man Who Discovered Quality*. New York: Viking Penguin.

Goodman, L., and W. Kruskal. 1954. Measures of association for cross classifications. *Journal of the American Statistical Association* 49: 732–64.

Greenwood, M., and G. U. Yule. 1915. The statistics of anti-typhoid and anti-cholera inoculations, and the interpretation of such statistics in general. *Proceedings of the Royal Society of Medicine* 8: 113–94.

Gross, P. R., and N. Levitt. 1994. *Higher Superstition: The Academic Left and Its Quarrels with Science*. Baltimore: Johns Hopkins University Press.

Grove, J. W. 1989. *In Defence of Science: Science, Technology and Politics in Modern Society*. Toronto: University of Toronto Press.

Harding, S. 1986. *The Science Question in Feminism*. Ithaca, N.Y.: Cornell University Press.

Hayles, N. K. 1992. Gender encoding in fluid mechanics: masculine channels and feminine flows. *differences: A Journal of Feminist Cultural Studies* 4: 16–44.

Heron, D. 1911. The danger of certain formulae suggested as substitutes for the correlation coefficient. *Biometrika* 8: 109–22.

Hughes, W. F., and J. A. Brighton. 1991. *Schaum's Outline of Theory and Problems of Fluid Dynamics*. 2d ed. New York: McGraw-Hill.

Irigaray, L. 1985. *This Sex Which Is Not One*. Trans. Catharine Porter. Ithaca, N.Y.: Cornell University Press.

Johnson, P. 1984. *A History of the Modern World*. London: Weidenfeld & Nicolson.

Joule, J. P. 1884. *The Scientific Papers of James Prescott Joule*. London: Physical Society.

Kevles, D. J. 1985. *In the Name of Eugenics*. Berkeley and Los Angeles: University of California Press.

Le Conte, J. 1864. On the adequacy of Laplace's explanation to account for the discrepancy between the computed and observed velocity of sound in air and gases. *Philosophical Magazine* 27: 1–32.

Lefkowitz, M. 1996. *Not out of Africa: How Afrocentrism Became an Excuse to Teach Myth as History*. New York: Basic Books.

Levin, M. 1988. Caring new world: feminism and science. *American Scholar*, Winter, 100–6.

MacKenzie, D. 1978. Statistical theory and social interests: a case study. *Social Studies of Science* 8: 35–83.

Nie, N. H., C. H. Hull, J. G. Jenkins, K. Steinbrenner, and H. H. Bent. 1975. *SPSS: Statistical Package for the Social Sciences*. 2d ed. New York: McGraw-Hill.

Pearson, K. 1900. Mathematical contributions to the theory of evolution—VII. On the correlation of characters not quantitatively measurable. *Philosophical Transactions of the Royal Society*, series A, 195: 1–47.

Pearson, K., and D. Heron. 1913. On theories of association. *Biometrika* 9: 159–315.

Press, W. H., B. P. Flannery, S. A. Teukolsky, and W. T. Vetterling. 1989. *Numerical Recipes*. Cambridge: Cambridge University Press.

Sparkes, A. W. 1991. *Talking Philosophy*. London: Routledge.

Spiegel, M. R. 1967. *Schaum's Outline of Theory and Problems of Theoretical Mechanics*. New York: McGraw-Hill.

———. 1988. *Schaum's Outline of Theory and Problems of Statistics*. 2d ed. New York: McGraw-Hill.

Stigler, S. M. 1986. *The History of Statistics: The Measurement of Uncertainty before 1900*. Cambridge, Mass.: Harvard University Press.

Stokes, G. 1851. An examination of the possible effect of the radiation of heat on the propagation of sound. *Philosophical Magazine* 1: 301–17.

Timoshenko, S. P. 1983. *History of Strength of Materials*. New York: Dover.

Truesdell, C. A. 1954. Introduction to *L. Euleri opera omnia*. Series II, vol 13. Lausanne: Societatus scientiarum naturalium helveticae.

———. 1980. *The Tragicomical History of Thermodynamics 1822–1854*. New York: Springer-Verlag.

Westfall, R. S. 1980. *Never at Rest: A Biography of Isaac Newton*. Cambridge: Cambridge University Press.

Wolpert, L. 1993. *The Unnatural Nature of Science*. London: Faber & Faber.

Yule, G. U. 1900. On the association of attributes in statistics. *Philosophical Transactions of the Royal Society*, series A, 194: 257–319.

———. 1912. On the methods of measuring association between two attributes. *Journal of the Royal Statistical Society* 75: 579–642.

———. 1924. *An Introduction to the Theory of Statistics*. London: Charles Griffin and Company.

Yule, G. U., and M. G. Kendall. 1950. *An Introduction to the Theory of Statistics*. London: Charles Griffin.

6

Evidence-Free Forensics
and Enemies of Objectivity

Paul R. Gross

Dr. Summers has not been able to explain to the readers of The
New York Review why Pasteur's discoveries . . . were multifaceted
and relative. Nor do I think he could name a single Nobel Prize–
winning discovery in physics, chemistry, or medicine where the
discoverers acted as he charges, by trying to navigate a safe passage between the constraints of empirical evidence on the one hand
and their personal or social interests on the other. If neither
Dr. Summers nor social historians of science can support their views
by such concrete examples, then the entire approach is clearly
humbug.

M. F. Perutz, letter, *New York Review
of Books*, February 6, 1997

One of the more absurd essays in *Science Wars* goes so far as to
examine figures of speech in Gross and Levitt's book itself in order
to prove that they are just male sexists and can consequently be
ignored as unreliable. But . . . if one were forced to put oneself into
the hands of Sarah Franklin, its author, or those of Levitt and Gross,
no sane person would choose to be at the mercy of what passes
for "reason" in Franklin.

Harold Fromm, "My Science Wars,"
Hudson Review 49 (Winter 1997)

Melancholy Fact

Writing in a new volume on feminism and science, philosopher Elisabeth A. Lloyd
does not mince words: her title is *Science and Anti-Science: Objectivity and Its Real
Enemies* (1996). Seven "real" enemies are named: Gerald Holton, Lewis Wolpert, (the
above quoted Max Perutz, Paul Gross, Norman Levitt, Margarita Levin, and (for lesser
violations) Clifford Geertz. Two others—Alan Sokal and Steven Weinberg—have been
charged elsewhere with related offenses, presumably after Lloyd's indictment went
to the printer. Such accusations pop up regularly these days, even in the pages of
Le Monde (see Dickson 1997, Duclos 1997). Lloyd's version is long and densely
endnoted, and italicized. Readers who think that smoke must imply fire might imagine that the accused are guilty of something. I would like to examine each case to show

that they are not, but even with the editor's concessions, my page limit is too small. Therefore, although I deal mainly with Lloyd's charges against Gross and Levitt (1994) (because here I am best able to match Lloyd's account with true texts), this rebuttal must stand for all seven. I do give some attention, however, to charges against Gerald Holton, as they refer to a book that I have read with special care. I also try to match Lloyd's methods with those of other accusers. They all follow a pattern, which is the subject of this chapter: smoke without fire.

At the moment, the tactics used to suggest fire are too common in academic writing. They are not new and antedate the so-called science wars, to which Lloyd's paper is addressed. It was a long struggle to reduce the frequency of such tactics in scholarly work, but they have returned—the equivalent of a stage conjuror's prattle. And a melancholy fact is that those who use these tactics are doing precisely what they say (with, as will emerge, no evidence) the "real enemies of objectivity" do. A stage illusion is best exposed by identifying the method of the trick and then revealing it in the performance. I have adapted this method by (1) identifying and naming (old) prototype devices (henceforth PD) used to construct false charges, giving examples; (2) finding these devices in Lloyd's bill of indictment; and (3) calling, finally, for the verdict.

Lloyd's most insistent claim is that feminist science critics are not hostile to science, that they see scientific knowledge as objective, that they have made fundamental contributions to science, and that the accused ignore the evidence to that effect. Neither Lloyd here, however, nor the other prosecutors identified later in this chapter provide any convincing evidence to support these claims. It is true that a few feminist science critics proclaim friendship toward science and the ideal of objectivity, but what they write and teach is often hostile and always influential. As they define it, "objectivity" is something vaguely akin to multiculturalism. Lloyd's commitments on these matters are summarized by her comment in an earlier paper:

> I would adopt Helen Longino's general approach, in which she characterizes objectivity in science as resulting from the critical interaction of different groups and individuals with different social and cultural assumptions, and different stakes. Under this view, the irreducibility of the social components of the scientific situation is accounted for. (1993, 150)

As we shall see, even this nebulous definition is abandoned when necessary; at need, Lloyd will insist not on "irreducibility" but that scientific results be separable from the politics of the scientist. This style of doing feminist epistemology and feminist science studies resembles what Paul Cantor sees in the new historicism: "As opposed to the despairing voice in T. S. Eliot's *The Waste Land* that says 'I can connect/Nothing with nothing,' the motto . . . seems to be 'I can connect anything with anything'" (1993, 24).

Intimations of Absurdity, or the Rhetorical Feint

The rhetorical feint is a debating move in which opposition points are dismissed by restating them in such a way as to imply their absurdity. The user indicates that proofs of absurdity will follow, but they never do. Instead, the attack shifts abruptly (as in fencing or boxing) to a different target. With luck, the judges remember the intimation of absurdity but not the absence of proof. PD: Lloyd's use of this device is so deft, so early, and so important to her argument that it can serve as its own paradigm. She introduces the accused enemies by listing their "pronouncements that materialize with mystifying but strict regularity whenever 'feminism' and 'science' are used in the same breath." There follows a long list of "mystifying" pronouncements: for example, "feminists—like many historians, sociologists, and anthropologists of science—wish to replace explanations of scientific success that are based on following the methods of science, with explanations purely in terms of power struggles, dominance, and oppression, and to ignore the role of evidence about the real world" (1996, 217). Such assertions, she implies, are absurd; no important feminists wish that. And, she adds, "there is a well-known body of feminist work which has systematically articulated the *negation* of each of the above beliefs in black and white" (217; italics in original).

The judge of this debate who remains alert through what follows awaits answers to these questions: Do some feminist science critics "wish to replace explanations of scientific success" (and so on)? If so, what fraction of feminist writing is of that kind? What is the "well-known body" of negation, and what fraction of feminist writing is of that kind? Lloyd's rhetorical feint drops the subject of "mystifying" pronouncements like a hot potato; the questions are never raised or answered. Instead, she alludes to negations that she does not, however, display. Thus is planted the idea that feminist "relativism" (Lloyd uses such sneer quotation marks)[1] and irrationalism are fictions or utterly insignificant in feminist science critique.

But, compare social ecologist Janet Biehl, whose feminist credentials seem sound, on the state of ecofeminism in 1990 (Lloyd's article was published in 1996):

> Ecofeminism's sweeping but highly confused cosmology introduces magic, goddesses, witchcraft, privileged quasi-biological traits, irrationalities, neolithic atavisms, and mysticism into a movement that once tried to gain the best benefits of the Enlightenment and the most valuable features of civilization for women, on a par with thinking and humane men. (Biehl 1991, 6)

And from a distinguished physician and feminist, Marcia Angell, the following:

> A particularly influential group to turn against science is a segment of the feminist movement. . . . To them, science is inherently "androcentric," because it was largely developed by men in patriarchal societies. . . . The 1986 book *Women's Ways of Knowing* suggests that women find the scientific method uncongenial because it emphasizes logical, linear thinking at the expense of intuitive, multi-faceted thinking. . . . But logical analysis is the most useful tool, and all scientific hypotheses must ultimately stand the test of objective evidence. Some feminists do not accept these constraints, *and many other women are to some extent influenced by their ideas.* (1996, 180; italics added)

To support her claim that feminists do not say the things about science that, as I have observed, every student in women's studies learns, Lloyd offers the irrelevant proposal that some feminists say things that in her view negate such statements. But of course, lots of feminists do say those "mystifying" things. They teach them, too.

Making Up Motives

This variant of ad hominem argument is used to obfuscate unwelcome disclosures. The obfuscator ignores specifics and proposes, instead, foul motives for the disclosers. If the motive is elaborate, accompanied by what looks like documentation, and pleasing to the audience, the disclosures can be camouflaged. A scan of the journals of political opinion, both left and right, shows any doubter how common this tactic is.

PD: Dorothy Nelkin, defending such *bonnes oeuvres* of science criticism as *Social Text*[2] and the game of "epistemological chicken" (Collins and Yearley 1992), attacks Alan Sokal and also the contributors (not just "scientists," as she calls them, but many humanists and social scientists) to *The Flight from Science and Reason* (Gross, Levitt, and Lewis 1995), by inventing a motive for their behavior: "The moral outrage of scientists [*sic*] stems from changes occurring in the field of science itself and in its relationship to the sources of financial support" (Nelkin 1996, A52). This is an elementary Marxist notion, that consciousness is social and social is economic, but there is no evidence for it. To be sure, there is a funding problem, but it began long before a few scientists—and others—took notice of what to them were fashionable distortions in science criticism. To the extent that scientists are nervous about funding, they delegate to their disciplinary societies the job of lobbying, not of arguing cultural questions. Indeed, they avoid, almost at any cost, the latter, and the nonscientist contributors to the "Flight" conference did not seem overly worried about science funding. Thus to attribute "moral outrage" and vulgar anxiety to the few scientists who have responded to some science studies shibboleths is grasping at straws.[3] Nelkin's variations on scientific angst are motive making (accompanied in print by a drawing of scientists in hard hats, attacking with blunt instruments critics shielding themselves with paper).

Lloyd: She invents, at inordinate length, a multilayered motive. It begins: "I propose *a way of thinking about the perceived conflicts* between these scientists and those who analyze the sciences within more inclusive social frameworks. . . . I suggest and develop a sympathetic interpretation which addresses the *social and political concerns* that are visible" (1996, 220; italics in original). That done, Lloyd offers a catalog of putative misdeeds of the "scientists" thus motivated. "Gross and Levitt conclude," she reports, "that the views of science they consider constitute a dire social threat" (220). This sets the tone for motive: these guys are just politicking and therefore simply crying wolf.

Nothing in Gross and Levitt's book, *Higher Superstition*, calls the threat "dire" (adj.: dreadful, desperate, urgent) or anything like it. Lloyd quotes from the book: "What is threatened is the capability of the larger culture, which embraces the mass media as well as the more serious processes of education, to interact fruitfully with the sciences . . . and above all, to evaluate science intelligently" (4). In what looks like (but isn't) a documenting endnote, she refers to the word *pernicious*, found elsewhere in the book. By a novel arithmetic, she adds all this up to "dire." Here is the "pernicious" passage from *Higher Superstition*:

> Although the criticisms we have examined amount, individually and collectively, to very little in strictly intellectual terms, it is nonetheless important for scientists and

fair-minded intellectuals—and this includes many left-wing thinkers—to take them very seriously. It is not without historical precedent that incoherent or simply incomprehensible opinions have had great and pernicious social effect. (15)

To deny any urgency, any political intent, or any desire to exclude (more about this later), our book also asks:

> Why should the doctrinal narrowness of black or women's studies departments be more objectionable than that of some schools of business administration, quite a few military science departments, or even athletic departments? No fire-breathing feminist zealot has ever had the power over the lives and minds of her charges that is exercised routinely by the football or basketball coach at a school with a major "program"—that is, one aspiring to be a significant NFL or NBA farm team. (1994, 37)

So the book calls no threat of science studies "dire" or of feminism "pernicious." It denies any such intention.[4] But this first layer of Lloyd's motive making insinuates an inquisitorial urgency. To this, she adds images of oligarchic conspiracy. In her fantasy, "some special, authoritative, even mythic stature is psychologically necessary" to science;[5] therefore (generously), "I [she] will not argue that no reason or no argument could be given that might justify a ban on the pursuit or dissemination of social studies of the sciences" (1996, 221).

Has anyone proposed "a ban on the pursuit or dissemination of social studies?" No. At its very beginning, the following statement appears in *Higher Superstition*: "We recognize that it is necessary for science patiently to abide social scrutiny, since science and its uses affect the prospects of the entire society" (7). And among other assurances, this:

> Apart from the most arrogant, they [natural scientists] concede that the psychological quirks and modes of personal interaction characteristic of working scientists are not entitled to special immunity from the scrutiny of social science. If bricklayers or insurance salesmen are to be the objects of vocational studies by academics, there is no reason why mathematicians or molecular biologists shouldn't sit still for the same treatment. (42)

But Lloyd must have her motive: "I conclude . . . that the efforts to demonize and dismiss feminist analyses of and contributions to the sciences, amount to illegitimate and unfounded attempts *to exclude participation by fully-qualified colleagues in these aspects of intellectual life solely on the basis of their feminist politics*" (1996, 218; italics in original). "Let us assume," she writes, "that social stability, the protection of the social *authority* of science is in everyone's best interest. . . . It could therefore appear that feminists and radical critics, when they expose and detail the scientific weaknesses and self-serving interests . . . of some scientific approaches and some scientists, are actually destabilizing" (1996, 221; italics in original).

Let's point out the implications: (1) "Scientific weaknesses" are exposed, not through the operations (and epistemic rules) of science, but by "feminists and radical critics." Nobody acquainted with the history of science can believe this. Scientific weaknesses were regularly exposed before there were feminists and radical critics. (2) The real issue is this: "inside information" must be "controlled" and the archetype of control is military intelligence! (Lloyd 1996, 222 ff). Objections to feminist science critique, according to Lloyd, protect the authority of "particular religious principles," just as military intelligence protects social stability in time of war. The reaction of a few scientists to feminist epistemology (it is this, not feminism, that they react to) is a military intelligence operation, mounted to maintain oligarchic power and stability, to keep the masses at heel. Invoking militarism is the rhetorical trick we called "moral intimidation" (*Higher Superstition*, 53).

Lloyd, a philosopher, cannot be unaware of the objections of other philosophers (leaving out Margarita Levin, who is among the accused)—including feminists Susan Haack, Noretta Koertge, and Janet Radcliffe Richards—to "feminist epistemology."[6] Rather, she must have chosen, in aid of her invented motive, to ignore them,[7] to ignore social scientists who see nihilism in fashionable versions of science studies (Schmaus, Segerstråle, and Jesseph 1992), to ignore even Thomas Kuhn[8]—so as to associate an odious motive with "scientists." It is as though Susan Haack had never argued in 1993, in her "Epistemological Reflections of an Old Feminist,"[9] that "there is no such connection between feminism and epistemology as the rubric 'feminist epistemology' requires" (1993, 55), as though there are not scholarly journals, even an entire issue of the *Monist* (1994), examining feminist epistemology, in which at least some distinguished thinkers express, to put it mildly, grave doubts (see esp. B. Gross 1994, Pinnick 1994). Thus Lloyd's indictment rests on an invented motive, a false analogy, and no facts. Recent arguments against "feminist epistemology" applied to science have nothing to do with oligarchy, exclusion, or social control.

Designed Quotations, Readings, and Paraphrases

By a *designed quotation* I mean something more sophisticated than simple misquotation; rather, a fabricated statement, enclosed wholly or partly in quotation marks or sometimes not at all, that is meant to be understood as a quotation, even though the intended victim never said or wrote it. Designed quotations flower among cultural studies adepts reacting to the notice lately taken of their more sensational declarations.

PD: Andrew Ross, writing in *the Nation* about the "Science Backlash on Techno-skeptics," has Gerald Holton identifying "the ill-educated masses" with "the Beast that slumbers below" (1995, 348). This reinforces Ross's contention that this backlash is a conspiracy of right-wing elitists. But as Holton used it,"the Beast that slumbers below" has nothing to do with any sort of masses, which appear nowhere in his text. Instead, Ross appropriated the "Beast" phrase from chapter 6 of Holton's 1993 book, *Science and Anti-Science*, giving it a new context. Holton is in fact discussing the "committed and politically ambitious parts of the anti-science phenomenon"—ideologies of "rabid ethnic and nationalistic passions" (184). Those ideologies, not "the . . . masses," are the slumbering beast. Ross does a like job on E. O. Wilson in the same piece (words attributed to Wilson were never uttered), and again, shamelessly, on Holton in the unfortunate issue of *Social Text* of which Ross was an organizer. Ross wrote, in the original introduction to *Science Wars*, that "apocalyptics like Holton see only science 'falling on its own sword'" (1996, 10). The passage in single quotation marks is not from Holton; it is from Oswald Spengler, to whose *refutation* Holton devotes his chapter 5. After complaints, the false quotation was changed in the book version of *Science Wars* by adding "Spengler" to "Holton" and so making both Spengler and Holton "apocalyptics"![10]

PD: Brian Siano (1996), in *Skeptic*, argues that political correctness (PC) is a myth and that its propagation is a right-wing conspiracy. He includes Gross and Levitt among the conspirators, despite their repeated—some say annoying—dismissals of PC as the issue. Siano quotes from an extract of their book in *Skeptical Inquirer*, a long quotation that is, however, discontinuous, with words enclosed in quotation marks interspersed with Siano's own. Together, all these words turn Gross and Levitt into enemies of the left. My coauthor, sorely vexed (he is a leftist), found the quoted words in the article from which they were taken. To form the quotation, Siano had cobbled together phrases from five different, unrelated, widely separated sentences, with the result that the designed quotation means the opposite of what the original intended. Levitt's dissection appears in a subsequent issue of the same journal (1997).

Lloyd: please distinguish again, where they occur, nested (single-mark) quotation marks from external (normal, double) ones: nested words are Lloyd quoting an "enemy,"; normal marks enclose quotations of Lloyd. Lloyd's treatment of the first listed enemy, Gerald Holton, has him approving, in *Science and Anti-Science*, the proposed agenda of the 1989 Nobel Conference, including "we have begun to think of science as a more subjective and relativistic project . . . Marxism and feminism, for example. This leads to grave epistemological concerns" (quoted in Lloyd, 1996, 218). Lloyd presents this as Holton's opinion, but it is not. First, the passage is in an endnote to Holton's chapter 5, "The Controversy over the End of Science" (Holton, 1993, 142–143), which is not the chapter (6) on the antiscience phenomenon. Second, Holton's point is that most scientists do *not* take such proposals seriously. Lloyd follows this quotation-of-a-quotation with the assertion that Holton's book is "about an 'antiscience movement.'" It isn't. Five of its six chapters are about arguments within science. Only the last chapter addresses antiscience, but not as "a movement."

Lloyd uses the phrase "intellectually least serious" (1996, 219), coupling it with "the feminist [a word Holton didn't use] view," thereby making Holton an antifeminist and thus—somehow—an enemy of objectivity. True, Holton does list four manifestations of contemporary antiscience, "starting from the intellectually most serious end" (1993, 153), which he identifies as "a type of modern philosopher who asserts that science can now claim no more than the status of one of the 'social myths'—the term used by Mary Hesse—not to speak of the new wing of sociologists of science who wish, in Bruno Latour's words, to 'abolish the distinction between science and fiction'" (quoted in Holton 1993, 153). Second are alienated intellectuals (such as Arthur Koestler); third are what Holton calls "the Dionysians, with their dedication ranging from New Age thinking to wishful parallelism with Eastern mysticism" (154). And finally, there is " a fourth group, again very different . . . a radical wing of the movement represented by such writers as Sandra Harding" (154). To convert this to a claim by Holton that feminism is intellectually trivial is the purpose of Lloyd's designed quotations.

Holton specifies his worries. In regard to feminism, they are clearly limited to the "radical wing . . . represented by such writers as—." They are not about feminism. Moreover, this is a minor concern among the big ones of chapter 6, which are this century's dangerous political movements wedded to antiscience. Holton's examples—and the only ones the book identifies—are "Creationism," "premillennialist Fundamentalism," "Fascist Germany," and "Stalin's USSR." For readers not aware of it, I note that this accused enemy of feminism is the coauthor, with Gerhard Sonnert, of a long-awaited, systematic study (in two volumes) of workplace and access problems for women in science. In her review, Sheila Tobias wrote, "Together, the two books constitute essential reading for those who would like to see more of our promising women achieve a life in science that they appear to want quite as much as they deserve" (1996, 13; see also Sonnert and Holton 1995). To make Gerald Holton into an antifeminist, as Lloyd does, is ludicrous.

Lloyd: "They embody the strategy of exclusion. Consider Gross and Levitt's description of feminists' 'acute and apocalyptic oppositionism': 'The announced goal, upon which feminists of the most disparate schools agree, is a science transformed. . . . [This is an enormously] ambitious project: to refashion the epistemology of science from the roots up'" (quoted by Lloyd, 1996, 750). Got it? The quoted words are arranged to suggest that Gross and Levitt construct a straw man, an "[enormously] ambitious project," so as, presumably, to ridicule feminist hubris. But (1) this specific sequence of phrases, referenced by Lloyd's endnote 127, is absent from the indicated pages (32, 108). More important, (2) the claim of reinventing science is not theirs at all. It is an announcement by feminist philosopher Sandra Harding: "I doubt that in our wildest dreams we ever imagined we would have to reinvent both science and theorizing itself" (see *Higher Superstition*, 1994, 132, for the full quotation and Margarita Levin's comment on it).

Lloyd: "Gross and Levitt spare us from having to speculate about what they're really after: they want to *exclude* feminists from university and intellectual life, period" (quoted by Lloyd, 1996, 251; italics in original). Placing an endnote (n. 134) after "what they're really after" makes it look like a quotation or an accurate reading—that is, like evidence. But no: the reader who looks up endnote 134 finds the following: "They argue, for example, that the high academic standing of these feminist researches *itself* 'raises serious questions about the presumed intellectual meritocracy of the academy'" (259; italics in original). Nothing more; no quotation, no call for exclusion. The suggestion (elsewhere in *Higher Superstition*) that scientists pay attention to work in other disciplines that purports to analyze the content of science is just that; the language is plain. We add that nonscience faculty would, and should, do the same for work in a science or mathematics department that analyzes, say, rhetoric or literary style (*Higher Superstition*, 256). Lloyd's endnote 134 is a false paraphrase, offering a thought as though quoted when in fact it is invented.

Lloyd: "Gross and Levitt threaten to eject humanities and social studies from the university unless they become more hospitable to being judged by natural scientists" (1996, 226). Again, she indicates by an endnote that this is a quotation or a proper reading. The endnote refers to pages 245–47 of *Higher Superstition*, but there is nothing on those pages (or anywhere else) that is sanely interpretable as a threat to eject anyone. It seems to me ironic that humanists, for whom texts should be of great importance, have recently found it acceptable to handle them with such insouciance. And, incidentally, anybody who does not suspect that the intellectual meritocracy of the academy is sometimes merely presumed must be, well, away from the office.

Concocting Conclusions

PD: In "The Loves of the Plants," in the February 1996 issue of *Scientific American*, Londa Schiebinger offers a lively account of Linnaeus's plant taxonomy based on sexual reproduction. She samples his overheated (by twentieth-century standards: he wrote in the eighteenth) prose: "Linnaeus emphasized the 'nuptials' of living plants as much as their sexual relations. Before their 'lawful marriage,' trees and shrubs donned 'wedding gowns'" (113). For a popular article, this one has laudable scholarly content, but it is narrow. The focus is on sexual metaphor, rather than on the problems, practical and philosophical, of Linnaeus's taxonomic venture. Schiebinger wants to show that botany was a patriarchal enterprise, pruriently interested in sex. Even if that was so (I do not concede it), it isn't the whole story or even the main one. There was a good reason to focus on plant reproduction: it had just been shown to be sexual, an astonishing discovery. Knowledge of fertilization was still far in the future (George Newport, 1824, and Oskar Hertwig, 1876). Because of the vast range of sexual differentiations, Linnaeus saw that these could be used to classify plants in a way that habitat, foliage, and color could not. He thereby classified and named all the plants then known and becoming known and attempted an account of all creation. His students and agents scoured the globe. "Linnaeus," as Daniel Boorstin observed, "made a world community of naturalists" (1985). Although elements of the Linnean systematics remain, most of them faded long ago.

Did Linnaeus's contemporaries dote on the sex? No. Even his admirers were embarrassed by it or rejected it. Boorstin focuses on this argument, but he does not dwell on the bitterest and most telling struggles, which were with Linnaeus's noble rivals, Buffon and Haller. A full and fair account, as in Larson (1994), shows that despite the success of the sex-based Linnean taxonomy—it worked for classification and Buffon's philosophical objections led nowhere—it was never automatically accepted.

Schiebinger writes: "researchers read nature through the lens of social relations in such a way that the new language of botany, among other natural sciences, incorporated fundamental aspects of the social world as much as those of the natural world" (1996). Maybe, but it depends on what she means by "incorporated" and "fundamental." Of course, botanical names are socially derived; where else are names to be found? But names ceased long ago to be critical to botany. Had there been many women botanists in the eighteenth century, nomenclature and categories might have been different, but it would have made little difference to what is known today about plant life. Today there are scores of women botanists, some of the best. The forgotten Linnean raptures don't seem to bother them; rather, they benefit from the knowledge that began with him.

Schiebinger is entitled to her emphasis, just as others are entitled to dispute it. But what she wants her readers to take away has nothing to do with her text:

> During the past few decades, the feminist critique of science combined with the process of more and more women becoming engaged as makers of knowledge has had a tremendous impact in the humanities, social sciences, and many of the sciences. We are just beginning to unravel how deeply gender has been worked into nature's body. Historical exposé, of course, is not enough, for what we unravel by night is often rewoven by day in the institutions of science. (1996, 115)

Gender worked deeply into nature's body? A rapture, like Linnaeus's "bridal beds," and more explicitly sexual. What follows also is explicit: crucial discoveries like these (Schiebinger's?) are being concealed "in the institutions of science." No, not by day or night. Publication in *Scientific American* is not concealment or reweaving. There is a flood of claims such as Schiebinger makes, which appear not only in books and popular magazines but also in scientific journals (e.g., *Science*). Has, then, the feminist critique had a "tremendous impact" on "many of the sciences?" If "impact" means an effect on scientific knowledge, then Schiebinger gives no evidence of it. Is it true, nevertheless? No. Have women made an impact on science? Of course, but not because they are feminist epistemologists or historians; rather because they do good science. Schiebinger's conclusion is a concoction.

Lloyd: From a hallmark work in constructivist sociology of science, *Leviathan and the Air Pump*, Lloyd takes a sentence that is discussed in *Higher Superstition*: "As we come to recognize the conventional and artificial status of our forms of knowing, we put ourselves in a position to realize that it is ourselves and not reality that is responsible for what we know" (Shapin and Schaffer 1985, 324). She provides an exegesis, arguing that (1) Shapin and Schaffer did not really mean it;(2) Gross and Levitt know that they didn't, and (3) yet they exploit it, together with a minor (?) flaw they say is in *Air Pump*'s account of Thomas Hobbes's conflicts with the Royal Society, in order to pin on Shapin and Schaffer a fundamentalism (social forces are decisive for scientific knowledge) that they do not espouse. Then Lloyd concocts two conclusions:

1. "After Gross and Levitt legislate that *no social explanation is compatible with a [potential]* 'internal' *one*, they go on to condemn Shapin and Schaffer for 'insisting that all such disputes are ideological'" (1996, 233; italics in original).

This statement has an endnote (n. 57), which just refers to a page number (68) in *Higher Superstition*. The conclusion Lloyd wants is the attributed Gross and Levitt fundamentalism: "No social explanation is compatible." But it doesn't exist: there are no such words on page 68, and those in single quotation marks (beginning with "insisting") have nothing to do with any "legislation." This fabricated paraphrase is contradicted by many statements in the book, including some on page 68.

2. Gross and Levitt "miss all the textual evidence supporting a more moderate reading."

This doesn't follow from the evidence, and not simply because the insistent mathematical follies of Hobbes, which Shapin and Schaffer discount, did, as Gross and Levitt (and others, specialists; see Bird 1996) argue, affect his relations with the Royal Society. It doesn't follow from Lloyd's lucubrations (1996, 231–34), which add little to the rising debate on Shapin and Schaffer's historiography. It is false on the basis of painstaking reexaminations of this—and related—stories of the scientific revolution that philosophers and historians are now publishing (see, for example, Cassandra Pinnick's chapter in this volume and, as background, her "Cognitive Commitment and the Strong Program," 1992).

A recent, comprehensive paper on the general subject, Alan Shapiro's "The Gradual Acceptance of Newton's Theory of Light and Color, 1672–1727" (1996), concerns a later work of Simon Schaffer, his 1989 "Glass Works: Newton's Prisms and the Use of

Experiment," but it is a constructivist analysis of Newton's theory of color, employing and extending the key arguments of *Leviathan and the Air Pump*. Shapiro's conclusion:

> [Schaffer's] historical explanation fails on several accounts. First, it does not agree with the actual sequence of events. *It cannot explain why Newton's theory was gradually established in Great Britain well before he had so much authority over "London experimenters,"* and why—independent of his power—it was beginning to take hold on the Continent within a few years of the *Opticks* . . . *Newton's "control over the social institutions of experimental philosophy" in London is too late in Great Britain and of minor consequence on the Continent*, especially among the Leibnizians in Germany. (1996, 132; italics added)

The role of social context in the acceptance of scientific knowledge is overemphasized, and the internals—especially scientific theory—are minimized or discounted. This argument will doubtless continue, but Lloyd's portrayal of Gross and Levitt as military intelligence zealots who "miss all the textual evidence" is, despite the plethoric endnoting, unjustified.

Crying Prejudice to Change the Subject

PD: The *Faculty Forum* is a Duke University periodical. Physicist Lawrence Evans is aware of the debate over "science studies." In the *Faculty Forum*, he dismantles some assertions of *The Golem: What Everyone Should Know about Science*, a popular sociology-of-science book arguing, among other things, that experiments scientists call critical to Einstein's relativity theory were nothing of the sort (Collins and Pinch 1994, Evans 1996). This props up the key notion of authors H. M. Collins and Trevor Pinch that contrary to the "myths of science," experiments never settle anything, but social arrangements do. *The Golem* contains, its title declares, everything you need to know about science.

Evans shows that what one supposedly needs to know includes mistakes and distortions (a detailed example can be found in Allan Franklin's chapter in this volume). Focusing on one of the book's "case studies"—relativity, the Michelson–Morley experiment, and the 1919 solar eclipse observations—he shows that *The Golem*'s version of those findings and the acceptance of relativity is wrong. The next issue of the *Forum* contains the responses from Collins and Pinch (1996) and Duke professor Arkady Plotnitsky (1996). They do not really dispute Evans's points but, instead, make new ones. They accuse him of prejudice and "a depressingly low tone for a debate over serious issues." Plotnitsky asseverates: Evans neglects subtleties in Jacques Derrida's comments on "the Einsteinian constant." Collins and Pinch complain that Evans uses the whole of relativity physics, rather than events of the short period of their interest (to which Evans replies later, in effect, Is *that* what everybody needs to know about science? [1996]). A reader who hasn't followed Evans's exposure of *The Golem* might be impressed; after all, prejudice is repellent. But he has simply touched the rawest nerve: these worst-case case studies of science are wrong about the science and are inadequate as history.

Lloyd: If prejudice were demonstrable, Lloyd's cry would be the most serious charge. It is, specifically, that the accused turn a blind eye to big mistakes in current science discovered, and to profound transformations wrought in science, by feminism. The putative prejudices of Gross and Levitt's *Higher Superstition* are prime targets: To effect *Exclusion at Any Price* (this is a subtitle of Lloyd's paper), they misrepresent and ignore this feminist achievement. The chapter "Auspicating Gender," one among nine, 13% of the pages, receives most of Lloyd's grapeshot. She has no comment on what the book says about literary theory, ecoradicalism, animal rights activism, the AIDS-is-a-social-disease movement, or other questions with which it is concerned; the "Gender" chapter and social constructivism are the insults. Of what does it consist, this *Discrediting and Exclusion of Feminists*? (another Lloyd subhead). The proposal that some feminists favor ideology over truth; that they tend to reject, or more commonly to redefine all meaning out of, "objectivity" (as Lloyd does here and as relativists do); and that they are often hostile to rationality.[11] But more viciously, in her view, these "real" enemies of objectivity deny a major feminist contribution so far to scientific knowledge. For Lloyd that is prejudice, ipso facto.

Lloyd argues: because the legend of science is that politics don't matter, only results do, feminists' contributions to science should not be judged by their politics. I say *Amen*. And note: Lloyd is arguing that politics and results are separable! But, she says, the enemies use a double standard: they ignore the politics of nonfeminists while attacking feminists for theirs. She demands that the breakthrough results (so she regards them) of Anne Fausto-Sterling and Ruth Bleier be given the same courtesy as Kekulé's discovery of the ring structure of benzene, because "science" doesn't cavil at his generative dream of a snake swallowing its tail (Lloyd 1996, 237 ff).[12] A peculiar invocation follows, of the distinction between the contexts of discovery and justification (a distinction that some feminist epistemologists Lloyd admires reject). But Fausto-Sterling's (1993) style of generative cognition—for example, that biology has gone astray because it views nature falsely—as a capitalist enterprise "red in tooth and claw," rather than as a socialist collective, seems to me not comparable with Kekulé's ophidian revery, nor are her "results" comparable as scientific knowledge.[13]

This section of Lloyd's paper is decorated with ringing claims from Fausto-Sterling and Bleier (such as: feminism has provided a new vision of science and has detected "gross procedural errors," and science has an "inaccurate understanding of biology's role in human development"), claims whose science Lloyd neither presents nor evaluates. She ends with what is supposed to be a feminist insight: "it makes for *better science*, to encourage the training and full participation of informed researchers with a variety of background experiences, preconceptions, and viewpoints" (1996, 238; italics in original).

Why, yes, but why is this a feminist science insight and not, say, the view of contemporary liberal society? Thousands of U.S. laboratories, most of them surely directed by non (not anti) feminists, are populated by students and coworkers of all genders (I accept the spectrum) from the four corners of the world. "Europeans" routinely go to Asia for postdoctoral research. Whatever exclusionary impulses there were in science of the past (they were certainly there; I have experienced a few of them), they do seem to be in the past—at least in the West. To their elimination, those of the seven "enemies of objectivity" I know well enough to judge have long been dedicated. To their elimination, every scientific society, including the 143,000-member American Association for the Advancement of Science, is officially committed on the only sound basis:

that brains and scientific talent are neither gender nor culture specific and that different points of view, when relevant to the research in progress, are worth having. Feminism, in an earlier and unproblematic form, had much to do with the growth of this attitude, but what the current feminist science criticism does is false. At the extreme, it goes in the opposite direction, of claiming superior "objectivity" for the oppressed.

Lloyd: "Thus, it seems that the threatening aspect of feminist and social analyses of the sciences is the revealing of the interdependence of social and cultural context, scientific practices and products, and the actions of individuals. . . . They [the critics] seem willing to acknowledge that, in *general*, scientists are social, human, beings, but unwilling to admit that the social, cultural context played any significant role in any *particular* case" (1996, 242; italics in original). This is false. For a paradigmatic argument about the role of social context in the content of science, the reader could follow back from the quotation from Max Perutz (1995) at the beginning of this chapter to the original exchange on Gerald Geison's biography of Louis Pasteur (1995). There, despite claims to the contrary, no convincing case is made for a significant role of Pasteur's sociocultural interests (of which he had plenty) in the content and acceptance of his scientific results, but the possibility is admitted by all parties. And remember, Lloyd herself argues elsewhere that results and politics are separable.

Lloyd: Gross and Levitt misinterpret Helen Longino's views—with prejudice—as "an overriding commitment to a specific scientific conclusion for ideological reasons" (1996, 245). It is impossible to dissect this claim without reprinting the six pages devoted in *Higher Superstition* to Longino's argument, including her views on theories of neurological development. It is placed at the end of a chapter that was designed to move up from the preposterous ("feminist algebra") to the serious (Fox Keller and Longino). Nevertheless, the intelligent Longino makes, in the end, this argument: Scientific content is inextricable from context; therefore we can choose the context we like, but it will not be possible to do science in a feminist context "until we change present conditions." And present conditions (social, academic-political, epistemological?) are bad. Is that not ideology?[14]

A recurrent theme in Lloyd's indictment is that because of their prejudice, Gross and Levitt (and others like them) appropriate feminist science (hide it) by calling it "just good science" or that they ignore it in "their non-engagement with her [Fausto-Sterling's] detailed scientific objections" or render Ruth Bleier's work "invisible" (1996, 246). (Never mind that Lloyd does nothing in her essay to make visible the scientific content of any such works.) Again, the analysis of feminist claims about scientific knowledge, in one chapter among nine, required selection. We devoted most of the available space to a few influential (outside, as well as inside, women's studies), well-qualified feminist writers. In our view, three of those were Sandra Harding, Evelyn Fox Keller, and Helen Longino. We made no attempt to render anyone, scientifically trained or not, "invisible"; all Lloyd's favorites are mentioned. I had written, and we refer in the book to, an earlier paper (1992) that discusses Fox Keller's early scientific research on slime-mold aggregation. It was sound but no more feminist than Platonist or Martian.

We wanted to consider the most recent, scientifically substantive feminist claims we could find of new scientific discoveries due to feminism, or corrections to existing science required by it. Those we found in a manifesto of the Biology and Gender Study Group (BGSG), whose leading spirit is an active biologist and author: Scott Gilbert (I have taught from his textbook). It was clear that this group had searched the literature for the most telling examples of feminist science they could find. In *Higher Superstition*, we gave the BGSG's claims (centered on embryology, including the fabricated "passive egg" issues) a detailed analysis. We found nothing new, certainly nothing feminist, except for some metaphor mongering, and our judgment has not been refuted.[15] Lloyd doesn't mention it. Space did not permit a digression to women primatologists (one of those categories we are accused of hiding), and some of them have been important. I think, for example, of Sarah Blaffer Hrdy, whose status among feminists varies, however, because she is also an advocate of sociobiology. And I grant that those women (not necessarily their epistemologies) may have changed primatology, although not nearly so much as proclaimed; nor has that change moved primatology or "science" toward the postmodern drama that Donna Haraway seems to equate with feminist thought.[16]

Several years have passed since we completed the manuscript of *Higher Superstition*. What now are the scientific revelations to which Lloyd alludes so fulsomely? In the libraries at Harvard University and the University of Virginia, I found only one recent book that looked as though it might have answers: *Women Changing Science*, by Mary Morse (1995). But unfortunately, it is not about changes in scientific knowledge: it is about women's workplace problems. Those are important, but not what was wanted. This book sports the obligatory subheads: "Science as a Male Pursuit," "Knowledge as Male," "The Contextual Nature of Scientific Inquiry," "Women's Scientific Styles," "Chaos Theory as Holist and Feminine."[17] Morse's examples of "illustrious female scientists" are Marie Curie and Jane Goodall. A more impressive list could have been offered had the author inquired outside feminist circles.[18] In this book, however, I did eventually find one claim of breakthrough science "practiced in a feminist context." This is an article by Margie Profet, of which Morse writes:

> Profet's *ground-breaking look* at an essential biological process *turns on its head the accepted tenet* that menstruation occurs *only* to prepare the uterus for fertilization. Profet's work, for which she was awarded a $225,000 MacArthur Foundation grant, is of the type that *may never have come out of nonfeminist science*. (1995; italics added)

Profet makes no such explicit claims. Her "Menstruation as a Defense against Pathogens Transported by Sperm" is an article in the *Quarterly Review of Biology* (1993), a hypothesis with a lengthy review of the literature of menstruation. Its impetus is adaptationist evolutionary biology, and it includes summaries of adaptationist principles and nods to leading exponents (G. C. Williams, Ernst Mayr). The hypothesis is that the function of menstruation is to purge the reproductive tract of sperm-borne pathogens. I don't know whether Profet is a feminist, although she may be: her abstract begins with "Sperm are vectors of disease" (1993, 335). Indeed they are, but as Profet notes later, the female reproductive tract is a reservoir; that's where the pathogens usually come from. No matter. This is a good study in "adaptationist" biology, a method whose goal is to understand, via the analysis of function, how structures and systems evolve.

Scientific speculation on the functions of menstruation has a long history, and it is true that the modern literature has been especially concerned with the remarkable endocrine control system and the cell biology of the endometrium. Still, the idea of a cleansing function of menstruation is not new. The literature contains speculations (just as Profet's is a speculation) that go far beyond "preparation for fertilization." Profet cites them. S. Thomas Shaw and Patrick C. Roche, for example, considered the cleansing function in their chapter for Baird's textbook (1985) on menstrual bleeding. If I say that this is "good scientific thinking" but not "feminist science," I shall earn Lloyd's (and perhaps Morse's) rancor. But this work would be "feminist science" only if the question, What is menstruation for?, had never been asked by nonfeminists or, more specifically, if a feminist-inspired answer were both true and unprecedented. The vast literature that Profet surveys—competently—denies both. Questions are raised (as by all such hypotheses), some of which Profet asks (and seeks answers to in the literature).

There is, however, a subtler issue. Can it be that antipathogenesis is the sole function of menstruation? No. No organ system or physiological process has only one function. Functions are like Linnaeus's taxons: they are conceptual subdivisions of a continuum. Physiological processes are products of the history of life and thus of manifold, selectively relevant contingencies. As a simple example, mammalian renal function eliminates liquid wastes. It might have been said, 80 years ago, that the only function of the kidney is micturition. But the kidney also controls water balance; it is a primary osmoregulator. It conserves, rather than eliminates, critical electrolytes and metabolites (e.g., Na^+, glucose, creatine, ascorbic acid). It is an endocrine organ; its hormones include renin and erythropoietin, which regulate blood pressure and hematopoiesis.

In short, menstruation surely came along, in ancestral mammals, as an advantageous modification of existing functions, in the context of internal fertilization and live bearing. Whatever selective advantages it might have conferred in its first manifestations, it would be amazing if it didn't now have a range of functions and adaptive values. So we were taught in first-year physiology at the University of Pennsylvania's School of Medicine back in 1951. What is newer about Profet's take is her neo-Darwinian emphasis on biological costs (loss of blood, energy spent in cycling endometrium) versus benefits. Her emphasis is on the cleansing benefit. It is a good thesis, done as others in the field do and have written. Perhaps feminism pointed her toward this problem, but she could just as well have been motivated by interest in the biochemistry of blood clotting. If there is any "-ism" salient in the inquiry, it is neo-Darwinism, not feminism.

Summation

In the *New York Review of Books*, Fiona McCarthy recently published (1997) a learned review of Olwen Hufton's *The Prospect before Her: A History of Women in Western Europe*. In it, she offers what seem to me important thoughts on women's history. I am not a historian, so this judgment may be wrong. Nevertheless, McCarthy's thoughts should commend themselves to the attention of anyone who believes in the value of uncovering what was important and was not so treated when history was just kings and politics. McCarthy's discomfort with some women's historiography is applicable to feminist science criticism:

> It is also important to acknowledge that the unprecedented growth of women's studies in the last twenty-five years has produced a great deal of very dull writing and—more dangerous—doubtful scholarship. Women's history has been vulnerable to the slipshodness of judgment and stridency of language that almost inevitably develop when an academic subject declines into a cause.

Yes, stridency is to be avoided, and doubtful scholarship for a good cause hurts it. Perhaps in the current argument over radical science studies, including the feminist versions, there has been too much stridency. Perhaps not: when questions of knowing are at stake, energetic debate is unavoidable, and "strident" often echoes "shrill." Recently, those questions have been addressed ostentatiously and one-sidedly in the academy; there has been insufficient argument until now. To argue against those ostentatious redefinitions of "knowledge" and "objectivity" has been to mark oneself a "traditionalist" or worse. This much, however, I know: Lloyd's indictment (like others by defenders of her clients) is an elaborate search for *excuses* to dismiss the arguments of her targets. Fair-minded observers should return the verdict "Not Proven."

Notes

Writing this paper has been difficult because I was not sure that it should be written. Discussions of it with Susan Haack, Gerald Holton, Norman Levitt, Janet Radcliffe Richards, and, of course, editor Noretta Koertge have been helpful, although the decision to proceed might prove in the end not to have been. In any case, the flaws in this text are mine alone. I acknowledge assistance for this work from the Klingenstein Fund.

1. One wonders about whose painstaking definitions and identifications of relativism are being sneered at. Laudan's in 1990? Boghossian's in 1996?

2. *Social Text* is a journal of cultural studies edited by Bruce Robbins and Andrew Ross and published by Duke University Press. "Science Wars," to which the epigraphic quotation from Harold Fromm refers, was the Spring–Summer 1996 issue of this journal.

3. Nothing like the "surprising number" Nelkin invokes. What is a "surprising" number? As far as I know, the number of scientists doing what Nelkin disapproves is fewer than a dozen.

4. In this section of her paper, Lloyd also provides a paraphrase in which Gross and Levitt denounce "social scientific and feminist authors" as "genocidal" (220). Nothing like that appears anywhere in the book. It is a pure invention.

5. Not, presumably, just necessary but psychologically necessary. Mythic stature and authority are, in other words, a deep, perhaps neurotic need, a hang-up of the "scientists" Lloyd is writing about.

6. Indeed, Susan Haack's analysis (1996) is in the same volume as Lloyd's.

7. I am aware of the danger: If I suggest that Lloyd has turned a blind eye to evidence unfavorable to her case, then she can return the compliment. And indeed she does. Among the few testable accusations Lloyd makes is that the accused systematically ignore great femi-

nist science and scientists. This is, however, false on several counts; see the section of this chapter "Crying Prejudice to Change the Subject."

8. See Kuhn 1983, 30: "The belated emergence during the last two decades of an internal sociology of science, one that takes scientific communities and their products as objects of sociological scrutiny, has shifted sociological concerns without altering the model used to pursue them. Interests remain the dominant factor that practitioners of the new field employ in explanation, and the interests they deploy remain predominantly socio-economic. To me the result often seems disaster."

9. This essay was published in shortened form as "Knowledge and Propaganda: Reflections of an Old Feminist," in *Partisan Review*, Fall 1993, 556-64, and reprinted in *Our Country, Our Culture*, ed. Edith Kurzweil and William Phillips (Boston: Partisan Review Press, 1994), 54–66. It was originally read at the 1992 APA meeting.

10. The matter of Ross's crude design-quoting of Holton and others remains unsettled as I write. There is no evidence that Ross or his publishers will do anything about it. For a depressing glimpse of damage done to "the masses"—Third World peasant women in this case—by the posturing of Western intellectuals (such as Ross) in empathy with the masses, see Nanda 1994.

11. For the judgment of another philosopher, Lloyd's Berkeley colleague John R. Searle, on the truth of this proposal, see his "The Mission of the University: Intellectual Discovery or Social Transformation?" (1994).

12. Encountering this, a sophisticated friend called it "a breathtaking non sequitur."

13. Compare Rocke, 1992: "In three papers published in 1865 and 1866, August Kekulé, professor of chemistry at the University of Ghent, proposed a theory of the structure of benzene that provided the basis for the first satisfactory understanding of aromatic compounds, a very large and important class of organic substances. It would be difficult to overestimate the impact of these papers. Within a decade most chemists accepted the theory as empirically verified and heuristically invaluable."

14. In case it's not clear, consider (1) the definition of ideology: a body of doctrine, myth, and so on, with reference to a social objective, design, or plan, together with strategies for its implementation; and (2) Sandra Harding on scientific rationality, back in 1986: "In our culture, reflecting on an appropriate model of rationality may well seem a luxury for the few, but it is a project with immense potential consequences: it could produce a politics of knowledge-seeking that would show us the conditions necessary to transfer control from the 'haves' to the 'have nots'" (Harding 1986, 20).

15. There is a newer example of feminist metaphor mining in the same scientific area (cell and molecular biology, development): Spanier's *Im/Partial Science*, 1995. Ripe examples of the genre turn up in the sections headed "As the Male Signifier, the Plasmid Is Dubbed an Essential Tool of Recombinant DNA," and "Distortions of Reproductive Cells and Fertilization." Reproductive cells and fertilization being my own specialty, I exercise self-restraint by limiting this note to the reference.

16. Anthropologist Matt Cartmill, on Donna Haraway's *Primate Visions* (New York: Routledge, 1989), which may be (one cannot be sure) a celebration of women's contributions to primatology: "This is a book that contradicts itself a hundred times; but that is not a criticism of it, because the author thinks contradictions are a sign of intellectual ferment and vitality. This is a book that systematically distorts and selects historical evidence; but that is not a criticism, because its author thinks that all interpretations are biased, and she regards it as her duty to pick and choose her facts to favor her own brand of politics . . . the author likes that sort of prose and has taken lessons in how to write it, and she thinks that plain, homely speech is part of a conspiracy to oppress the poor. . . . Despite Haraway's protestations to the contrary, Primate Visions strikes me as an expression of hostility and contempt, to the scientific enterprise in general and to primatologists in particular" (1991).

17. "Chaos theory's focus on the nonlinear qualities of nature have [sic] been described by such terms as curvaceousness, fluidity, complexity, and attraction—all characteristics attributed to the Western concept of femininity" (Morse 1995, 14).

18. One example: the extraordinary physicist Lise Meitner, about whom a new biography has appeared, which gets, as it happens, an admirable review/memoir from one of the accused—M. F. Perutz (1997).

References

Angell, M. *Science on Trial: The Clash of Medical Evidence and the Law in the Breast Implant Case*. New York: W. W. Norton and Company, 1996.

Biehl, J. *Rethinking Ecofeminist Politics*. Boston: South End Press, 1991

Bird, A. Squaring the circle: Hobbes on philosophy and geometry. *Journal of the History of Ideas* 57(2), 217–23, 1996

Boghossian, P. What the Sokal hoax ought to teach us. *Times Literary Supplement*, December 13, 14–15, 1996.

Boorstin, D. J. *The Discoverers*. New York: Vintage Books, 1985.

Cantor, P. A. Stephen Greenblatt's new historicist vision. *Academic Questions*, Fall, 21–36, 1993.

Cartmill, M. Review of *Primate Visions*, by Donna Haraway. *International Journal of Primatology* 12(1): 67–75, 1991.

Collins, H. M., and T. Pinch. *The Golem: What Everyone Should Know about Science*. Cambridge: Cambridge University Press (Canto ed.), 1994.

———. Golem authors respond to Lawrence Evans. *Faculty Forum* (Duke University), November, 1–2, 1996.

Collins, H. M., and S. Yearley. Epistemological chicken. In *Science as Practice and Culture*, ed. A. Pickering. Chicago: University of Chicago Press, 1992.

Dickson, D. The "Sokal Affair" takes a transatlantic turn. *Nature*, 305 (30 January), 381, 1997.

Duclos, D. 1997. Sokal n'est pas Socrate. *Le Monde*, 3 January, 1997.

Evans, L. Should we care about science "studies?" *Faculty Forum* (Duke University), October, 1–3, 1996.

Fausto-Sterling, A. Is nature really red in tooth and claw? *Discover*, 24 April, 24–27, 1993.

Fromm, H. My science wars. *Hudson Review* 49(4), winter: 599–609, 1997.

Geison, G. L. *The Private Science of Louis Pasteur*. Princeton: Princeton University Press, 1995.

Gross, B. R. 1994. What could a feminist science be? *The Monist*, 77(4), October, 434–43, 1994.

———. On the 'gendering' of science. *Academic Questions*, 5(2), 10–23, 1993.

Gross, P. R., and N. Levitt. *Higher Superstition: The Academic Left and Its Quarrels with Science*. Baltimore: Johns Hopkins University Press, 1994.

Gross, P. R., N. Levitt, and M. W. Lewis, (editors), *The Flight from Science and Reason*. Vol. 775 of *Annals of the New York Academy of Sciences*. New York: New York Academy of Sciences, 1995.

Haack, S. Epistemological reflections of an old feminist. *Reason Papers*, Fall, 31–43, 1993.

———. Science as social?—yes and no. In *Feminism, Science, and the Philosophy of Science*, ed. L. H. Nelson & J. Nelson, pp. 79–93. Dordrecht: Kluwer Academic Publishers, 1996.

Harding, S. *The Science Question in Feminism*. Ithaca: Cornell University Press, 1986.

Holton, G. *Science and Anti-Science*. Cambridge, MA: Harvard University Press, 1993.

Kuhn, T. S. (1983). Reflections on receiving the John Desmond Bernal award. *4S Review*, 1(4), 26–30.

Larson, J. L. *Interpreting Nature: The Science of Living Form from Linnaeus to Kant*. Baltimore: Johns Hopkins University Press, 1994.

Laudan, L. *Science and Relativism: Some Key Controversies in the Philosophy of Science*. Chicago: University of Chicago Press, 1990.

Levitt, Norman. No co-conspirator. *Skeptic* 5(1), 1997, 26–28.

Lloyd, E. A. Pre-theoretical assumptions in evolutionary explanations of female sexuality. *Philosophical Studies* 69, 139–53, 1993.

———. Science and anti-science: objectivity and its real enemies. In *Feminism, Science, and the Philosophy of Science*, ed. L. H. Nelson & J. Nelson, pp. 217–259. Dordrecht: Kluwer, 1996.

McCarthy, F. How the other half lived. *The New York Review of Books*, 6 February, 4–8, 1997.

Morse, M. *Women Changing Science: Voices from a Field in Transition*. New York: Insight Books/Plenum Press, 1995.

Nanda. M. The science wars in India. *Dissent*, Winter, 1997, 78–83.

Nelkin, D. What are the science wars really about? *The Chronicle of Higher Education*, 26 July, 1996, A52.

Perutz, M. F. The pioneer defended. *The New York Review of Books*, 25 December, 1995, 54–8.

———. Letter. *The New York Review of Books*, 6 February, 1997, 41–2.

———. Passion for science. *The New York Review of Books*, 20 February, 39–42, 1997.

Pinnick, C. Cognitive commitment and the strong program. *Social Epistemology*, 6(3), 289–98,1992.

———. Feminist epistemology: implications for philosophy of science. *Philosophy of Science*, 61, 646–57, 1994.

Plotnitsky, A. Another response to Lawrence Evans. *The Faculty Forum (Duke University)*, November, 2–6, 1996.

Profet, M. Menstruation as a defense against pathogens transported by sperm. *The Quarterly Review of Biology*, 68(3), September, 1993, 335–96.

Rocke, A. J. Kekule's benzene theory and the appraisal of scientific theories. In *Scrutinizing Science: Empirical Studies of Scientific Change*, ed. A. Donovan, L. Laudan, & R. Laudan. Baltimore: Johns Hopkins University Press, 1992.

Ross, A. Science backlash on technoskeptics. *The Nation*, 2 October, 1995, 346–50.

———. Introduction. *Social Text*, 46–47, 1996, 1–13.

Schaffer, S. Glass works: Newton's prisms and the use of experiment. In *The Use of Experiment: Studies in Natural Science*, ed. D. Gooding, T. Pinch, & S. Schaffer, pp. 67–104. Cambridge: Cambridge University Press, 1989.

Schiebinger, L. The loves of the plants. *Scientific American*, February, 110–5, 1996.

Schmaus, W., U. Segerstråle, and D. A. Jesseph. A manifesto. *Social Epistemology*, 6(3), 243–65, 1992.

Searle, J. R. The mission of the university: intellectual discovery or social transformation? *Academic Questions*, 7(1), 80–85, 1994.

Shapin, S., and S. Schaffer. *Leviathan and the Air Pump: Hobbes, Boyle, and the Experimental Life*. Princeton: Princeton University Press, 1985.

Shapiro, A. E. The gradual acceptance of Newton's theory of light and color, 1672–1727. *Perspectives on Science*, 4(1), 59–140.

Shaw, S. T., and P. C. Roche. The endometrial cycle: aspects of hemostasis. In *Mechanism of Menstrual Bleeding*, ed. D. T. Baird & E. A. Michie, pp. 7–26. New York: Raven Press, 1985.

Siano, B. The great political correctness conspiracy hoax. *Skeptic*, 4(3), 52–61, 1996.

Sonnert, G., and G. Holton. *Gender Differences in Science Careers*. New Brunswick, N.J.: Rutgers University Press, 1995.

————. *Who Succeeds in Science? The Gender Dimension.* New Brunswick, N.J.: Rutgers University Press, 1995.

Spanier, B. B. 1995. *Im/Partial Science.* Bloomington: Indiana University Press, 1995.

Tobias, S. Review. *Physics and Society*, 25(3), 13, 1996.

Weil, N. La Mystification pedagogique du professeur Alan Sokal. *Le Monde*, 20 December, 1996.

7

Is Darwinism Sexist?
(And if It Is, So What?)

Michael Ruse

> Darwin's theories were conditioned by the patriarchal culture in
> which they were elaborated: he did not invent the concept of sexual
> difference. The late nineteenth-century elaboration of that concept
> took place in the context of the growing demands of women. Its
> success owed much to women's own ambivalence about their role
> in society and the nature of womanhood. The defeat of the move-
> ment did not depend on the intervention of the men of science.
> But the support offered by a science that was for the most part
> accepted in its day as objective and value-free immeasurably
> strengthened patriarchy for decades to come. The *Origin* provided
> a mechanism for converting culturally entrenched ideas of female
> hierarchy into permanent, biologically determined, sexual hierarchy.
> Erskine, "The Science of Female Inferiority," 118

To say that evolutionary theory—Darwinian evolutionary theory in particular—
comes saddled with a bad reputation among feminists is akin to saying that Hitler
had a thing about the Jews. For about 20 years now, since the human sociobiological
controversy of the 1970s, Darwinism—the theory of organic evolution through natural
selection—has been right at the top of the feminist hate list (Ruse 1979b). Both the
work of Charles Darwin himself and the work of his modern supporters have been
held to scorn and ridicule, judged to be gross chauvinistic prescriptions masquerad-
ing as genuine science. Or perhaps they are genuine science, in which case the rot is
even worse than seems on first sight, for the moral corruption then clearly lies right
at the heart of the whole enterprise (see, e.g., Birke 1986; Fausto-Sterling 1985, 1993;
Haraway 1989; Harding 1986; Hubbard 1983; Keller 1984; Richards 1983; Russett 1989;
Sleeth Mosedale 1978; and many more).

In fact, as is often the case when evangelicals seize hold of a theme, things are rather
more complex than they seem on first sight. They are also rather more interesting—
about both Darwinism in particular and science in general.

Is Darwinism Sexist?

Well yes, you can make a case for the prosecution. Indeed, we did not need feminists
to tell us this, for—if we mean by "sexist" the "valuing of one sex (usually males)

above the other, without good grounds"—scholars of the subject were there at least as quickly (see my *Darwinian Revolution* [Ruse 1979a], a work of deliberate synthesis). Certainly, if you look at Darwin's own writings, the *Descent of Man* in particular, you will find some pretty conventional Victorian sentiments about the sexes and their relative worths. "Man is more courageous, pugnacious, and energetic than woman, and has a more inventive genius" (Darwin 1871, 2: 316). In compensation, woman has "greater tenderness and less selfishness" (326). You could be reading a novel by Charles Dickens.

And if you skip over to the present, ignoring all the non-Darwinian evolutionists who were saying very similar things, you will find the same refrain. On reading Edward O. Wilson's *Sociobiology: The New Synthesis* (1975), you will be little surprised that the critics' ire was raised. Wilson's later book, *On Human Nature* (1978), is even worse.

> Males [in the majority of animal species] are characteristically aggressive, especially toward one another and most intensely during the breeding season. . . . It pays males to be aggressive, hasty, fickle, and undiscriminating. In theory it is more profitable for females to be coy, to hold back until they can identify males with best genes. In species that rear young, it is also important for the females to select males who are more likely to stay with them after insemination.
>
> Human beings obey this biological principle faithfully. (Wilson 1978, 125)

Or what about Richard Dawkins in the *Selfish Gene* (1976)? Females, we learn, have two options open to them. There is the "he man" strategy in which they align themselves with the biggest hulk in the hope that, in turn, their sons will be hulklike and thus sufficiently sexy to have lots of mates. Or there is the "domestic bliss" strategy, in which females play the little woman, offering grateful males a taste of home-life nirvana.

Say what you like. The picture in the Darwinian world of the female is hardly inspiring or enviable. Which would you rather be under the Darwin or the Wilson or the Dawkins scenarios? A male or a female? The feminists are right. Darwinism is sexist.

Is Darwinism Feminist?

Although the story usually stops here, it should not. What the critics ignore—perhaps prefer not to know—is the historical fact that just as Darwinism was used as a vehicle for sexist views, so also it was, and is, used as a vehicle for feminist views. In the past, consider Alfred Russel Wallace, the codiscoverer of natural selection. He held the view that human progress depends ultimately on female sexual selection, that men cannot be trusted with the future of our race. Fortunately, in the future, young women will take over the reins, choosing as mates only those males with the highest moral and intellectual properties. Thus upward progress is guaranteed:

> In such a reformed society the vicious man, the man of degraded taste or of feeble intellect, will have little chance of finding a wife, and his bad qualities will die out with himself. The most perfect and beautiful in body and mind will, on the other hand, be most sought and therefore be most likely to marry early, the less highly endowed later,

and the least gifted in any way the latest of all, and this will be the case with both sexes. From this varying age of marriage, . . . there will result a more rapid increase of the former than of the latter, and this cause continuing at work for successive generations will at length bring the average man to be the equal of those who are among the more advanced of the race. (Wallace 1900, 2: 507)

You may object that if Wallace truly thought any of this to be remotely possible— that young women in the future would freely mate with only the best of the male crop—then his knowledge of human nature was about on a par with his touching beliefs in the integrity of spiritualists. But the fact remains that Wallace made his claims in the name of evolution. And, after Darwin, if any nineteenth-century evolutionist deserves the label "Darwinian," it is he.

Likewise today, Wilson's student, the anthropologist Sarah Blaffer Hrdy, in *The Woman That Never Evolved* (1981), has presented a feminist picture of human evolution and resultant nature that would not be out of place in *Ms.* magazine. She argues that the reason human females uniquely conceal their ovulation—they do not come into heat—is to keep the poor males on a parental care-giving string. Because from one brief sexual encounter, a male does not know, he cannot know, whether or not he is the father of the resultant child; he must stay around—keeping out rivals, but incidentally also helping with child rearing. If he does not, the woman will cut him off. Lysistrata in the genes, one might say. Feminism made flesh, one might also say. And, like Wallace, Hrdy cannot be dismissed as a side phenomenon, a kind of scientific freak. She is one of the most successful evolutionists active today and a member of the National Academy of Sciences, no less.

Popular Science

We have a veritable Hegelian contradiction. Darwinism is sexist. Darwinism is feminist. How can this be? The obvious answer is that, in some sense, Darwinism is simply a clotheshorse on which people hang any ideology that they find comforting. You are a sexist? Darwinism will accommodate you. You are a feminist? Darwinism will accommodate you, too. This suggests that there is either something distinctively cultural (or culture welcoming) about Darwinism in particular or (assuming Darwinism to be typical) there is something cultural (or culture welcoming) about the scientific enterprise itself.

The evolutionist's answer will be that we are failing to distinguish real or professional or mature evolutionary work from that in the popular domain (Ruse 1996). We will be told that we must draw a distinction between the evolution of the full-time researcher in the field and the evolution given to the public in museums, magazine articles, and popular works like those written by Stephen Jay Gould. The real standard against which we should make our judgments is professional science—the science of quantification, of measurement, of predictive accuracy, of unificatory power, of consistency, of fertility (in the sense of pointing to new problems and fertile fields of inquiry), of elegance or simplicity. This is the science of "epistemic standards" (McMullin 1983, 15). Popular science is the area in which values and judgments and ideologies are not only expected but welcomed. And it is in this area exclusively that we find judgments about the relative worths of males and females, wherever you come down on the divide. Professional or mature science is neutral in value and culture. It is trying to give—and to give only—a disinterested view of objective reality.

Is this response effective, judged by the examples of evolutionary reasoning that we have just considered? Certainly, some of our instances do seem to be cases of popular science rather than mature professional science, particularly in regard to contemporary instances. Dawkins's *Selfish Gene* is obviously a work of the popular domain. With its catchy title and racy examples, it is a paradigm of work produced for the general reader—who properly responded by making it a best-seller. Hrdy's *Woman That Never Evolved* had the same fate. Here, too, the title is an indicator, echoing the thrilling story of World War II: *The Man That Never Was*. Likewise, the content is for the nonexpert; no mathematics is a good indicator of that. Even Wilson's work, especially *On Human Nature*, belongs more in the public domain than in the fully professional realm. As Wilson himself wrote at the beginning of that particular work: "*On Human Nature* is not a work of science; it is a work about science" (1978, x).

What about our nineteenth-century authors? Wallace is easy to categorize. He was certainly capable of writing in-the-trenches professional science—a piece on butterfly mimicry, for instance (Wallace 1866)—but most of the time he was writing commercially for the public: to make a living and to spread the gospel for the latest cause with which he had become enthused—spiritualism, socialism, antivaccinationism, nolife-on-Mars-ism (Wallace 1900, 1903, 1905, 1907). The things he wrote on human evolution, late in his life, owed more to his selective reading, in this case, the futuristic novel *Looking Backward 2000–1887* (1887) by the American Edward Bellamy (who had precisely the same ideas that Wallace adopted)—than to anything resembling genuine scientific research.

Darwin is perhaps more problematical, but he, too, hardly qualifies as fully professional when he is writing on our species, especially *The Descent of Man*. Like Wallace, Darwin could write fully professional science—the barnacle taxonomy, for instance (Darwin 1851a, b; 1854a, b)—and one might even include here the material (two-thirds of the work) in the *Descent* on sexual selection. But his material on our own species was all taken at second hand and was not at the cutting edge of the science, as is to be found in his followers like Huxley (Greene 1977). It was more an old man putting together his reflections because they were expected of him. Since the work was clearly directed at the general reader, it is not surprising that Darwin included all the prejudices of his sex—not to mention his class and his race—as when he turned to such subjects as the virtues of capitalism and the burdens of the white man.

Professional/Mature Science

I think there is considerable merit in this response. As scholars of the subject have shown, even though Darwin is rightly called the "father of evolutionary theory," there is something deeply troubling about the level of most of his writings on the subject. He tends to address the audience of his patrons—that of his father and his uncle Josh—rather than his fellow scientists (Desmond and Moore 1992; Ruse 1979a, 1996). The same is true of the other evolutionists just discussed. But if we grant this, what about the other side to the case? Is the real scientific work—the mature work produced by professionals for professionals—as value free, in particular as sexism-value free, as the evolutionist would claim?

It is certainly the case that when thinking now at this professional level, evolutionists have been very interested in sex. Indeed, the actual solution to the very existence of sexuality has been labeled the biggest outstanding problem in need of solution. But is the writing sexist (here or elsewhere)? Look at the classic discussion on sex ratios by R. A. Fisher, a man who in his private life treated his wife like chattel (Box 1978). Talking about the care of offspring by their parents, he writes:

> Let us consider the reproductive value of these offspring at the moment when this parental expenditure on their behalf has just ceased. If we consider the aggregate of an entire generation of such offspring it is clear that the total reproductive value of all the males in this group is exactly equal to the total value of all the females, because each sex must supply half the ancestry of all future generations of the species. From this it follows that the sex ratio will so adjust itself, under the influence of Natural Selection, that the total parental expenditure incurred in respect of children of each sex, shall be equal; for if this were not so and the total expenditure incurred in producing males, for instance, were less than the total expenditure incurred in producing females, then since the total reproductive value of the males is equal to that of the females, it would follow that those parents, the innate tendencies of which caused them to produce males in excess, would, for the same expenditure, produce a greater amount of reproductive value; and in consequence would be the progenitors of a larger fraction of future generations than would parents having a congenital bias towards the production of females. Selection would thus raise the sex-ratio until the expenditure upon males became equal to that upon females. (Fisher 1930, 142–43)

The whole point of the discussion is that if there is an imbalance between males and females, then natural selection will act to right the order. Unless you insist that the very fact of drawing attention to sex is sexist, there is nothing particularly sexist here. If you are redefining the notion of "sexism" to mean something like "simply talking about sex," then at the very least, you must provide reasons why anyone should be concerned. After all, are not the feminists as guilty of this sin (if such it be) as are the male chauvinists?

This is not the end of the discussion, however. The feminist critic will argue that although it is certainly the case that evolutionary biology (at this professional level) is not necessarily sexist, it has a tendency to be so. For a start, there is the disproportionate attention paid to males. The problems and the solutions may not themselves be sexist, but in the unequal choice of topics, sexism shows through. Nevertheless, although I certainly concede that this is a possibility, I have yet to see hard empirical evidence that this is in fact so. It is true that if you look at certain areas of evolutionary research, you are likely to see more interest in males. I suspect that sexual selection work qualifies here, because by and large (though not inevitably), it is the males who are the more gaudy and striking in appearance and behavior. But if you look in other areas, the interest is in females. No one could say that the topic of hymenopteran sociality is a minor pursuit—it has always been one of the chief fascinations of evolutionists from the *Origin* to (especially to) the present—but generally it is the females, the queens and the workers, who are the focus of attention, rather than the males, the drones. Everyone expects an organism to be dronelike if it can get away with it. But why would an organism ostensibly give up its reproduction for the benefit of others? That is a real problem (see Hamilton 1996).

My suspicion regarding the choice of topics is that a Fisherian balance obtains. If evolutionists start to specialize disproportionately on one sex, bright graduate students will quickly spot open academic ecological niches for work on the other sex. So if not topics, then what about motives? Could it not be that evolutionists, being predominantly male, are more interested in males, whatever sex they may end up working on? Again, I am not sure of the empirical evidence. Indeed, I am not sure how one might even set about testing such a claim.

Whatever the case, surely here that much-derided distinction between the context of discovery and that of justification still has some bite. Whatever the interests, the question is, What is produced, and how good is it? Today's leading evolutionist William Hamilton, who solved the hymenopteran sociality problem, has an intriguing solution to the existence-of-sex problem. He argues that thanks to sexuality and the consequent genetic shuffling, higher organisms like mammals and birds present a moving target to their biggest threat, parasites. That is, they can stay one step ahead of the more rapidly evolving microorganisms that prey on them (Hamilton 1980; Hamilton, Axelrod, and Tanese 1990; Hamilton and Zuk 1982, 1989). Hamilton admits candidly that he first devised this hypothesis because otherwise he could see no evolutionary point to males—the precise negative position taken by other evolutionists, like George Williams (1975) who thinks that sexuality in higher organisms is maladaptive (but for physiological and morphological reasons, it can no longer be eliminated). As Hamilton put it in an interview:

> There are several reasons for not liking Williams's position on sex. One is that things as close to us as lizards do sometimes abandon sex, and also it's annoying in ordinary human terms that human reproductive technology is going ahead so fast that soon it will probably be possible for us to abandon sex—I don't like it, and since the sex to be abandoned will likely be the male and being a man myself, I don't like to think we have to be in this position of regarding males as really useless. And so I produced a reason of why we are here and why we should be kept going. (Interview with author, spring 1991)

In fact, I am not even sure that this motive is really that sexist. Hamilton's starting position is that females are indispensable and that males prima facie are not. But even if it is sexist, the finished product does not show the motives. And it must in any case stand the test of experience, not simply the approval of those who share Hamilton's evolutionary wish list. As it is, the general opinion at the moment on Hamilton's hypothesis is probably "not proven." But the proving or not proving will come through epistemic criteria like predictive fertility and the ability to withstand falsification attempts. It will not be a matter of cultural values, sexist or otherwise.

Values through Metaphor

Feminist critics will respond that although there may be some truth in the "context of discovery/context of justification" dichotomy, nevertheless much science—especially evolutionary science—comes through analogies and similes and metaphors based on our cultural experience, and with this come values including, if one lives in a sexist society, as we certainly do, values prizing males over females. So whether or not evolutionists intend it, even in their most professional work sexism can and does appear.

Let me say that I am sympathetic to this line of argument. I agree that cultural metaphors do play a major role in evolutionary theorizing and that with them can come society's values. From the eighteenth century on, the "division of labor" has played a starring role in evolutionary discussions. It is thought that within the individual organism there is an apportioning of different roles by different organs and that at the group level there is a utilizing of different niches by different species. It is silly to deny that with this thinking has come approval—both for what happens in nature and in reflecting on what we think should happen in human affairs. Listen, for instance, to Darwin himself, in the *Origin* (sixth edition of 1872) on the subject of progress. He sees the division as the very peak of organic perfection:

> Natural Selection acts exclusively by the preservation and accumulation of variations, which are beneficial under the organic and inorganic conditions to which each creature is exposed at all periods of life. The ultimate result is that each creature tends to become more and more improved in relation to its conditions. This improvement inevitably leads to the gradual advancement of the organisation of the greater number of living beings throughout the world. But here we enter on a very intricate subject, for naturalists have not defined to each other's satisfaction what is meant by an advance in organisation. . . . Von Baer's standard seems the most widely applicable and the best, namely, the amount of differentiation of the parts of the same organic being, in the adult state, as I should be inclined to add, and their specialisation for different function; or, as Milne Edwards would express it, the completeness of the division of labour. (1959, 221)

Despite all this approval, it does not mean that evolutionary biology suddenly becomes a matter of whim, an epiphenomenon on societal practices and norms. In the end, the division of labor is cherished in biology for its epistemic virtues. Darwin again, later in the *Origin*, on the social insects:

> I have now explained how, I believe, the wonderful fact of two distinctly defined castes of sterile workers existing in the same nest, both widely different from each other and from their parents, has originated. We can see how useful their production may have been to a social community of ants, on the same principle that the division of labour is useful to civilised man. (1959, 421)

Despite the continued approval ("useful to *civilised* man"), had the division-of-labor metaphor not led to epistemic triumphs, in this case explanatory power, it would have been discredited—as have group selection hypotheses, even though evolutionists from Wallace (1900) and Prince Peter Kropotkin (1902/1955) onward have been drawn to such hypotheses because they want desperately to believe that they are natural, hence the way that human society should be modeled (Todes 1989).

Nevertheless, the Darwin case does show that cultural values can slip between the cracks of the epistemological virtues of a theory. From Karl Marx on, people have rightly seen Darwin's work as supportive of Victorian industrialism and capitalism (see the letter from Marx to Engels, June 18, 1862, reprinted in Marx and Engels 1965,

128). But has this insertion of values happened, does this still happen, in mature evolutionary biological practice and theorizing—with respect to sexism, that is? I do not deny the possibility or that it may well have happened or indeed may still be happening. Yet before feminists start to celebrate, I want firm evidence that in practice it happens in a major way and that there are no countereffective moves.

One case in which the infiltration of sexism strikes me as plausible is in the work of the leading English evolutionist Geoffrey Parker, who first made his mark through a stunning series of papers on the mating habits of the dung fly (Parker 1969, 1970a, b, c, d, e, f, g). This somewhat unattractive organism performs its mating rituals on the tops of newly deposited pats of cow feces. The males hang around the pats, waiting for females to arrive. When they do, the males rush to grab the females and to copulate with them, after which fertilized eggs are laid in the pats, thus providing shelter and food for the next generation.

Parker's early papers are models of epistemic brilliance as they explain such anomalies as the males having expended much energy finding free females and then ceasing to copulate before all the eggs are fertilized. (This happens because time spent searching for new mates is more valuable from an evolutionary perspective than time spent getting a sperm into every last egg.) But one might also say that Parker's early papers are also pretty sexist in that they concentrate almost exclusively on the males' practices, regarding the females as archetypal Victorian females waiting there to be ravished. And certainly, by his own admission, this was not much more than Parker's own scale of values transferred down to the insect world.

> Q. The 1960s was the generation of sexual freedom. Does this tie in with interest in sexual selection? Is this a crazy idea?
>
> A. No. I don't think it's crazy. It was a time of liberation. It was a time of sexual freedom. I think it is impossible if you are interested in biology, and if you are interested in behavior, not to think about human behavior. And we certainly did. (Interview with author, spring 1991)

Before long, however, Parker himself started to change his thinking about the relative practices of males and females. (He did his earliest work when he was a young man in the late 1960s, a time of free sexual behavior for all, whereas he did his later work in the 1970s, when he was married with a daughter and when the feminist movement was starting to make its case with some vigor.) Parker's scientific work, no less epistemical rigorous and successful, started to show an increased sensitivity to females and to their role in the evolutionary scheme of things. Now his dung-fly females no longer passively await their fate. They, too, are actively involved in the choice of mates, and their behaviors become as crucial to the overall picture as are the behaviors of the males. "The manner in which females arrive at the dropping will be an important determinant of male search strategy" (Parker 1974, 102).

I would not say that this is overtly feminist work—the males still play a pretty significant role—but it is certainly a lot less sexist. This conclusion leads me to suspect that although the feminist today has every right, obligation even, to keep up the scrutiny, she should be wary of assuming that what may have happened in the past is necessarily the pattern for what happens today.

Conclusion

Let me sum up. In dealing with issues like sexism and Darwinism, professional evolutionists would urge us—with good reason, I think—to distinguish between popular science (the science of the magazine, the museum, the television program) and the work that they do as trained scientists (the science of the university, of the graduate student and supervisor, of the learned journal or monograph). In popular science, cultural or nonepistemic values are to be found openly; they are expected. These values can and have included values demeaning to women, but they can and have included values giving women preferential status. Quite apart from any reservations one might have about making too many judgments about the values of the past against the values of the present, my impression is that writers and other practitioners of popular science today are particularly sensitive to gender issues, as they are to related topics like racism. Paradigmatic is the so-called Tower of Time at the Smithsonian Institution in Washington D.C., a columnlike panorama of life's history topped by the trio of a black man, a Chinese woman, and an aged white male.

In the science of the professionals, cultural values are much less prominent. Indeed, part of the culture of professional scene is that it be culture free! This is not to say that such cultural values cannot slip in, maybe in the choice of topics, perhaps slopping over from intentions and interests, possibly from the metaphors and analogies drawn on from the culture of the day. With respect to sexist ideas, particularly given the history of our culture, one cannot say that Darwinian evolutionism has never been influenced—tainted, if you will—in this way. However, this is certainly not universally so, and I have given reasons to suggest that there are built-in factors in the culture of science (like the need to find underexploited areas of study), as well as external factors from the culture of the day, that keep (and will continue to keep) sexism at bay.

Epistemic factors—predictive fertility, consistency, elegance, and so forth—do count in professional science. They count overwhelmingly, whatever the cultural values. They provide a check not only on the possibility of gross sexism but also on those who would prescribe revisions of evolutionary biology in directions that are feminism friendly. Science—and I refer now to the whole of science and not just evolutionary biology—is a product of culture, and as such, it surely shows its origins. But it is silly to claim that it is simply a matter of fulfilling personal agendas. There are controls and guidelines—epistemic values—aimed at bringing the products of science into correspondence with external nature. This may be an ideal never achieved, but it does bring an objectivity to science, professional science in particular (Ruse 1996, 1998). Darwinian evolutionism shows this, which is why the time has now come to move on from such knee-jerk accusations of gross and unbridled sexism as I quoted at the beginning of this essay.

References

Bellamy, E. 1887. *Looking Backward 2000–1887*. New York: Tricknor.
Birke, L. 1986. *Women, Feminism and Biology*. Brighton: Wheatsheaf.
Box, J. F. 1978. *R. A. Fisher: The Life of a Scientist*. New York: Wiley.

Darwin, C. 1851a. *A Monograph of the Fossil Lepadidae; or, Pedunculated Cirripedes of Great Britain.* London: Palaeontographical Society.

———. 1851b. *A Monograph of the Sub-Class Cirripedia, with Figures of All the Species. The Lepadidae; or Pedunculated Cirripedes.* London: Ray Society.

———. 1854a. *A Monograph of the Fossil Balanidae and Verrucidae of Great Britain.* London: Palaeontographical Society.

———. 1854b. *A Monograph of the Sub-Class Cirripedia, with Figures of All the Species. The Balanidge (or Sessile Cirripedes); the Verrucidae, and C.* London: Ray Society.

———. 1859. *On the Origin of Species.* London: Murray.

———. 1871. *The Descent of Man.* London: Murray.

———. 1959. *The Origin of Species by Charles Darwin: A Variorum Text,* ed. Morse Peckham. Philadelphia: University of Pennsylvania Press.

Dawkins, R. 1976. *The Selfish Gene.* Oxford: Oxford University Press.

Desmond, A., and J. Moore. 1992. *Darwin: The Life of a Tormented Evolutionist.* New York: Warner.

Erskine, F. 1995. "The Origin of Species" and the science of female inferiority. In *Charles Darwin's "The Origin of Species": New Interdisciplinary Essays,* ed. D. Amigoni and J. Wallace, 95–121. Manchester: Manchester University Press.

Fausto-Sterling, A. 1985. *Myths of Gender.* New York: Basic Books.

———. 1993. Genetics and male sexual orientation. *Science* 261: 1257.

Fisher, R. A. 1930. *The Genetical Theory of Natural Selection.* Oxford: Clarendon Press.

Greene, J. 1977. Darwin as a social evolutionist. *Journal of the History of Biology* 10: 1–27.

Hamilton, W. D. 1980. Sex versus non-sex versus parasite. *Oikos* 35: 282–90.

———. 1996. *The Narrow Roads of Gene Land.* San Francisco: Freeman.

Hamilton, W. D., R. Axelrod, and R. Tanese. 1990. Sexual reproduction as an adaptation to resist parasites. *Proceedings of the National Academy of Science, USA* 87: 3566–73.

Hamilton, W. D., and M. Zuk. 1982. Heritable true fitness and bright birds: a role for parasites. *Science* 218: 384–87.

———. 1989. Letter to editor. *Nature* 341: 289–90.

Haraway, D. 1989. *Primate Visions.* London: Routledge.

Harding, S. 1986. *The Science Question in Feminism.* Ithaca, N.Y.: Cornell University Press.

Hrdy, S. B. 1981. *The Woman That Never Evolved.* Cambridge, Mass.: Harvard University Press.

Hubbard, R. 1983. Have only men evolved? In *Discovering Reality,* ed. S. Harding and M. B. Hintikka, 45–69. Dordrecht: Reidel.

Huxley, L., ed. 1900. *The Life and Letters of Thomas Henry Huxley.* London: Macmillan.

———. 1918. *Life and Letters of Sir Joseph Dalton Hooker.* London: Murray.

Keller, E. F. 1984. *Reflections on Gender and Science.* New Haven, Conn.: Yale University Press.

Kropotkin, P. 1902/1955. *Mutual Aid,* ed. A. Montague. Boston: Extending Horizons Books.

Marx, K., and F. Engels. 1965. *Selected Correspondence.* Moscow: Progress.

McMullin, E. 1983. Values in science. In *PSA 1982,* ed. P. D. Asquith and T. Nickles, 3–28. East Lansing, Mich.: Philosophy of Science Association.

Mill, J. S. 1869/1975. The subjugation of women. In J. S. Mill, *Three Essays.* Oxford: Oxford University Press.

Parker, G. A. 1969. The reproductive behaviour and the nature of sexual selection in *Scatophaga stercovaria* L. (Diptera: Scatophagidae)—III. Apparent intersex individuals and their evolutionary cost to normal searching males. *Transactions of the Royal Entomology Society of London,* 305–23.

———. 1970a. The reproductive behaviour and the nature of sexual selection in *Scatophaga stercovaria* L. (Diptera: Scatophagidae)—I. Diurnal and seasonal changes in population density around the site of mating and position. *Journal of Animal Ecology* 39: 185–204.

————. 1970b. The reproductive behaviour and the nature of sexual selection in *Scatophaga stercovaria* L. (Diptera: Scatophagidae)—II. The fertilization rate and the spatial and temporal relationships of each sex around the site of mating and oviposition. *Journal of Animal Ecology* 39: 205–28.

————. 1970c. The reproductive behaviour and the nature of sexual selection in *Scatophaga stercovaria* L. (Diptera: Scatophagidae)—IV. Epigamic recognition and competition between males for the possession of females. *Behaviour* 37: 113–39.

————. 1970d. The reproductive behaviour and the nature of sexual selection in *Scatophaga stercovaria* L. (Diptera: Scatophagidae)—V. The female's behaviour at the oviposition site. *Behaviour* 37: 140–68.

————. 1970e. The reproductive behaviour and the nature of sexual selection in *Scatophaga stercovaria* L. (Diptera: Scatophagidae)—VIII. The origin and evolution of the passive phase. *Evolution* 24: 744–88.

————. 1970f. Sperm competition and its evolutionary consequences in the insects. *Biological Reviews* 45: 525–67.

————. 1970g. Sperm competition and its evolutionary effect on copula duration in the fly *Scatophaga stercovaria. Journal of Insect Physiology* 16: 1301–28.

————. 1974. The reproductive behaviour and the nature of sexual selection in *Scatophaga stercovaria* L. (Diptera: Scatophagidae)—IX. Spatial distribution of fertilization rates and evolution of male search strategy within the reproductive area. *Evolution* 28: 93–108.

Richards, E. 1983. Darwin and the descent of woman. In *The Wider Domain of Evolutionary Thought*, ed. D. Oldroyd and I. Langham. Dordrecht: Reidel.

Ruse, M. 1979a. *The Darwinian Revolution: Science Red in Tooth and Claw.* Chicago: University of Chicago Press.

————. 1979b. *Sociobiology: Sense or Nonsense?* Dordrecht: Reidel.

————. 1996. *Monad to Man: The Concept of Progress in Evolutionary Biology.* Cambridge, Mass.: Harvard University Press.

————. 1998. *Evolution and Its Values: An Essay on the Nature of Science.* Cambridge, Mass.: Harvard University Press.

Russett, C. E. 1989. *Sexual Science.* Cambridge, Mass.: Harvard University Press.

Sleeth Mosedale, S. 1978. Science corrupted: Victorian biologists consider "the woman question." *Journal of the History of Biology* 11: 1–55.

Todes, D. P. 1989. *Darwin without Malthus: The Struggle for Existence in Russian Evolutionary Thought.* New York: Oxford University Press.

Wallace, A. R. 1866. On the phenomena of variation and geographical distribution as illustrated by the *Papilionidae* of the Malayan region. *Transactions of the Linnean Society of London* 25: 1–27.

————. 1900. *Studies: Scientific and Social.* London: Macmillan.

————. 1903. *Man's Place in the Universe.* London.

————. 1905. *My Life: A Record of Events and Opinions.* London: Chapman & Hall.

————. 1907. *Is Mars Habitable?* London.

Williams, G. C. 1975. *Sex and Evolution.* Princeton, N.J.: Princeton University Press.

Wilson, E. O. 1975. *Sociobiology: The New Synthesis.* Cambridge, Mass.: Harvard University Press.

————. 1978. *On Human Nature.* Cambridge, Mass.: Harvard University Press.

Part III

Interests, Ideology, and the Neglect of Experiments

A popular genre in contemporary STS writings deals with so-called controversy studies. One narrates various scientific debates in terms of the professional, disciplinary, and/or ideological interests of the participants, claiming that the frequent appeals to experimental results function primarily as rhetorical devices. The resolution of scientific controversies is said to come about as a result of the "negotiation" of competing interests and is not shaped by the probative force of evidential considerations.

It is indeed the case that recent work on the "philosophy of experiment" emphasizes how difficult it can be to construct and execute good experiments. To standardize and calibrate new apparatus and experimental techniques requires industry, ingenuity, and creativity. In a book called *The Neglect of Experiment*, Franklin chastises traditional philosophy of science for taking the empirical base of science too much for granted and for neglecting the epistemological subtleties of good experimental design.

Unfortunately, many authors of controversy studies seem to have concluded that all experimental results are inconclusive and that experiments never end. Their narratives of negotiation tend to neglect the epistemic role of experiments altogether. To carry out such an interpretation, they must also discount what scientists themselves say about the role of evidence in scientific decision making and reinterpret such remarks as self-serving.

There is no question that scientists are, in general, fiercely protective of their own reputations and hardly adverse to self-promotion. But as Merton and Popper pointed out long ago and as Hull more recently elaborated, science is a case par excellence in which it is possible to do well by doing good! As long as scientists are rewarded for performing experiments that are accurate, reproducible, well controlled, and elegant, it will be in their "interest" to design such experiments, and it will be in the "interest" of other scientists to dramatize any weaknesses in the results of their colleagues. But this kind of "friendly–hostile competition," as Popper called it, is hardly a threat to the epistemic integrity of science. Quite the contrary. If properly institutionalized, it provides the critical impetus for scientific progress.

In this section, we revisit several popular controversy studies in which it is alleged that ideology and interests carried the day. As the following essays demonstrate, in each case the STS authors have either neglected or misunderstood the relevance of the experimental data or mathematical arguments. As Lakatos used to say, there may be no instant rationality in science, but this does not mean there are never good evidential reasons for scientific choices.

The cold fusion episode with its media circus, disputes between chemists and physicists, and lingering adherents despite the seeming triumph of the elite research institutes might seem to be a case in which the sociology of scientific knowledge approach (SSK) would provide a good explanation of what happened. Indeed, Collins and Pinch have set it up as a paradigm of "science as usual." William McKinney argues, however, that despite the initial hype and frenzy, the determining factors in the outcome were rather mundane methodological considerations having to do with bad thermal data, as well as an inconclusive gamma ray spectrum. It is indeed science as usual to shelve or throw out irreproducible results. There is no reason to think cold fusion would be alive if the original experiments had been done by Boston physicists instead of Utah chemists.

Alan Franklin takes on two showpieces in the SSK corpus: Collins's analysis of the attempt to detect gravity waves as an example of the so-called experimenter's regress and Pickering's much discussed book claiming that quarks were socially constructed. The two case studies differ enormously in depth and detail, but they both arrive at the same conclusion: although the scientific community chose to reject Weber's gravity waves in the first case and to accept Weinberg–Salaam's weak neutral currents in the second, there was no compelling reason for either choice. Like Robert Frost's traveler in the woods, the path not taken could have been just as fair. But Franklin shows otherwise. By reconstructing the sequence of experimental results and the inferences drawn from them, he argues that scientists had strong evidential reasons for their conclusions.

John Huth wrestles with Latour's attempt to claim Relativity Theory as an example of the incorporation of sociological factors into the content of physics. Neither Latour's prose nor his ideas are easily accessible, and we might have anticipated that the move from relativitistic physics to relativistic epistemology would turn out to be fallacious. It is fascinating, however, to follow Huth's clear diagnosis of Latour's muddled ruminations on semiotics, abstractions, synchronizing of clocks, and the role of the observer in Relativity Theory. Latour concludes that Einstein did not fully understand his own theory because it is really about the delegation of power. Huth respectfully submits that Latour did not understand Einstein.

8

When Experiments Fail
Is "Cold Fusion" Science as Normal?

William J. McKinney

The Competing Accounts

The history and philosophy of science (HPS, STS, or whatever currently fashionable name you choose) has enough difficulty figuring how to handle cases of scientific success (however defined)—for example, Newton, Lavoisier, and Einstein—that it should come as no surprise that more ambiguous cases are dealt with, well, more ambiguously. The curious case of the 1989 claim to have initiated a sustained nuclear reaction in a bench-scale experiment, so-called cold fusion, is one such case.

At this early point, a disclaimer is in order. I will not attempt to act as a judge on the final scientific or technological value of the results discussed here. Rather, my focus is on the early days of the dispute, from 1989 to 1991, and the methodological issues. I begin with a brief synopsis.

On March 23, 1989, Martin Fleischmann and B. Stanley Pons announced to the media that they had induced a nuclear reaction in an electrochemical cell (figure 8.1). In experiments conducted over the preceding five years in a basement laboratory at the University of Utah, the two electrochemists claimed to have detected the telltale signs of deuteron–deuteron (d–d) fusion: a 2.45 MeV neutron flux (from γ-ray evidence), ^{3}He, tritium, and, most significantly, excess heat generation that, they maintained, could be used for practical ends. Nearly simultaneously, Steven Jones, a Brigham Young University physicist, announced similar but less grandiose results, reporting a relatively small neutron flux and little to no excess heat, consistent with d–d fusion on a very small, but not commercially viable, scale. To be sure, most of the controversy in this case surrounded the grand claims made by the Utah group, and for the sake of brevity, the Utah claims will occupy most of this chapter as well.

In brief, the early days of the case evolved as follows:

1. With respect to accepted theory regarding d–d fusion, both teams reported interesting experimental results, with Jones's results being consistent with existing theory and the Fleischmann and Pons's results not.

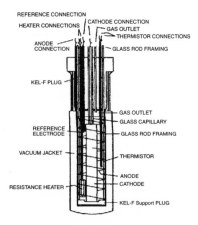

Figure 8.1 The Fleischmann and Pons apparatus. (Fleischmann et al. 1990, 295).

2. Controversy over Fleischmann and Pons's nuclear data soon emerged against a background of accepted theory and practice.
3. Against this background, attempts to discredit Fleischmann and Pons's nuclear data began shortly after the initial press conference.
4. Various attempts to embed Fleischmann and Pons's results in a new theoretical context were proposed and published, largely in response to anomalous thermal data.

To be sure, this rough description of the early story ignores several essential factors in painting an honest picture of cold fusion. The personal stories of the scientists, especially of Fleischmann, a Fellow of the Royal Society, and his former pupil, Pons, offer insight into the motivations behind the experimental program. There are institutional issues as well, including the fact that the relationship of the Utah and Brigham Young teams was alternately competitive and cooperative, as well as the availability of governmental funding from the state of Utah for the Utah team and from the U.S. Department of Energy for the Brigham Young team. Finally, there is the question of power, disciplinary turf guarding, and influence. Jones, a physicist, was using the tools of the chemist in a domain generally inhabited only by physicists, whereas Fleischmann and Pons, both electrochemists, used the tools of their own trade in the domain of the physicists. It is the latter issues that dominate the story of cold fusion told in *The Golem: What Everyone Should Know about Science.*

Harry Collins and Trevor Pinch's *Golem* presents a compelling and interesting rendition of the early days of cold fusion. By their own admission, they focus exclusively on "human matters" (Collins and Pinch 1993, 2). Indeed, such "human matters" are the stock-in-trade of the science and technology studies (STS) approach, with its emphasis on "interest models" of science and science as a social institution. Collins and Pinch maintain that cold fusion, among the other cases in their book, is a "piece of beautiful scientific work," emblematic of "science as normal," and that this beautiful normalcy is revealed by their approach to the case (Collins and Pinch 1993, 2, 78). Their approach to the case embraces all the factors that my brief enumeration misses. It stresses competition between chemists and physicists, indeed jealousy on the part of physicists; haste on the part of all interested parties to find a safe, inexpensive, and plentiful energy source; and an unsavory kind of scientific orthodoxy that unnecessarily protects established theory and practice in the face of potentially damaging competition and anomalous experimental data.

Such claims echo in part the assertions and arguments in Gieryn's important sociorhetorical analysis of the cold fusion case (see Gieryn 1992). In this telling of the tale, cold fusion becomes "more interesting and less real" owing to the interplay between the media and professional jealousy and competition, revealing that scientific experimentation alone is insufficiently powerful to decide scientific disputes. I argue here that the STS approach to cold fusion, though certainly important, misses key epistemological factors that explain much of the response of the scientific community in this case.

The STS approach should be applauded for its attack on "naively scientistic" versions of the cold fusion tale (see, e.g., Close 1991, Mallove 1991), yet by intentionally ignoring methodological issues and claiming to relate "what actually happened" in this case, Collins and Pinch tell an incomplete, and occasionally misleading, tale (Collins and Pinch 1993, 2).

In *The Golem*, Collins and Pinch wish to de-mystify science, stripping it of the trappings often furnished by philosophers of science and their rational reconstructions. What remains in their version of cold fusion is a self-proclaimed "science in the raw"—science the way it really happened. Only in this way can we see what "normal" science really is. Well, consider that if cold fusion is "science as normal," then such science would be defined by continual disciplinary border wars such as those between chemists and physicists, and the practice of science would be dominated by theoretical presuppositions and narrow professional and broad socioeconomic interests. To be sure, these factors enter into any accurate telling of a scientific case. To deny them is to be both naive and historically dishonest. But Collins and Pinch present them as if they are the hallmarks of science, for after all, cold fusion is "science as normal."

I wish to paint a different picture. Initially, cold fusion was anything but normal, as its study was beset with excessive posturing in the media, and the attempts to replicate the Fleischmann–Pons results relied solely on media reports rather than accurate published experimental descriptions. Eventually, however, cold fusion became normal, but not for the reasons of the human issues discussed in *The Golem* and "The Ballad." Rather, it happened for the very reasons that these accounts ignore: It became normal through explicit appeals to evidence and the canons of scientific method. The interest model used by Collins and Pinch and their ilk is an important means by which we can understand science, but it cannot stand alone. Unless it is augmented with a detailed discussion of the methodological issues in each scientific episode, it will emerge as a story as incomplete as the scientistic versions it, so correctly, criticizes.

To show the inadequacies of the STS approach, I offer a detailed analysis of the methodological issues in this case and then contrast the explanation of scientific behavior provided by my account with the narratives offered by Collins and Pinch, and Gieryn. I begin with the controversy over the nuclear data.

Nuclear Data

As Collins and Pinch tell the story, the bulk of the cold fusion controversy surrounded the nuclear data, particularly the γ-ray spectrum, which Fleischmann and Pons used to argue for the presence of a nuclear reaction. Collins and Pinch certainly are correct about the importance of this issue, but they do not correctly describe how Pons and Fleischmann's claims were evaluated. I argue that close analysis of their neutron and γ-ray data reveals errors and anomalies that even the most junior physicists in this field would detect. In fact, these problems were detected quite early by nuclear physicists, who offered the most resistance to Fleischmann and Pons's claims. The controversy surrounding the γ-ray data was certainly the most damaging. I maintain that it is in this very sense that, contrary to Collins and Pinch, cold fusion is anything but normal science.

Fleischmann and Pons reported that a deuteron–deuteron (d–d) reaction of some kind was induced by the electrolytic implantation of deuterium nuclei into the lattice structure of palladium electrodes. Their claim was that the experimental evidence pointed to a heretofore unknown nuclear reaction. They argued that the following reactions, referred to as reactions (v) and (vi) in their 1989 paper, could not account for the bulk of energy released in their experiment.

$$\text{(v) } d + d \rightarrow {}^3T + {}^1H$$
$$\text{(vi) } d + d \rightarrow {}^3He + n \text{ (2.45 MeV)}$$

They claimed to have detected excess enthalpy, which varies with the applied current density in the cell and the electrode volume, and a γ-ray spectrum, which points to a 2.45 MeV neutron flux in the vicinity of 4×10^4 s^{-1}. Since the original 1989 publication, it has become apparent that Fleischmann and Pons's γ-ray spectrum was artifactual at best. Let us now look in detail at the story of the mysteriously mobile γ-ray peak.[1]

Fleischmann and Pons argued for their neutron flux based on their γ-ray data and the following neutron capture reaction: ${}^1H + n \,(2.45 \text{ MeV}) = {}^2D + \gamma\,(2.225 \text{ MeV})$. They claimed to have detected a 2.225 MeV γ-ray peak on their spectrum, a plot of which appears in figure 8.2. They announced that their spectrum confirms that 2.45 MeV neutrons are indeed generated in the electrodes by reaction (vi). These γ-rays are generated by reaction (vii): "We note that the intensities of these spectra are weak and, in agreement with this, the neutron flux calculated from measurements with the dosimeter is of the order of 4×10^4 s^{-1} for a 0.4 cm $\times$ 10 cm rod electrode polarized at 64 m A cm^{-2}" (Fleischmann and Pons 1989, 305–6). Although Fleischmann and Pons insisted that their γ-ray spectra confirmed the existence of 2.45 MeV neutrons produced in reaction (vi), it is here that the first, and most vocal, of their critics pointed to critical experimental problems.

The first of these problems emerged in the "Errata" published by the *Journal of Electroanalytical Chemistry* shortly after its initial 1989 publication. Inexplicably, the Utah team changed the scale on the y-axis of their γ-ray spectrum: "Fig. 1A: The y-axis should be labeled from 0 counts (minimum) to 1000 counts (maximum), i.e. replace 0, 500/0, 10000, 15000, 20000 and 25000 counts by 0, 200, 400, 600, 800 and 1000 counts respectively" (Fleischmann et al. 1989a, 187).[2] The inexplicably revised axis, which points to neutron flux reduced by a factor of 25, is shown in figure 8.3.

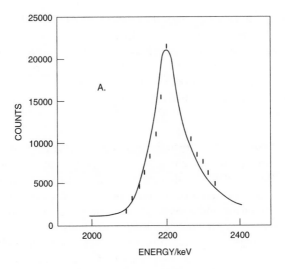

Figure 8.2   The Fleischmann-Pons γ-ray spectrum. (Fleischmann and Pons 1989, 306).

In addition to the sudden change in the vertical scale, critics noted that the shape of the peak was atypical. Such spectra do not display single peaks but multiple peaks that correspond to the range of energies characteristic of natural background. Finally, critics such as Abriola and, most notably, Petrasso, pointed out that Fleischmann and Pons's peak was too narrow, and in fact, the neutron capture peak had been moved between Fleischmann's first public presentation on March 28, 1989, at Great Britain's

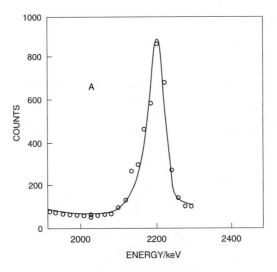

Figure 8.3   The revised Fleischmann-Pons γ-ray spectrum. (Fleischmann et al. 1989a, 188).

Harwell Laboratory, and the first published data (1989a) in April 1989. At the Harwell presentation, Fleischmann pointed to a *2.5 MeV* γ-ray peak (see figure 8.4). A revised spectrum does appear in the first published paper (1989), but although the peak appears at 2.2 MeV (see figure 8.2), the equation that the authors cited as describing 2.45 MeV neutron capture is incorrect, indicating an erroneous 2.45 MeV γ-ray peak. "The spectrum of γ-rays emitted from the water bath's due to the (n,γ) reaction:

(vii) $^1H + n$ (2.45 MeV) $= {}^2D + \gamma$ (2.45 MeV)" (Fleischmann and Pons 1989, 302).

Although the electrochemists corrected equation (vii) in the published "Errata," they gave no real explanation for the changes. At this point, the informed skeptic must ask, "Which are the real data?" and of course, "Can this be 'science as normal?'"

Abriola et al. reported that a consistent upper limit for both γ-ray and neutron production in Fleischmann and Pons–type cells was detected at three orders of magnitude lower than that reported by Fleischmann and Pons. In addition, Abriola noted the following: "The high energy resolution measurements of the emitted γ-rays performed in the present work reveal the presence of a background line of 2204.5 keV which may be difficult to unfold in a low resolution experiment" (Abriola et al. 1989, 360). Keep in mind that Fleischmann and Pons reported that they had detected a peak in their own, controversial, γ-ray spectrum at 2.225 MeV (2225 keV). The Abriola team employed a detector with a resolution of about 2.2 keV at 1332 keV. In the range of this experiment, that is, ~2.2 MeV, Abriola argued that the resolution for their detector was about "25 times better than that of the NaI counter employed by [Fleischmann and Pons] as judged from the γ-ray spectrum shown there" (Abriola et al. 1989, 357).

This information is crucial, for Abriola demonstrated that a 2.2045 MeV background line coincident with ^{214}Bi decay is a scant 20 keV from the 2.225 MeV peak shown in

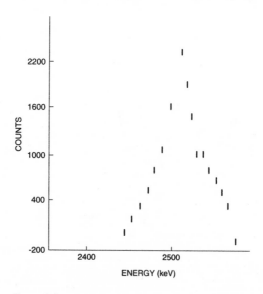

Figure 8.4 Erroneous γ-ray spectrum as presented at Harwell. (Close 1991, 284).

the Fleischmann and Pons spectrum. Since these two peaks are only 20 keV apart, it seems highly unlikely that a detector with a resolution 25 times lower than the 2.2 keV resolution of the Abriola detector could separate the true peak from background. In other words, Fleischmann and Pons were actually operating with a detector that could distinguish only those peaks with a separation greater than ~60 keV. This, of course, does not assume that Fleischmann and Pons's lab was subject to background sources identical to those present in Abriola's lab. But it does mean that their apparatus might not have been able to distinguish *any* background from true signal if the two were within the instrument's relatively low resolving power.

Abriola demonstrated the danger of this possibility. If background such as his instrument was able to detect is subtracted from the 2.225 MeV peak that Fleischmann and Pons detected, a neutron flux is detected at three orders of magnitude below the Fleischmann and Pons flux, 13 s^{-1} compared with 4×10^4 s^{-1}. Abriola's upper limit of total neutron production agreed with the γ-ray spectrum, confirming that the Abriola experiment detected neutron production at three orders of magnitude lower than did the Fleischmann and Pons experiment (Abriola et al. 1989, 359).[3]

In a more damaging commentary, Petrasso et al. (1989a) pointed to inconsistencies in the Fleischmann and Pons data. They reported the following:

- The line width of the Fleischmann and Pons 2.225 MeV signal is a factor of two below the predicted value for a 2.45 MeV neutron source.
- Given the resolution of the Fleischmann and Pons apparatus, a Compton edge at 1.99 MeV should have been reported.[4] This would seem to negate the claim that Fleischmann and Pons did indeed measure 2.225 MeV γ-rays from reaction (vii) (figure 8.5).

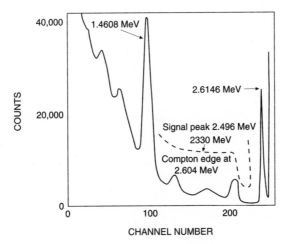

Figure 8.5 Complete γ-ray spectrum published in *Nature* by Fleischmann et al. Note the explicit reference to a 2.496MeV peak, as opposed to the 2.2MeV peak reported in their 1989a and b. (Source: Fleischmann et al. 1989c, p. 667)

- Given the resolving power of their detector and measured background, the Fleischmann and Pons 2.225 MeV peak actually occurred at 2.5 MeV.

Each of these critiques points to the ambiguous determination of the correct energy of the Fleischmann and Pons spectrum.[5]

The most damaging of these three critiques is the last, which points to the misidentification of the energy of the 2.2 MeV peak. Recall that Fleischmann and Pons had inexplicably altered their spectrum, first moving the peak from 2.5 MeV to the expected 2.2 MeV level and then adjusting the vertical axis on the plot. Petrasso observed that based on the Utah background measurements, up to 4,000 counts per channel were detected in the vicinity of 2.2 MeV. According to Petrasso, "The only relevant part of the entire spectrum (between 1.46 and 2.61 MeV) in which the background was as low as 400 counts was at an energy in the vicinity of 2.5 MeV, not at 2.22 MeV" (Petrasso et al. 1989b, 184). Thus Petrasso concluded that the Fleischmann and Pons spectrum contained instrumental artifacts, not genuine evidence of the capture of 2.5 MeV neutrons.

In response to this critique, Fleischmann et al. finally published their full γ-ray spectrum. Inexplicably, however, their questionable signal peak was moved once again, this time from 2.2 MeV back to their original 2.5 MeV (see figure 8.6). "The peak under discussion is centered at 2.496 MeV" (Fleischmann et al. 1989b, 667). In this published reply to Petrasso, the Utah team contended that rather than using their radiation data to argue for the occurrence of a d–d reaction, they measured these spectra as a safety precaution. This claim was bolstered by the fact that the person who made the very first radiation measurements in the Fleischmann and Pons lab—University of Utah radiation safety officer R. J. Hoffman—appears for the first time as a coau-

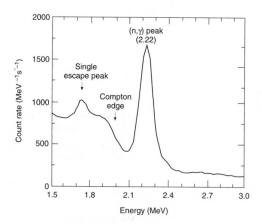

Figure 8.6 The Petrasso γ-ray spectrum from the capture of 2.45MeV neutrons showing the predicted Compton edge. Note that their peak is indeed centered at 2.2MeV, not at 2.5, as Fleischmann and Pons had reported originally. (Source: Petrasso et al. 1989a, p. 184)

thor. But unfortunately, as Petrasso writes in his simultaneously published reply to the Utah group, Fleischmann et al. never really address the original Petrasso et al. criticisms. In fact, Petrasso maintains in the reply that due to a miscalibration in the Fleischmann et al. spectrometer, their true signal peak does not even rest at 2.496 MeV but, instead, at about 2.8 MeV.[6] Clearly, the Utah claim to have generated d–d fusion neutrons was to be doubted, for even Fleischmann et al. were not clear about the identity of their results.

The Abriola and Petrasso critiques were confirmed by other laboratories investigating the reported neutron flux. For example, Gai et al. (1989) reported a neutron flux of approximately 1 ± 1.5 counts per hour.[7] Salamon et al. also found little to no evidence of fusion activity in their 1990 paper. This particular study was accorded great significance by the scientific community, for Salamon is a physicist at the University of Utah and was permitted to conduct his experiments on the actual Fleischmann and Pons cells in the University of Utah lab where the cold fusion effects were allegedly detected.[8] This study was also the center of much controversy, as its release in *Nature* appeared, along with two scathing editorials declaring cold fusion to be dead, on the one-year anniversary of Fleischmann and Pons's monumental announcement. Lawyers for Fleischmann and Pons (on the University of Utah payroll) had threatened that any damage to their clients' reputations precipitated by the Salamon paper would be met with civil action (Salamon et al. 1990, 405). Yet again, this brings to mind the question, "Is this 'science as normal?'" It may be business as normal, but it does not license, even in the face of the other studies in *The Golem*, the inference that it is science as normal.

The importance of the preceding critiques cannot be stressed enough. If it is true that Fleischmann and Pons did not obtain valid experimental results for their γ-ray and neutron data, then their claim that fusion reaction (vi) occurred comes into question. However, none of these criticisms addresses the anomalous heat generation, and by Salamon's own admission, any such generation might point to a yet unknown fusion reaction, as the Utah group speculated in their first published paper. This is underscored by Petrasso who, in a lecture at the MIT Plasma Fusion Center, remarked, "We may turn out to be the big allies of Fleischmann and Pons if they can now prove that they have fusion, because what we have demonstrated is that they basically didn't have any neutrons at all coming from their heat producing cell. . . . So now they can claim that they are having neutronless heat generation" (Mallove 1991, 147). At this point, let us turn our attention to that most curious of the data—the thermal data claimed by Fleischmann and Pons.

Thermal Data

Collins and Pinch also agree that controversies about thermal phenomena are central to the thermal data, but they understate the problems involved by claiming that testing for thermal data in this experiment is tantamount to practicing high school physics![9] If this were indeed the case, then surely we would not see the kind of work to be described here. From the beginning, the most dramatic aspect of the cold fusion announcements were the claims that the cells produced large quantities of heat energy.

But exactly how much heat quickly became the issue. Was it more than the electrical energy put into the cell? In their original paper, Fleischmann and Pons reported that enthalpy generation could exceed 10 W cm^{-3} of the Pd cathode. Note that their claims for most excess heat are *projected* results, not *measured* results.[10]

In their table 1, all data for rod cathodes at current densities of 512 mA cm^{-2} are based not on measured results but on an extrapolation from a 1.25-cm cathode length to a 10-cm cathode length. Unfortunately, the authors do not account for the means by which this extrapolation is calculated. This, of course, means that the figures in their table 2 are simply projections: that the first column expresses the percentage of breakeven based on total joule heat applied to the cell, that the second column expresses the percentage of breakeven based on the total energy applied to the cell, and that the third column expresses the percentage of breakeven based on total energy applied to the cell with a cell potential of 0.5 V. All these figures are projections, with no indication whatsoever as to how they were made.

These figures highlight, yet again, the unusual and aberrant nature of Fleischmann and Pons's rendition of cold fusion. To offer estimations and extrapolations without showing how these were obtained is grounds for failing an undergraduate lab report. By anyone's definition, this is neither science as normal nor beautiful science.

Excess enthalpy in the 1989 paper was calculated by first estimating the upper and lower bounds on the joule heating rates based on a balancing of the following electrochemical reactions. Again, I adopt the authors' numerical conventions for convenience (Fleischmann and Pons 1989, 301, 303).

(i) $D_2O + e^- \rightarrow D_{ads} + OD^-$
(ii) $D_{ads} + D_2O + e^- \rightarrow D_2 + OD^-$
(iv) $D_{ads} + D_{ads} \rightarrow D_2$
(viii) $4OD^- \leftrightarrow D_2O + 4e^- + O_2$

Finally, the authors argued (Fleischmann et al. 1990) that their enthalpy data represent lower limits owing to their chosen method of calculation. They used a "black box" approximation of the calorimeter heat and mass transfer. In short, a black box model depicts complex experimental instruments as a network of inputs and outputs, in which the input parameters are manipulated by equations inside the apparatus—the "black box." Figure 8.7 depicts Fleischmann and Pons's "black box" model.[11]

Approximately 1,000 measurements allowed Fleischmann et al. to estimate that their experimental cells were capable of producing energy generation rates as high as 100 W cm^{-3}. This value is up to five times greater than that reported in Fleischmann and Pons 1989, although the authors do not explain why this is the case. In the 1989 paper, Fleischmann and Pons reported excess enthalpy in the range of 1–20 W cm^{-3}. Note that this was only a *projected* excess. As I will show, experimental confirmations reported a much lower *measured* heat excess.

Finally, Fleischmann et al. (1990) reported that their excess enthalpy increased with increased current density, although they admitted that more detailed experiments were needed to determine whether or not there was a lower "threshold density" at which the enthalpy generation began (319).

So what is the cause of these sporadic heat effects? Clearly, not all the experimen-

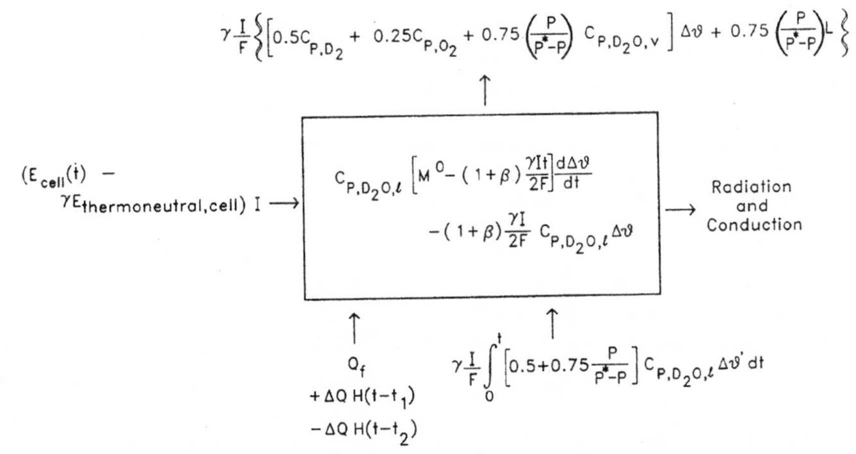

$$\gamma \frac{I}{F} \left\{ \left[0.5 C_{P,D_2} + 0.25 C_{P,O_2} + 0.75 \left(\frac{P}{P^*-P} \right) C_{P,D_2O,v} \right] \Delta \vartheta + 0.75 \left(\frac{P}{P^*-P} \right) L \right\}$$

$\uparrow$

$$(E_{cell}(t) - \gamma E_{thermoneutral,cell}) I \rightarrow \quad \boxed{ \begin{array}{c} C_{P,D_2O,\ell} \left[M^0 - (1+\beta) \frac{\gamma I t}{2F} \right] \frac{d\Delta\vartheta}{dt} \\[2mm] - (1+\beta) \frac{\gamma I}{2F} C_{P,D_2O,\ell} \Delta\vartheta \end{array} } \quad \rightarrow \begin{array}{c} \text{Radiation} \\ \text{and} \\ \text{Conduction} \end{array}$$

$\uparrow \qquad\qquad\qquad \uparrow$

$$\begin{array}{c} Q_f \\ + \Delta Q \, H(t-t_1) \\ - \Delta Q \, H(t-t_2) \end{array} \qquad \gamma \frac{I}{F} \int_0^t \left[0.5 + 0.75 \frac{P}{P^*-P} \right] C_{P,D_2O,\ell} \, \Delta\vartheta' \, dt$$

Figure 8.7 Fleischmann et al. "black box" model. (Fleischmann et al. 1990, 330).

tal runs generated excess heat. In fact, only six of 19 cells generated greater than 1 W of excess heat. In addition, the Fleischmann and Pons heat effect was only qualitatively confirmed.

Interestingly, the Utah calorimetry was largely ignored by the physics community, which chose to concentrate on radiation measurements. In this respect, accounts of cold fusion such as those in *The Golem* are right on the mark. Such an observation would be consistent with the claim that physicists attacked Fleischmann and Pons only when the two electrochemists strayed out of their own backyard and into the highly specialized backyard of neutron and γ-ray detection. What such a generalization misses, however, is the fact that not only did Fleischmann and Pons stray from their own electrochemical turf but also they did it incompetently. For this reason alone, we can find fault in their work, whether we are physicists, sociologists, historians, or philosophers.

It is the excess heat that was the most puzzling, and robust, of the Utah results, however, and this excess was not without its own puzzles as well. As Bockris, et al. note in their review of laboratory work attempting to confirm the Fleischmann and Pons results, "The excess heat is rarely a constant phenomenon; rather, it occurs in bursts of activity, sometimes, but not always, associated with a change of system parameters (a change in current density, for example)" (1990, 13). Most published results were contained in preliminary discussion notes that did not speculate on the cause of the anomalous heat generation, and in no case was the excess heat a recurrent phenomenon. In fact, excess heat was reported in only a small fraction of electrolytic cells in any of the experiments that I will describe later.

As will be seen as typical of heat confirmations, a Texas A&M team, led by Kainthla, reported in their brief discussion note that excess heat was generated in only three of their 10 experimental cells. In what has become, for the most part, standard procedure in such calorimetry experiments, a heat transfer–type calorimeter immersed in a constant temperature water bath was calibrated by passing increasing amounts of direct current and noting the potential drop across it.

The calorimeter was calibrated by passing different amounts of direct current through the electric heater (resistance 245 W) in the solution of 0.1 M LiOD (made by dissolution of Li in D_2O) and noting the potential drop across it. For each value of electrical power put in, steady state temperature difference (monitored on a y-t recorder) was noted and plotted to yield a calibration curve as shown in Fig. 1. This curve was repeatedly reestablished by several of the authors of this paper and found to be stable to within ±0.05 watts. (Kainthla et al. 1989, 1315).

Three experimental trials yielded anomalous thermal output—that is, heat excess beyond that shown in the calibration curve.[12] One of these cells, known as "cell B8," produced excess heat for a period of approximately 33 hours. Cells B9 and A9 also produced excess heat output, but for much shorter durations. In addition, once the current was shut off for even short periods and then reapplied, the heat excess did not return. Using similar techniques of data analysis, Yun et al. (1991) found an excess of 25–128 W cm^{-3}, amounting to only about 25% to 75% excess heat beyond their calibration curve. This group reported heat bursts in only five of their 22 experimental cells. Similarly, Srinivasan and Appleby (1989) also reported a low-level heat excess, estimated at 10%.

The heat excess in the Kainthla experiments was estimated to be at a level of approximately 2 to 5 W cm^{-3} of the palladium cathode, significantly lower than the 10 to 100 W cm^{-3} reported by Fleischmann and Pons. In fact, the heat excess of the Kainthla group amounts to 10% to 30%, whereas Fleischmann and Pons reported excess heat ranging from 100% to 600%. As Bockris notes, most of the reports of excess heat fall into the category of low-level heat excess—that is, excess heat less than 100%.

Similarly low heat generation was reported by Scott et al. (1990), who wrote in a preliminary technical note that 5% to 10% heat excess was detected in their experiments at the Oak Ridge National Laboratory. These experiments underscore the need to stabilize the heat effects, for in experimental cells run for more than 1,000 hours, excess heat was produced only in those cases in which system perturbations, accidental or purposeful, were initiated.

The Scott experiment was of some significance because the experiments were performed on both open and closed cells. One of the most common sources of error in the calorimetry of electrolytic cells is the possible recombination of electrolytically separated compounds, in this case the exothermic recombination of oxygen and deuterium. In their open cells, where oxygen and deuterium gases were vented off, heat excesses were reported only in those cases in which there were disturbances in the cell's operating parameters, such as decreases in electrolyte temperature. In closed cells, similar heat excesses were reported when system parameters were perturbed. The Scott team noted the significance of the closed cell results and pointed out that, by not requiring the assumption that electrolysis gases *do not* recombine, there is less inherent uncertainty in their experimental results.

Finally, it appears that one stinging criticism of the original 1989 calorimetry experiment—that the calorimetry cells were not well stirred and that the temperatures measured were those of local "hot spots"—was successfully dismissed in the 1990 Fleischmann et al. paper. The most damaging report for the 1989 calorimetry data came from the 1989 American Physical Society meeting in Baltimore. Nathan Lewis demonstrated that he was able to replicate the Fleischmann and Pons data by not stirring the liquid in his cells, but he could not replicate the results while the electrolytic solution was stirred (Close 1991, 214). This was confirmed by Schultze et al.: "Since there are surface inhomogeneities there may be local hot spots at the electrode surface" (Schultze et al. 1989, 1304). In their 1990 paper, Fleischmann et al. guarded against this attack: "The Dewar cells were maintained in specially constructed water baths stirred with Techne Tempunit TU-16A stirrer/regulators. Since March 1989, we have maintained two or three water baths; each bath contained up to five Dewar cells so that we were able to run 8 to 15 cells simultaneously" (296). In addition, they noted the difficulty in designing a calorimetrically accurate cell:

> For the particular case under consideration, where vigorous gas evolution takes place at both electrodes, the transport of heat from the electrode surfaces and equalization of the temperature differences will be dominated by convective mixing by eddies induced in the boundary layers and their decay by eddy diffusion into the bulk of the liquid. The use of long narrow calorimeters ensures rapid radial mixing. (Fleischmann et al. 1990, 322)

Note that similar to the criticisms of Fleischmann and Pons's radiation data, what we see here are appeals to method. To be sure, this is science as normal, but it is normal because it anticipates the methodological issues—namely, calibration, quantitative analysis, and repeatability—and not because of professional jealousy or the drive to power Salt Lake City and Provo with heavy water and palladium.[13]

Discussion

Collins and Pinch make a single broad claim regarding the cold fusion controversy, that much of the controversy surrounding Fleischmann and Pons resulted from jealous physicists guarding their turf:

> Part of the skepticism came from fusion researchers being only too familiar with grandiose claims which shortly afterwards turn out to be incorrect. . . . For them, solving the world's energy problems via a breakthrough in fusion had about as much a chance of being true as the perennial claims to have superseded Einstein's theory of relativity. Though fusion researchers, well-used to spectacular claims, and with their own billion-dollar research programs to protect, were incredulous, other scientists were more willing to take the work seriously. Pons and Fleischmann have fared much better with their colleagues in chemistry where, after all, they are acknowledged experts. (Collins and Pinch 1993, 66)

As I have illustrated, the controversy over Fleischmann and Pons was more than just guarding turf. Yes, what initially made the attack on Fleischmann and Pons worth pursuing may well have been a desire to protect multibillion-dollar research programs, but, in the end, it was methodological claims regarding experimental care and instrumental competence that carried the day, and these criticisms came from chemists as well as physicists.

Consider another claim from The Golem:

> This is not to say that fusion physicists simply rejected the claims out of hand (although a few did), or that it was merely a matter of wanting to maintain billion-dollar investments (although with the Department of Energy threatening to transfer hot fusion funding to cold fusion research there was a direct threat to their interests), or that this was a matter of the blind prejudice of physicists over chemists (although some individuals may have been so prejudiced); it was to imply that no scientist could hope to challenge such a powerfully established group without having his or her credibility put on the line. (Collins and Pinch 1993, 74)

Note the conflation of reasons for examining Fleischmann and Pons's results in the first place and the reasons for rejecting those results. Indeed, the credibility of any scientist is put on the line every time he or she makes a public claim. Fleischmann and Pons were no different. Putting one's credibility on the line is one matter; having it damaged is another. Fleischmann and Pons may well have had their credibility put on the line because they quite publicly strayed into the hot fusion physicists' multibillion-dollar backyard. Unfortunately for them, they were unequipped to do so. It was carelessness, not the challenging of the fusion establishment, that led to the criticisms of Fleischmann and Pons.

In the end, Collins and Pinch make two simple mistakes. On the one hand, they are prepared to claim that cold fusion is science as normal based on one case study—cold fusion itself. Although we might charitably maintain that they make this claim based on the other six case studies in The Golem, seven cases of aberrant science hardly warrant the claim that science is a "bumbling giant" and that this is "normal" (Collins and Pinch 1993, 2). I could easily point to seven careless colleagues, yet I would never be willing to make the weak inductive claim that all my colleagues are careless. They may very well be, but I lack the evidence to support such a claim. This is a methodological pronouncement, the kind that Collins and Pinch are unwilling to make and the kind that I insist must be made to tell an accurate history of science.

Finally, Collins and Pinch commit sins of factual omission. In addition to ignoring methodological concerns of the kind that I just described, they make far too many broadly unsubstantiated claims in their telling of the cold fusion tale. For the sake of brevity, I point to one example.[14]

Recall Lewis's claim that Fleischmann and Pons had reported artifactual heat data by not stirring their electrochemical cells. Collins and Pinch write, "However, it seems that Lewis' charges were misplaced. Pons and Fleischmann claimed that there was no need to stir the electrolyte because the deuterium bubbles produced by the reaction did the job sufficiently well" (1993, 70). Although this was indeed the case in 1989, recall from the preceding discussion that, by the time of their 1990 paper, Fleischmann et al. had indeed changed their tune and that they reported using stirring devices for their experimental runs. Such a difference in our accounts is more than simple attention to minutiae, considering Collins and Pinch's conclusion from their report of the incident:

> As in other controversies what was taken by most people to be a "knockdown" negative result turns out, on closer examination, to be itself subject to the same kinds of ambiguities as the results it claims to demolish. If Lewis' measurements had been un-

packed in the same kind of detail reserved for Pons and Fleischmann they might not have seemed as compelling as they did at the time. In the atmosphere of the Baltimore meeting, where physicists were baying for the blood of the two chemists, and where a whole series of pieces of negative evidence was presented, Lewis was able to deliver the knock-out blow. (1993, 70–71)

First, it is clearly the case that Lewis's data were not the knockout blow; rather, it was the Abriola and Petrasso reports, which showed the artifactual nature of the Utah neutron data. In addition, Collins and Pinch do not account for the fact that Fleischmann and Pons eventually resorted to stirred water baths. One wonders whether this was a case of Fleischmann and Pons's bowing to the pressure of the big, bad physicists or the realization that they had neglected a simple procedure that every undergraduate chemistry student knows to follow. Based on Collins and Pinch's telling in *The Golem*, we will never know. Finally, speculation with respect to Lewis's own measurements being ignored in the feeding frenzy over Fleischmann and Pons does a disservice to the history of science. Maybe Lewis's results would have remained robust or maybe not. Either way, Collins and Pinch do not give their readers the scientific detail necessary to decide. We are left with an ambiguous telling of an ambiguous case. This, in the end, is not an improvement on the current state of science studies.

In conclusion, Collins and Pinch and their colleagues have done science studies a great service in choosing to focus on the broader, less methodological issues that naively scientistic accounts of the scientific process have ignored for far too long. But if Collins and Pinch wish to remain true to their title, *The Golem: What Everyone Should Know about Science*, they must tell the whole story, not just the parts that have been neglected. Replacing one incomplete story with another serves nobody's interests.

Notes

1. Neutron generation data was taken in cells, similar to the one in figure 8.1, employing 0.4 cm × 10 cm Pd-rod cathodes using a neutron dose equivalent rate monitor. Fleischmann and Pons used the resulting γ-ray data to bolster their claim that a neutron flux of $4 \times 10^4 \text{ s}^{-1}$ was generated.

2. Fleischmann and Pons note in the "Errata" that they were assisted throughout their research effort by Marvin Hawkins, a graduate student whose name they inadvertently left off their initial paper. His name appears on all subsequent work.

3. It is interesting to note that although this measurement is significantly lower than that reported for Fleischmann and Pons, it is still somewhat larger than that reported by Jones (1989), who detected a neutron flux of $4.1 \pm 0.8 \times 10^{-3} \text{ s}^{-1}$. The Abriola paper refers to the Jones paper in its introduction, but it singles out the Fleischmann and Pons paper in the body of the paper and the critique.

4. The Compton edge is best understood as the end of the Compton spectrum of electron energies responsible for the continuous part of the spectrum. Since no spectrometer operates with perfect resolution, one would expect to see such effects on most spectra (Tsoulfanadis 1983, 355).

5. Petrasso's experiment uses a known neutron source (^{239}Pu) submerged in a water tank to produce the neutron thermalization reaction. Petrasso measured a resolving power of approximately 5% with a NaI scintillator similar to that used by Fleischmann and Pons (Petrasso et al. 1989b, 183). In addition, the resolution of such detectors can be estimated by the follow-

ing equation: $R(E) = DE/E \approx R(E_n)(E_n/E)^{1/2}$. Petrasso notes that $R(E_n)$, the measured resolution at energy En, can be determined by using either a ^{60}Co source (1.33 MeV peak), or a ^{40}K decay line (1.46 MeV). Based on Fleischmann and Pons's published 2.2 MeV peak, Petrasso notes that the line width indicates a resolving power of 2.5%, a factor of at least two *below Petrasso's estimates for the same instrument*. Such a discrepancy in resolving power indicates that Fleischmann and Pons should report a more sharply defined Compton edge at 1.99 MeV than that shown by Petrasso (see figure 8.6). Instead, Fleischmann and Pons report none at all.

6. "Although they inexplicably no longer claim to have observed the 2.2 MeV neutron capture γ-ray, they now contend that their γ-ray signal line is a true γ-ray energy 2.496 MeV and, most importantly, that this signal is evidence for nuclear reactions in their cell. They make this claim despite their inability to identify the nuclear process associated with their purported 2.496 MeV γ-ray or to account for its distinctly unphysical line shape. Furthermore, after correcting their spectral data for miscalibration in energy, we argue that their signal line (with a correct energy of 2.8 MeV rather than 2.496 MeV) as well as all lines above their (true) ^{208}Tl peak are instrumental artefacts in the upper channels of their spectrum analyzer" (Petrasso et al. 1989a, 668).

7. "The estimated neutron flux in this experiment is at least a factor of 50 times smaller than that reported by Jones et al. and about one million times smaller than that reported by Fleischmann et al. The results suggest that a significant fraction of the observed neutron events are associated with cosmic rays" (Gai et al. 1989, 29).

8. In addition to an upper limit of one 2.45 MeV neutron per hour from the experimental cells, significantly lower than that reported by Fleischmann and Pons, Salamon and his colleagues report the following: "During the 831h (live time) of monitoring γ-ray emissions from electrolytic cells in Pons' laboratory, no evidence was seen of radiation from any known d+d (or p+d) fusion reaction. The upper limits placed on power from these reactions range between 10^{-12} and 10^{-6} W, which are many orders of magnitude lower than the sensitivity of the calorimetric measurements made by Pons' group; therefore, if a heat excess were to have occurred during our period of observation, one could conclude that *no known fusion process contributed significantly to that excess*" (Salamon et al. 1990 404; italics added).

9. For the sake of brevity and owing to the scientific community's general lack of interest, I have elected not to discuss the fusion gas data.

10. The following summarizes their 1989 calorimetry results (Fleischmann and Pons 1989, 304–5; italics added):

- Excess enthalpy generation is a function of applied current density, which relates to the chemical potential of the cell, and the electrode volume.
- In experimental runs in excess of 120 h, heat in excess of 4 MJ cm^{-3} of electrode was liberated.
- In terms of the amount of energy released relative to that consumed, values exceeding 1000% of break-even *are projected*.
- *Projections* for experiments using a D2O, DTO, and T2O mixture, rather than solely D2O, place the energy output at 10^5–10^6% of break-even, and an enthalpy release of > 10 kW cm^{-3} of the electrode.

11. The energy and mass balances performed on the "black box" model allowed Fleischmann et al. to derive the following equation to describe the heat flow out of their cells in terms of heat transfer coefficients:

$$\Delta t \cong Q / (4 \, k_R'' \, t^3 \, bath - yI/t^0) \, (1 - [1 + 1] \, It/2FM^0)$$

where t is cell temperature, Q is the rate of heat transfer, t is time, and I is applied cell current. For a detailed derivation of this equation, please see Fleischmann et al. 1990, 323–37.

12. Heat excess is estimated as $W_{est} = (E - 1.54)I$, where E is the cell voltage and I is the applied current (Kainthla et al. 1989).

13. On a more detailed methodological note, much of the work on the calorimetry investigation is determining whether or not the excess heat is an artifact or is easily explained by electrochemical causes. Recall the list of known electrochemical processes that could generate heat in a typical Fleischmann and Pons–type cell. In total, these reactions account for a total of 2.11 W cm^{-3}, not nearly enough to account for the 10 to 100 W cm^{-3} reported by Fleischmann and Pons between 1989 and 1990 and not nearly enough to account for most of the subsequent work on excess heat. For instance, Oriani et al. (1990), who report a relatively high heat generation, in excess of 60 W cm^{-3}, note the examination of possible chemical reactions: "To examine possible chemical reactions that may be responsible for such magnitudes of energy generation, assume first the exothermic formation of a palladium deuteride that (surprisingly) does not have a corresponding palladium hydride. Assuming that the hypothetical PdD$_2$ has a DH of formation of ~200 kJ/mol, one calculates that at 1A the excess power would be 1.04W if all the D$_2$O reacted to form hydride. Since our work and that of many others demonstrates that essentially all the D$_2$ leaves the cell at steady state, the calculated excess power from this source decreases by a factor of at least 50. Furthermore, should such a hypothetical deuteride be formed at high current densities at which a high activity of dissolved deuterium is generated, at low current densities and correspondingly low activities the PdD$_2$ would decompose *endothermically*" (R. Oriani et al. 1990, 657; italics added). Oriani goes on to consider the analogous formation of a lithium deuteride, which is similarly dismissed as the cause of all the excess heat. The Oriani results were obtained in a fashion similar to that used in experiments discussed in this text.

14. So that I am not accused of refusing to practice what I preach, the interested reader should turn to Collins and Pinch 1993, 64–68 and 75–77, for their discussions of nuclear and thermal data and the lack of essential detail.

References

Abriola, D., et al. 1989. Examination of nuclear measurement conditions in cold fusion experiments. *Journal of Electroanalytical Chemistry* 265: 355–60.

Bockris, J. O., et al. 1990. A review of the explanations of the investigations of the Fleischmann-Pons phenomenon. *Fusion Technology* 18: 11–13.

Close, F. 1991. *Too Hot to Handle: The Race for Cold Fusion*. Princeton, N.J.: Princeton University Press.

Collins, H., and T. Pinch. 1993. *The Golem: What Everyone Should Know about Science*. Cambridge: Cambridge University Press.

Fleischmann, M., and S. Pons. 1989. Electrochemically induced nuclear fusion of deuterium. *Journal of Electroanalytical Chemistry* 261: 301–8.

Fleischmann, M., et al. 1989a. Errata. *Journal of Electroanalytical Chemistry* 263: 187–88.

———. 1989b. Measurement of γ-rays from cold fusion. Correspondence in *Nature* 339: 667.

———. 1990. Calorimetry of the palladium–deuterium–heavy water system. *Journal of Electroanalytical Chemistry* 287: 293–348.

Gai, M., et al. 1989. Upper limits on neutron and γ-ray emission from cold fusion. *Nature* 340: 29–34.

Gieryn, T. 1992. The ballad of Pons and Fleischmann: how cold fusion became more interesting, less real. In *The Social Dimension of Science*, ed. E. McMullin. South Bend, Ind.: Notre Dame University Press.

Jones, S. E., et al. 1989. Observation of cold nuclear fusion in condensed matter. *Nature* 338: 737–40.

Kainthla, R. C., et al. 1989. Sporadic observation of the Fleischmann–Pons heat effect. *Electrochimica Acta* 34: 1315–18.

Mallove, E. 1991. *Fire from Ice.* New York: Wiley.

Oriani, R. A., et. al. 1990. Calorimetric measurements of excess power output during the cathodic charging of deuterium into palladium. *Fusion Technology* 18: 652–58.

Petrasso, R. D., et al. 1989a. Measurement of γ-rays from cold fusion. Correspondence in *Nature* 339: 667–69.

———. 1989b. Problems with the γ-ray spectrum in the Fleischmann et al. experiment. *Nature* 339: 183–85.

Salamon, M. H., et al. 1990. Limits on the emission of neutrons, γ-rays, electrons and protons from Pons/Fleischmann electrolytic cells. *Nature* 344: 401–5.

Schultze, J. W., et al. 1989. Prospects and problems of electrochemically induced nuclear fusion. *Electrochimical Acta* 34: 1289–1313.

Scott, C. D., et al. 1990. Measurement of excess heat and apparent coincident increases in the neuron and gamma-ray count rates during the electrolysis of heavy water. *Fusion Technology* 18:103–14.

Tsoulfanadis, N. 1983. *Measurement and Detection of Radiation.* New York: McGraw-Hill.

Yun, K., et al. 1991. Calorimetric observations of heat production during electrolysis of 0.1M LiOD + D2O solution. *Journal of Electroanalytical Chemistry* 306: 279–85.

9

Avoiding the
Experimenters' Regress

Allan Franklin

Collins and the Experimenters' Regress

Harry Collins is well known for his skepticism concerning both experimental results and evidence. He develops an argument that he calls the "experimenters' regress" (Collins 1985, chapter 4, pp. 79–111): What scientists take to be a correct result is one obtained with a good, that is, properly functioning, experimental apparatus. But a good experimental apparatus is simply one that gives correct results. Collins claims that there are no formal criteria that one can apply to decide whether or not an experimental apparatus is working properly. In particular, he argues that calibrating an experimental apparatus by using a surrogate signal cannot provide an independent reason for considering the apparatus to be reliable.

In Collins's view, the regress is eventually broken by negotiation within the appropriate scientific community, a process driven by factors such as the career, social, and cognitive interests of the scientists, but one that is not decided by what we might call epistemological criteria, or reasoned judgment. Thus Collins concludes that his regress raises serious questions concerning both experimental evidence and its use in evaluating scientific hypotheses and theories. Indeed, if no way out of the regress can be found, then he has a point.

Collins's strongest candidate for an example of the experimenters' regress in actual scientific practice is presented in his history of the early attempts to detect gravitational radiation, or gravity waves. In this case, the physics community was forced to compare Weber's claims that he had observed gravitational waves with the reports from six other experiments that failed to detect them. On the one hand, Collins argues that the decision between these conflicting experimental results could not be made on epistemological or methodological grounds—He claims that the six negative experiments could not legitimately be regarded as replications[1] and hence become less impressive. On the other hand, Weber's apparatus, precisely because the experiments used a new type of apparatus to try to detect a hitherto unobserved phenomenon,[2] could not be subjected to standard calibration techniques.

In this essay, I examine Collins's account of the first attempts to detect gravitational radiation. I then present my own account of the episode, which differs substantially from his, and argue that his account is misleading and provides no grounds for belief in the experimenters' regress. On the contrary, even in this extremely complex and atypical case of experimental uncertainty, scientists had good orthodox epistemic reasons for arriving at their conclusions as to whether gravity waves had really been detected.

Collins's Account of Gravity Wave Detectors

Collins illustrates the experimenters' regress and his skepticism concerning experimental results with the early history of gravity wave detectors.[3] He begins with a discussion of the original apparatus developed by Joseph Weber (figure 9.1) which later became standard. Weber used a massive aluminum alloy bar,[4] or antenna, which was supposed to oscillate when struck by gravitational radiation.[5] The oscillation was to be detected by observing the amplified signal from piezo-electric crystals attached to the antenna. The signals were expected to be quite small—the gravitational force is much weaker than the electromagnetic force and the bar had to be well insulated from other sources of noise such as electrical, magnetic, thermal, acoustic, and seismic forces. Because the bar was at a temperature different from absolute zero, thermal noise could not be avoided, so Weber set a threshold for pulse acceptance that was in excess of those expected from thermal noise.[6] In 1969, Weber claimed to have detected approximately seven pulses/day due to gravitational radiation.

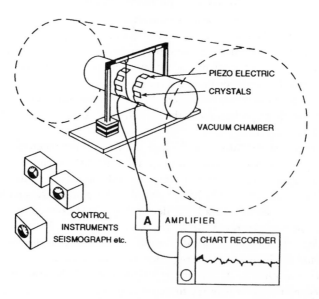

WEBER-TYPE GRAVITY WAVE ANTENNA

Figure 9.1 A Weber-type gravity wave detector (Collins 1985).

Because Weber's reported rate was far greater than that expected from calculations of cosmic events (by a factor of more than 1,000), his early claims were met with skepticism. During the late 1960s and early 1970s, however, Weber introduced several modifications and improvements that increased the credibility of his results. He claimed that above-threshold peaks had been observed simultaneously in two detectors separated by 1,000 miles. It was extremely unlikely that such coincidences were due to random thermal fluctuations. In addition, he reported a 24-hour periodicity in his peaks, a sidereal correlation that indicated a single source for the radiation, located perhaps near the center of our galaxy. These results increased the plausibility of his claims sufficiently so that by 1972 three other experimental groups had not only built detectors but had also reported results. However, none was in agreement with Weber.

Collins notes that after 1972, Weber's claims were less and less favored. During 1973, three different experimental groups reported negative results, and subsequently these groups, as well as three others, reported further negative results. No corroboration of Weber's results was reported during this period. Although in 1972 approximately a dozen groups were involved in experiments aimed at checking Weber's findings, by 1975 no one, except Weber himself, was still working on that particular problem. Weber's results were regarded as incorrect.

The reasons offered by different scientists for their rejection of Weber's claims were varied, and not all these scientists agreed on their relative importance. Between 1972 and 1975, it was discovered that Weber had made several serious errors in his analysis. His computer program for analyzing the data contained an error, and his statistical analysis of residual peaks and background was questioned and thought to be inadequate. Weber also claimed to find coincidences between his detector and another distant detector when, in fact, the tapes used to provide the coincidences were actually recorded more than 4 hours apart. Weber had found a positive result where even he did not expect one. Others cited the failure of Weber's signal to noise ratio to improve, despite his "improvements" to his apparatus. In addition, the sidereal correlation disappeared. Perhaps most striking were the uniformly negative results obtained by six other groups. Collins argues, however, that these reasons were insufficient to reject Weber's results.

If Collins is correct in arguing that the negative evidence provided by the replications of Weber's experiment—the application of what we might call epistemological criteria, combined with Weber's acknowledged errors—is insufficient to explain the rejection of Weber's results, then he must provide another explanation. At this point, Collins invokes the experimenters' regress cited earlier. He argues that if the regress is a real problem in science, then scientists should disagree about what constitutes a good detector, and this is exactly what his fieldwork shows.[7] Collins presents several excerpts from interviews with scientists working in the field that show their differing opinions on the quality of detectors. At the beginning, when scientists were responding positively to Weber's experiments, they offered different reasons for believing in Weber's claims. Some were impressed by the coincidences between two separated detectors; others cited the fact that the coincidence disappeared when one detector signal was delayed relative to the other or Weber's use of the computer for analysis.[8] But not everyone agreed, and three years later when the majority view had shifted against Weber, scientists again gave different reasons for rejecting his program. Collins argues that these differing opinions demonstrate the lack of any consensus on formal criteria for the adequacy of gravitational wave detectors.

Collins also notes that after 1975 scientists stopped pursuing other possible explanations of the conflict between Weber's results and those of his critics. These included not only prosaic differences in the detectors—that is, piezo-electric crystals versus other strain detectors, the antenna material, and the electronics—but also more theoretical hypotheses such as the invocation of a new, "fifth force." There was even speculation in some quarters that the gravity wave findings were the result of random mistakes, deliberate lies, or self-deception or could even be explained by psychic forces. Yet by 1975 all these alternative explanations of the discordant results of Weber and his critics had been dropped. It was simply accepted that Weber had made systematic errors. Collins suggests that this was not a necessary conclusion and that scientists might have reasonably investigated the other more radical possibilities. Why, according to Collins, did a premature consensus form? Collins proposes that the key factor was the social impact of the negative evidence provided by scientist Q.[9] Collins argues that it was not so much the intrinsic power of Q's experimental result but, rather, his forceful and persuasive presentation of that result and his careful analysis of thermal noise in an antenna that turned the tide.

In the course of presenting his case that there were insufficient epistemic grounds for rejecting Weber's reports, Collins deals in detail with the attempt to break the experimenters' regress by the use of calibration. In this episode, most experimenters calibrated their gravity wave detectors by injecting a pulse of known acoustical energy at one end of their antenna and measuring the output of their detector. This served to demonstrate that the apparatus could detect energy pulses and also provided a measure of the sensitivity of the apparatus. One might object, however, that the acoustic pulses were not an exact analogue of gravity waves. Another experimenter adopted a different method of calibration, using a local, rotating laboratory mass to more closely mimic gravity waves.[10]

According to Collins, Weber was initially reluctant to calibrate his own antenna acoustically but did eventually do so. However, his observations included quite a different method of analyzing the output pulses. He used a nonlinear energy algorithm, whereas his critics used a linear algorithm that was sensitive to both the phase and the amplitude of the signal. (For a discussion of this difference see Franklin 1994, app. 1.) The critics argued that one could show quite rigorously, and mathematically, that the linear algorithm was superior in detecting pulses. The issues of the calibration of the apparatus and the method of analysis used were inextricably tied together. When the calibration was done on Weber's apparatus, it was found that the linear algorithm was 20 times better at detecting the calibration signal than was Weber's nonlinear algorithm. For the critics, this established the superiority of their detectors. Weber did not agree. He argued that the analysis and calibration applied only to short pulses, those expected theoretically and used in the calibration, whereas the signal he was detecting had a length and shape that made his method superior.

Collins concludes,

> The anomalous outcome of Weber's experiments could have led toward a variety of heterodox interpretations with widespread consequences for physics. They could have

led to a schism in the scientific community or even a discontinuity in the progress of science. Making Weber calibrate his apparatus with the electrostatic pulses was one way in which his critics ensured that gravitational radiation remained a force that could be understood within the ambit of physics as we know it. They ensured physics' continuity—the maintenance of links between past and future. Calibration is not simply a technical procedure for closing debate by providing an external criterion of competence. In so far as it does work this way, it does so by controlling interpretive freedom. It is the control on interpretation which breaks the circle of the experimenters' regress, not the "test of a test" itself. (Collins 1985, 105–6)

Collins states that the purpose of his argument is to demonstrate that science is uncertain. He grants, however, "For all its fallibility, science is the best institution for generating knowledge about the natural world that we have" (1985, 165).

Although I agree with Collins concerning the fallibility of science and its status as "the best institution for generating knowledge about the natural world we have," I believe there are serious problems with his argument. These are particularly important because the regress argument, despite Collins's disclaimer, does seem to cast doubt on experimental evidence and its use in science and therefore on the status of science as knowledge.

Collins's argument can be briefly summarized as follows: There are no rigorous independent criteria for either a valid result or a good experimental apparatus; all attempts to evaluate the apparatus are dependent of the outcome of the experiment. This leads to the experimenters' regress in which a good detector can be defined only by its obtaining the correct outcome, whereas a correct outcome is one obtained using a good detector. This is illustrated by the discussion of gravity wave detectors. In practice, the regress is broken by negotiation in the scientific community, but the decision is not based on anything that one might call epistemological criteria. This casts doubt on not only the certainty of experimental evidence but also its very validity. Thus, experimental evidence cannot provide grounds for scientific knowledge.

An Alternative History of Gravity Wave Detection

Collins might correctly argue that the case of gravity wave detectors is a special case, one in which a new type of apparatus was being used to try to detect a hitherto unobserved quantity.[11] I agree.[12] I do not, however, agree that one could not present arguments concerning the validity of the results or that one could not evaluate the relative merits of two results, independent of the outcome of the two experiments. The regress can be broken by reasoned argument. I will also demonstrate that the published record gives the details of that reasoned argument.

Let us now reexamine the early history of attempts to observe gravity waves. As we shall see, in this controversy the crucial question was not about what constituted a good gravity wave detector; all the experiments did, in fact, use variants of the Weber antenna. Rather, the questions were whether or not the detector was operating properly and whether or not the data were being analyzed correctly. The discordant results reported by Weber and his critics and the disagreements about how best to resolve the differences are not unusual occurrences in the history of physics, particularly at the beginning of an experimental investigation of a phenomenon.[13] Collins correctly concludes on the basis of his interviews that Weber's critics did not always agree about the force of particular arguments, but this does not mean that by the end of the controversy, each did not have compelling reasons for believing that Weber was wrong. To understand how the scientists arrived at their conclusions, we must look at the full history of the episode as given in published papers, conference proceedings, and public letters. I believe that the picture these give is one of overwhelming evidence against Weber's result and that the final decision, although not governed by an algorithm, was unquestionably reasonable and based on epistemological criteria.

I begin with the debates about calibration and Weber's analysis procedure. The question of determining whether or not there is a signal in a gravitational wave detector or whether or not two such detectors have fired simultaneously is not easy to answer. There are several problems. One is that there are energy fluctuations in the bar due to thermal, acoustic, electrical, magnetic, and seismic noise. When a gravity wave strikes the antenna, its energy is added to the existing energy. This may change either the amplitude or the phase, or both, of the signal emerging from the bar. It is not just a simple case of observing a larger signal from the antenna after a gravitational wave strikes it. This difficulty informs the discussion of which analysis procedure was the best to use.

The nonlinear, or energy, algorithm preferred by Weber was sensitive only to changes in the amplitude of the signal. The linear algorithm, preferred by everyone else, was sensitive to changes in both the amplitude and the phase of the signal. Weber preferred the nonlinear procedure because it resulted in the output of several pulses exceeding the threshold for each input pulse to his detector. Weber admitted, however, that the linear algorithm, preferred by his critics, was more efficient at detecting calibration pulses. Results on the superiority of the linear algorithm for detecting calibration pulses were reported by both Kafka (Shaviv and Rosen 1975, 258–59) and Tyson (Shaviv and Rosen 1975, 281–82). Tyson's results for calibration pulse detection are shown for the linear algorithm in figure 9.2 and for the nonlinear algorithm in figure 9.3. The peak for the linear algorithm is clear, whereas the peak for the nonlinear procedure is not apparent.[14]

Nevertheless, Weber preferred the nonlinear algorithm, his reason being that this procedure gave a more significant signal than did the linear one. This is illustrated in figure 9.4, in which Weber's data analyzed with the nonlinear algorithm are presented in (a) and the same data analyzed with the linear procedure are presented in (b). To decide which analysis procedure was better, Weber was, in fact, using the method that gave a positive result. If anyone in this case was "regressing," it was Weber. However, contrary to what Collins's account would have us believe, the rest of the scientific community was sharply critical of this maneuver.

Weber's critics objected to his failure to calibrate his apparatus. Nonetheless, they accommodated Weber's position as much as possible by analyzing their own data using both algorithms. If it was the case that, unlike the calibration pulses, for which the linear algorithm was clearly superior, using the linear algorithm either masked or failed to detect a real signal, then using the nonlinear algorithm on their data should have produced a clear signal. But none appeared. Typical results are shown in figures 9.3 and 9.5. Figure 9.3 gives Tyson's data analyzed with the nonlinear algorithm and shows

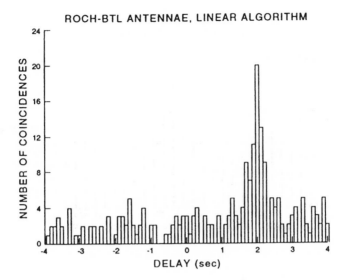

Figure 9.2 A plot showing the calibration pulses for the Rochester–Bell Laboratory collaboration. The peak due to the calibration pulses is clearly seen (Shaviv and Rosen 1975).

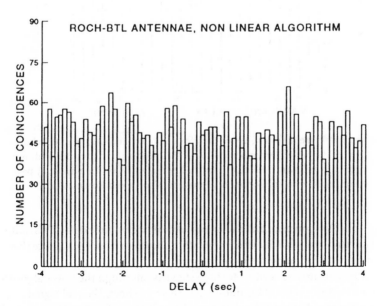

Figure 9.3 A time-delay plot for the Rochester–Bell Laboratory collaboration, using the nonlinear algorithm. No sign of any zero-delay peak can be seen (Shaviv and Rosen 1975).

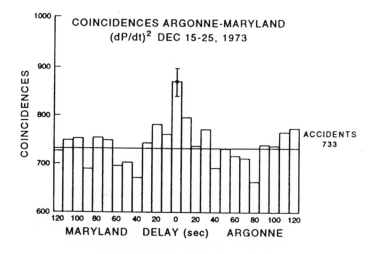

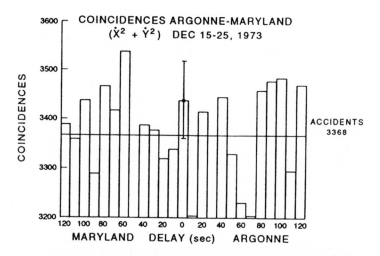

Figure 9.4 Weber's time delay-data for the Maryland–Argonne collaboration for the period December 15–25, 1973. The top graph uses the nonlinear algorithm, whereas the bottom uses the linear algorithm. The zero-delay peak is seen only with the nonlinear algorithm (Shaviv and Rosen 1975).

not only no calibration peak but also no signal peak at zero time delay. It is quite similar to the data analyzed with the linear algorithm shown in figure 9.5. (Note that for this data run, no calibration pulses were inserted.)[15]

Weber had an answer ready. He suggested that although the linear algorithm was better for detecting calibration pulses, which were short, the real signal of gravitational waves was a longer pulse than most investigators thought. He argued that the nonlinear algorithm that he used was better at detecting these longer pulses. The critics did think that gravitational radiation would be produced in short bursts. Still, if

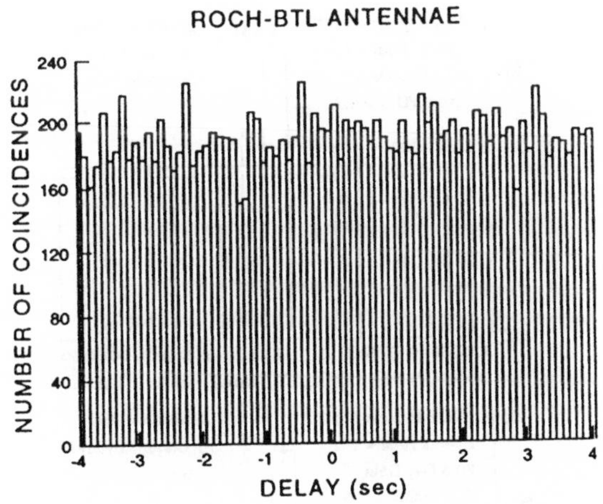

ROCH-BTL ANTENNAE

Figure 9.5 A time-delayplot for the Rochester–Bell Laboratory collaboration, using the linear algorithm. No sign of a zero-delay peak is seen (Shaviv and Rosen 1975).

the signal were longer, one would have expected a positive result to appear when the critics' data were processed with the nonlinear algorithm. It did not (see figure 9.3). Drever also reported that he had looked at the sensitivity of his apparatus with arbitrary waveforms and pulse lengths. Although he found a reduced sensitivity for longer pulses, he did analyze his data to look explicitly for such pulses. He found no effect. He also found no evidence for gravity waves using the short-pulse (linear) analysis.

Despite Collins's claim that the experiments performed by the critics could not be legitimately considered to be replications, one finds there was considerable cooperation among the various groups. They exchanged both data tapes and analysis programs, and this led to the first of several questions about possible serious errors in Weber's analysis of his data. Douglass first pointed out that there was an error in one of Weber's computer programs:

> The nature of the error was such that any above-threshold event in antenna A that occurred in the last or the first 0.1 sec time bin of a 1000 bin record is erroneously taken by the computer program as in coincidence with the next above-threshold event in channel B, and is ascribed to the time of the later event. Douglass showed that in a four-day tape available to him and included in the data of (Weber et al. 1973), nearly all of the so-called "real" coincidences of 1–5 June (within the 22 April to 5 June 1973 data) were created individually by this simple programming error. Thus not only some phenomenon besides gravity waves could, but in fact did cause the zero-delay excess coincidence rate. (Garwin 1974, 9)

Weber admitted the error but did not agree with the conclusion. He claimed that even after the error was corrected, there was a positive signal.[16] It is clear, however, that this error raised legitimate doubts about the correctness of Weber's results (figure 9.6).

Another serious question was raised concerning Weber's use of varying selection thresholds in the analysis of his data. Weber's critics used a single threshold to ana-

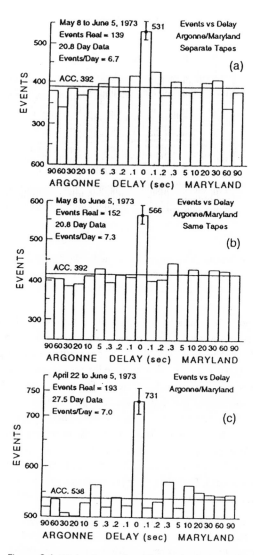

Figure 9.6 Weber's results. The peak at zero time delay can be clearly seen (Weber et al. 1973).

lyze all their data. They accused Weber of varying his threshold to maximize his final result. Garwin (1974), in fact, showed that such a procedure could indeed produce a positive result that was merely an artifact of the selection procedure. Kafka also showed, using his own data, that varying the selection procedure could, on occasion, create a positive signal (Shavin and Rosen 1975, 265). A further criticism was offered by Levine and Garwin (1974, Garwin 1974). They questioned whether or not Weber's apparatus could have produced the signal he reported. Using a computer simulation, they argued that the signal should have been far broader than the one Weber published.

Weber denied both charges. He did not specify his method of data selection for his histogram, however. In particular, he did not state that all the results presented in a particular histogram had the same threshold.
Weber reported another odd result.

> First, Weber has revealed at international meetings (Warsaw, 1973. etc.) that he had detected a 2.6-standard deviation excess in coincidence rate between a Maryland antenna [Weber's apparatus] and the antenna of David Douglass at the University of Rochester. Coincidence excess was located not at zero time delay but at "1.2 seconds," corresponding to a 1–sec intentional offset in the Rochester clock and a 150–millisecond clock error. At CCR-5, Douglass revealed, and Weber agreed, that the Maryland Group had mistakenly assumed that the two antennas used the same time reference, whereas one was on Eastern Daylight Time and the other on Greenwich Mean Time. Therefore, the "significant" 2.6 standard deviation excess referred to gravity waves that took four hours, zero minutes and 1.2 seconds to travel between Maryland and Rochester. (Garwin 1974, 9)

Weber answered that he had never claimed that the 2.6 standard-deviation effect he had reported was a positive result. But by producing a statistically significant result where none was expected, Weber had certainly cast doubt on his analysis procedures.

Let us summarize the evidential situation concerning gravity waves at the beginning of 1975. The results were discordant. Weber had reported positive results on gravitational radiation, whereas six other groups had reported no evidence for such radiation. The critics' results were not only more numerous but had also been carefully cross-checked. The groups had exchanged both data and analysis programs and confirmed their results. The critics had also investigated whether or not their use of a linear algorithm could account for their failure to observe Weber's reported results. They had used Weber's preferred procedure, a nonlinear algorithm, to analyze their own data, and still found no sign of an effect. They had also calibrated their experimental apparatuses by inserting acoustic pulses of known energy and finding that they could detect a signal. Weber, however, as well as his critics, could not detect such calibration pulses even when using his analysis procedure.

In addition, several other serious questions were raised about Weber's experimental procedures. These included an admitted programming error that generated spurious coincidences between Weber's two detectors, possible selection bias by Weber, Weber's report of coincidences between two detectors when the data had been taken 4 hours apart, and whether or not Weber's experimental apparatus could produce the narrow coincidences claimed.

The critics' results were clearly far more credible than Weber's. They had checked their results by independent confirmation, which included sharing data and analysis programs. They had also eliminated a plausible source of error, that of the pulses being longer than expected, by analyzing their results using the nonlinear algorithm and by looking for such long pulses. They had also calibrated their apparatuses by injecting pulses of known energy and observing the output. Contrary to Collins, I believe that the scientific community made a reasoned judgment when they rejected Weber's results and accepted those of his critics. Although no formal rules were applied—for example, if you make four errors rather than three, your results will lack credibility; or if there are five, but not six, conflicting results, your work will still be credible—the procedure clearly followed the norms of scientific experimentation.

I also question Collins's account of Garwin's role (Scientist Q). Although Garwin was a major figure in this episode and did present strong and forceful arguments against Weber's result, the same arguments were being made at the time by other scientists. At GR7, for example, the 1974 Conference on General Relativity for which we have transcriptions of a panel discussion on gravity waves, Garwin's experiment was mentioned only briefly, and although the arguments about Weber's errors and analysis were made, they were not attributed to the absent Garwin.[17]

A point that should be emphasized is that although calibration—and its success or failure—played a significant role in the dispute, it was not decisive. Other arguments were both needed and provided. In most cases, failure to detect a calibration signal would be a decisive reason for rejecting an experimental result.[18] In this case, however, it was not. The reason for this was that the scientists involved seriously considered whether or not an injected acoustic energy pulse was an adequate surrogate for a gravity wave. Their doubts about its adequacy led to the variation in analysis procedures and to the search for long pulses.

I have argued that Collins's claim that the experimenters' regress played a role in this episode is wrong. He conflates the difficulty of getting an experiment to work with the problem of demonstrating that it is working properly. I believe that I have shown that epistemological criteria were reasonably applied to decide between Weber's result and those of his critics. Although calibration was not decisive in the case of gravity wave detectors, nor should it have been, it is often a legitimate and important factor and may even be decisive in determining the validity of an experimental result. Both the argument about the impossibility of replication and the lack of criteria in deciding the validity of experimental results have failed. The history of gravity wave detectors has not established what Collins claims it does. There are no grounds for belief in the experimenters' regress.

Epilogue

At the present time, gravity waves have not been detected by either the use of Weber bar antennas or the newer technique of using an interferometer, in which the gravitational radiation has a differential effect on the two arms of the interferometer and thus change the observed interference pattern. The radiation has not been detected, even though current detectors are several orders of magnitude more sensitive than those in use in 1975.[19]

Gravity waves have been observed indirectly, however. They have been detected by measuring the change in orbital period of a binary pulsar. Such a binary system should emit gravitational radiation, thereby losing energy and decreasing the orbital period. This effect was initially measured using the results of Hulse and Taylor (1975), which provided the initial measurement of the period, and of Weisberg and Taylor (1984), which measured the period at a later time. The measured change in the period was $(-2.40 \pm 0.09) \times 10^{-12}$ s s^{-1}, in excellent agreement with the theoretical prediction of $(-2.403 \pm 0.002) \times 10^{-12}$ s s^{-1}. "As we have pointed out before most relativistic theories of gravity other than general relativity conflict strongly with our data, and would appear to be in serious trouble in this regard. It now seems inescapable that gravitational radiation exists as predicted by the general relativistic quadrupole formula" (Weisberg and Taylor 1984, 1350). More recent measurements and theoretical calculations give $(2.427 \pm 0.026) \times 10^{-12}$ s s^{-1} (measured) (Taylor and Weisberg 1989) and $(2.402576 \pm 0.000069) \times 10^{-12}$ s s^{-1} (theory) (Damour and Taylor 1991). If General Relativity Theory is correct, Weber should not have observed a positive result. Weber's original signal was 1,000 times larger than that predicted by General Relativity.

Notes

1. Collins offers two arguments concerning the difficulty, if not the virtual impossibility, of replication. The first is philosophical. What does it mean to replicate an experiment? In what way is the replication similar to the original experiment? A rough and ready answer is that the replication measures the same physical quantity. Whether or not it, in fact, does so can, I believe, be argued for on reasonable grounds, as I will do.

Collins's second argument is pragmatic. This is the fact that in practice it is often difficult to get an experimental apparatus, even one known to be similar to another, to work properly. Collins illustrates this with his account of Harrison's attempts to construct two versions of a TEA laser (Transverse Excited Atmospheric) (Collins 1985, 51–78). Even though Harrison had previous experience with such lasers and had excellent contacts with experts in the field, he had great difficulty building the lasers. Hence the difficulty of replication.

Ultimately Harrison found errors in his apparatus, and once these were corrected, the lasers operated properly. As Collins admits, "In the case of the TEA laser the circle was readily broken. The ability of the laser to vaporize concrete, or whatever, comprised a universally agreed criterion of experimental quality. There was never any doubt that the laser ought to be able to work and never any doubt about when one was working and when it was not" (Collins 1985, 84).

Although Collins seems to regard Harrison's problems with replication as casting light on the episode of gravity waves, as supporting the experimenters' regress, and as casting doubt on experimental evidence in general, it really doesn't work. As Collins admits (see the quotation in the preceding paragraph), the replication was clearly demonstrable. One may wonder what role Collins thinks this episode plays in his argument.

2. In my more detailed discussions of this episode (Franklin 1994, 1997), I argued that the gravity wave experiment is not at all typical of physics experiments. In most experiments, the adequacy of the surrogate signal used in calibrating the experimental apparatus is clear and unproblematical. In cases in which it is questionable, considerable effort is devoted to establishing the adequacy of that surrogate signal. Although Collins chose an atypical example, I believe that the questions he raises about calibration in general and this particular episode of gravity wave experiments should be answered.

3. One cannot examine Collins's sources in any detail because he uses interviews almost exclusively, and to maintain anonymity, he refers to scientists by a letter only. In addition, he makes no references to any of the published scientific papers involved, not even to those of Weber.

4. This device is often referred to as a *Weber bar*.

5. Gravity waves were predicted by Einstein's general theory of relativity. Just as an accelerated electrically charged particle produces electromagnetic radiation (light, radio waves, etc.), so does accelerated mass produce gravitational radiation (gravity waves). Such radiation can be detected by the oscillations produced in a large mass when it is struck by gravity waves. Because the gravitational force is far weaker than the electromagnetic force, a large mass must be accelerated to produce a detectable gravity wave signal. (The ratio of the gravi-

tational force between the electron and the proton in the hydrogen atom compared with the electrical force between them is 4.38×10^{-40}, a small number indeed). The difficulty of detecting a weak signal is at the heart of this episode.

6. Given any such threshold, there is a finite probability that a noise pulse will be larger than that threshold. The point is to show that there are pulses in excess of those expected statistically.

7. This might also be expected when a new detector is first proposed and there has been little experience in its use. Although one may think about sources of background in advance, it is the actual experience with the apparatus that often tells scientists which of them are present and important.

8. Weber originally analyzed the data using his own observation of the output tapes.

9. Any reader of the literature can easily identify Q as Richard Garwin.

10. A local oscillating mass is also not an exact analogue. Although it produces tidal gravitational forces in the antenna, it does not produce gravity waves. Only a distant source could do that. Such a mass would, however, have a gravitational coupling to the antenna, rather than an electromechanical one.

11. I rely primarily on a panel discussion on gravitational waves that took place at the Seventh International Conference on General Relativity and Gravitation (GR7), at Tel Aviv University, June 23–28, 1974. The panel included Weber and three of his critics, Tyson, Kafka, and Drever, and they presented papers and also permitted discussion, criticism, and questions. Their papers include almost all the important and relevant arguments concerning the discordant results. The proceedings were published as Shaviv and Rosen 1975. Unless otherwise indicated, all quotations in this section are from Shaviv and Rosen 1975. I give the author and the page numbers in the text.

12. One might then wonder why he uses such an atypical example as his illustration of the experimenters' regress.

13. For a discussion of other similar episodes, those of experiments on atomic parity violation, on the 17 keV neutrino, and on the fifth force in gravity, see Franklin 1990, 1993a and b, 1995.

14. The calibration pulses were inserted periodically during data-taking runs. The peak was displaced by 2 seconds by the insertion of a time delay, so that the calibration pulses would not mask any possible real signal, which was expected at zero time delay.

15. Collins does not discuss the fact that Weber's critics exchanged both data and analysis programs and that they analyzed their own data with Weber's preferred nonlinear analysis algorithm and failed to find a signal. This fact, as documented in the published record, seems to argue for using epistemological criteria in evaluating the discordant experimental results.

16. This claim was never corroborated in the published literature.

17. The panel discussion on gravitational waves covers 56 pages, 243 to 298, in Shaviv and Rosen 1975. But Tyson's discussion of Garwin's experiment occupies only one short paragraph (approximately one-quarter of a page), p. 290.

18. See the discussion of calibration in Franklin 1994, 1997.

19. An account of an experiment using such a detector appears in Astone et al. 1993. Using a very sensitive cryogenic antenna, they set a limit of no more than 0.5 events/day, in contrast to Weber's claim of approximately seven events/day.

References

Astone, P., M. Bassan, P. Bonifazi, et al. 1993. Long-term operation of the Rome "Explorer" cryogenic gravity wave detector. *Physical Review D* 47: 362–75.

Collins, H. 1985. *Changing Order*. London: Sage.

———. 1994. A strong confirmation of the experimenters' regress. *Studies in History and Philosophy of Modern Physics* 25: 493–503.

Damour, T., and J. H. Taylor. 1991. On the orbital period change of the binary pulsar PSR 1913 + 16. *Astrophysical Journal* 366: 501–11.

Franklin, A. 1990. *Experiment, Right or Wrong.* Cambridge: Cambridge University Press.

———. 1993a. Discovery, pursuit, and justification. *Perspectives on Science* 1: 252–84.

———. 1993b. The Rise and Fall of the Fifth Force: Discovery, Pursuit, and Justification in Modern Physics. New York: American Institute of Physics.

———. 1994. How to avoid the experimenters' regress. *Studies in the History and Philosophy of Science* 25: 97–121.

———. 1995. The resolution of discordant results. *Perspectives on Science* 3: 346–420.

———. 1997. Calibration. *Perspectives on Science* 5: 31–80.

Garwin, R. L. 1974. Detection of gravity waves challenged. *Physics Today* 27: 9–11.

Hulse, R. A., and J. H. Taylor. 1975. A deep sample of new pulsars and their spatial extent in the galaxy. *Astrophysical Journal* 201: L55–L59.

Levine, J. L., and R. L. Garwin. 1974. New negative result for gravitational wave detection, and comparison with reported detection. *Physical Review Letters* 33: 794–97.

Shaviv, G., and J. Rosen, eds. 1975. General Relativity and Gravitation: Proceedings of the Seventh International Conference (GR7), Tel-Aviv University, June 23–28, 1974. New York: Wiley.

Taylor, J. H., and J. M. Weisberg. 1989. Further experimental tests of relativistic gravity using the binary pulsar PSR 1913 + 16. *Astrophysical Journal* 345: 434–50.

Weber, J., M. Lee, D. J. Gretz, et al. 1973. New gravitational radiation experiments. *Physical Review Letters* 31: 779–83.

Weisberg, J. M., and J. L. Taylor. 1984. Observations of post-Newtonian timing effects in the binary pulsar PSR 1913 + 16. *Physical Review Letters* 52: 1348–350.

10

Do Mutants Die
of Natural Causes?

The Case of Atomic Parity Violation

Allan Franklin

A ndrew Pickering's *Constructing Quarks* (1984b) is an early product of the social constructivist attempt to portray scientific theories as culturally specific products of social negotiation.[1] Pickering says in the first chapter of this richly detailed history of high-energy physics that he will try to offer a mirror image of the standard realist accounts in which "experiment is seen as the supreme arbiter of theory" (Pickering 1984b, 5). Whereas "the scientist legitimates scientific judgments by reference to the state of nature; I attempt to understand them by reference to the cultural context in which they are made" (8). During much of the two decades covered in this book, fundamental particle theoreticians were casting about in many directions, and experimenters were correspondingly unsure as to which experiments to undertake and which parameters to measure. Pickering's account of these uncertain periods, based in part on extensive interviews with key figures, is both fascinating and unforced. It is only when the scientists feel that they are getting some answers and their views start to converge that Pickering's history starts to diverge from what his informants are reporting.

Pickering focuses on the period following the "November Revolution" of 1974 when the Weinberg–Salam theory quickly became the "Standard Model" of electroweak interactions. (Weinberg, Salam, and Glashow received the Nobel Prize in 1979.) Unimpressed by the widespread experimental confirmations of this theory, Pickering remarks, "Quite simply, particle physicists accepted the existence of the neutral current because they could see how to ply their trade more profitably in a world in which the neutral current was real" (Pickering 1984a, 87). However, Pickering does take very seriously the early experiments on atomic parity violation, which were anomalous for the Weinberg–Salam (W–S) unified theory of electroweak interactions. These experiments, performed at Oxford University and at the University of Washington and published in 1976 and 1977, measured the parity-nonconserving optical rotation in atomic bismuth. The results disagreed with the predictions of the Weinberg–Salam theory, whereas, by contrast, a different atomic parity experiment, performed at the Stanford Linear Accelerator Center in 1978, on the scattering of polarized electrons from deuterons confirmed the theory.[2]

In a section entitled "Slaying the Mutants," Pickering argues that the scientific community simply regarded the Oxford and Washington experiments as mutants and didn't take them seriously. He claims that by 1979 the high-energy physics community accepted the Weinberg–Salam theory as established, even though "there had been no *intrinsic* change in the status of the Washington–Oxford experiments" (1984b, 301). In Pickering's view,

> particle physicists *chose* to accept the results of the SLAC experiment, *chose* to interpret them in terms of the standard model (rather than some alternative which might reconcile them with the atomic physics results) and therefore *chose* to regard the Washington–Oxford experiments as somehow defective in performance or interpretation. (301)

The implication seems to be that these choices were made so that the experimental evidence would be consistent with the standard model[3] and that there were not good, independent reasons for them.[4] In short, Pickering argues that the mutants died not from natural causes but because they didn't fit in with the career interests of the opportunistic scientific community.

Pickering regards this case as supporting his view that "there is no obligation upon anyone framing a view of the world to take account of what twentieth century science has to say" (1984b, 413). He thus denies that science is a reasonable enterprise based on valid experimental or observational evidence.[5] In this essay, I reexamine the history of this episode, presenting both Pickering's interpretation and an alternative explanation of my own, arguing that there were good evidential reasons for the decision of the physics community.

The Early Experiments: Washington and Oxford

The search for atomic parity violations arose from the attempt to test the Weinberg–Salam unified theory of electromagnetic and weak interactions in a new domain. The model predicted that one would see weak neutral-current effects in the interactions of electrons with hadrons, the strongly interacting particles. The effect would be quite small when compared with that of the dominant electromagnetic interaction but could be distinguished from it by the fact that it violated parity conservation.[6] Thus a demonstration of such a parity-violating effect and a measurement of its magnitude would test the W–S theory.[7]

One such predicted effect was the rotation of the plane of polarization of polarized light when it passed through bismuth vapor. Such a rotation is possible only if parity is violated. This was the experiment performed by the Oxford and Washington groups. They both used bismuth vapor but used light corresponding to different transitions in bismuth, $\lambda = 648$ nm (Oxford) and $\lambda = 876$ nm (Washington). They published a joint preliminary report noting that "we feel that there is sufficient interest to justify an interim report" (Baird et al. 1976, 528). They reported values for R, the parity-violating parameter, of $R = (-8 \pm 3) \times 10^{-8}$ (Washington) and $R = (+10 \pm 8) \times 10^{-8}$ (Oxford). "We conclude from the two experiments that the optical rotation, if it exists, is smaller than the values -3×10^{-7} and -4×10^{-7}

predicted by the Weinberg–Salam model plus the atomic central field approximation" (529).

Pickering offers the following interpretation:

> The caveat to this conclusion was important. Bismuth had been chosen for the experi-
> ment because relatively large effects were expected for heavy atoms, but when the effect
> failed to materialise a drawback of the choice became apparent. To go from the calcula-
> tion of the primitive neutral-current interaction of electrons with nucleons to predic-
> tions of optical rotation in a real atomic system it was necessary to know the electron
> wave-functions, and in a multi-electron atom like bismuth these could only be calcu-
> lated approximately. Furthermore, these were novel experiments and it was hard to
> say in advance how adequate such approximations would be for the desired purpose.
> Thus in interpreting their results as a contradiction of the Weinberg–Salam model the
> experimenters were going out on a limb of atomic theory. Against this they noted that
> four independent calculations of the electron wavefunctions had been made, and that
> the results of these calculations agreed with one another to within twenty-five percent.
> This degree of agreement the experimenters found "very encouraging" although they
> conceded that "Lack of experience of this type of calculation means that more theoreti-
> cal work is required before we can say whether or not the neglected many-body effects
> in the atomic calculation would make R [the parity-violating parameter] consistent with
> the present experimental limits." (1984b, 295–96)

Pickering attributes all the uncertainty in the comparison between experiment and
theory to the theoretical calculations and none to the experimental results themselves.

The comparison was even more uncertain than Pickering implies and included
uncertainty in the experimental results. The experimenters reported that the "quoted
statistical error represents 2 s.d. [standard deviations]. There are, however, also systematic effects which we believe do not exceed $\pm 10 \times 10^{-8}$, but which are not yet fully
understood" (Baird et al. 1976, 529). Thus, there were possible systematic experimental uncertainties of the same order of magnitude as the expected effect. As Pickering
states, these were novel experiments, using new and previously untried techniques.
This also tended to make the experimental results uncertain.

The theoretical calculations of the expected effect were also uncertain. The Oxford–
Washington joint paper noted that Khriplovich had argued, in a soon-to-be-published
paper, that the approximate theory overestimated R by a factor of approximately 1.5.
In addition, the four calculations agreed with their mean only to within approximately
$\pm 25\%$. This made the largest and smallest calculated values of R differ by almost a
factor of two.

In September 1977, both the Washington and Oxford groups published more detailed accounts of their experiments, with somewhat revised results (Baird et al. 1977,
Lewis et al. 1977).[8] Both groups again reported results in substantial disagreement with
the predictions of the Weinberg–Salam theory, although the Washington group stated
that "more complete calculations that include many-particle effects are clearly desirable" (Lewis, 1977, 795). The Washington group now reported a value of $R = (-0.7 \pm
3.2) \times 10^{-8}$, which not only disagreed with the W–S prediction of approximately $-2.5
\times 10^{-7}$ but also was inconsistent with their own earlier result of $(-8 \pm 3) \times 10^{-8}$. This
inconsistency was not addressed in the published paper, but it was discussed in the
atomic physics community and lessened the credibility of the result.[9] The Oxford result

was $R = (+2.7 \pm 4.7) \times 10^{-8}$, again in disagreement with the Weinberg–Salam prediction of approximately -30×10^{-8}. They noted, however, that there was a systematic effect in their apparatus. They had found a change in ϕ_r, the rotation angle, due to slight misalignment of the polarizers, optical rotation in the windows, and so forth, of order 20×10^{-8} radians.

> Unfortunately, it varies with time over a period of minutes, and depends sensitively on the setting of the laser and the optical path through the polarizer. While we believe we understand this effect in terms of imperfections in the polarizers combined with changes in laser beam intensity distribution, we have been unable to reduce it significantly. (Baird et al. 1977, 800)

A systematic effect of this size certainly cast doubt on the result.

How were these results viewed by the physics community? In the same issue of *Nature* in which the joint Oxford–Washington paper was published, Frank Close, a particle theorist, summarized the situation:

> Is parity violated in atomic physics? According to experiments being performed independently at Oxford and the University of Washington the answer may well be no.... This is a very interesting result in light of last month's report ... claiming that parity is violated in high energy "neutral current" interactions between neutrinos and matter. (Close 1976, 505)[10]

Close noted that as the atomic physics results stood, they appeared to be inconsistent with the predictions of the Weinberg–Salam model supplemented by atomic physics calculations. He also remarked, "At present the discrepancy can conceivably be the combined effect of systematic effects in atomic physics calculations and systematic uncertainties in the experiments" (1976, 505–6). Pickering stated that "if one accepted the Washington–Oxford result, the obvious conclusion was that neutral current effects violated parity conservation in neutrino interactions and conserved parity in electron interactions" (1984b, 296). Close discussed this possibility along with another alternative that had an unexpected (on the basis of accepted theory) energy dependence, so that the high-energy experiments (the neutrino interactions) would be predicted to show parity nonconservation, whereas the low-energy atomic physics experiments would not. "Whether such a possibility could be incorporated into the unification ideas is not clear. It also isn't clear, yet, if we have to worry. However, the clear blue sky of summer now has a cloud in it. We wait to see if it heralds a storm" (Close 1976, 506).

In Pickering's view, the 1977 publication of the Oxford and the Washington results indicated that "the storm that Frank Close had glimpsed had materialised and was threatening to wash away the basic Weinberg–Salam model, although not the gauge-theory enterprise itself" (Pickering, 1984b, 298).[11] There is some support for this in Miller's summary in *Nature* of the Symposium on Lepton and Photon Interactions at High Energies, held in Hamburg, August 25–31, 1977 (Miller 1977). Miller noted that Sandars had reported that neither his group at Oxford nor the Washington group had seen any parity-violating effects and that "they have spent a great deal of time checking both their experimental sensitivity and the theory in order to be sure" (288). Miller went on to state (as Pickering also reported) that

> S. Weinberg and others discussed the meaning of these results. It seems that the SU(2) $\times$ U(1) is to the weak interaction what the naive quark-parton model has been to QCD,

a first approximation which has fitted a surprisingly large amount of data. Now it will be necessary to enlarge the model to accommodate the new quarks and leptons, the absence of atomic neutral currents, and perhaps also whatever it is that is causing trimuon events. (288)

Nevertheless, I believe that the uncertainty in these experimental results made the disagreement with the W–S theory only a worrisome situation and not a cause for epistemological crisis, as Pickering believes it should have been.[12] In any event, the monopoly of Washington and Oxford was soon broken.

Discordant Results

The experimental situation changed in 1978 when Barkov and Zolotorev (1978a, b), two Soviet scientists from Novosibirsk, reported measurements on the same transition in bismuth studied by the Oxford group. Their results agreed with the predictions of the W–S model. They gave a value for $\psi_{exp}/\psi_{W-S} = (+1.4 \pm 0.3)\,k$, where ψ was the angle of rotation of the plane of polarization by the bismuth vapor and k was a factor between 0.5 and 1.5, introduced because of inexact knowledge of the bismuth vapor. They concluded that their result did not contradict the predictions of the Weinberg–Salam model. A point to be emphasized here is that agreement with theoretical prediction depended (and still does depend) on which method of calculation one chose. A somewhat later paper (Barkov and Zolotorev 1978a) changed the result to $\psi_{exp}/\psi_{W-S} = 1.1 \pm 0.3$.

Subsequent papers in 1979 and 1980 (Barkov and Zolotorev 1979, 1980a,b) reported more extensive data and found a value for $R_{exp}/R_{theor} = 1.07 \pm 0.14$. They also reported that the latest unpublished results from the Washington and Oxford groups, which had been communicated to them privately, also now showed parity violation, although "the results of their new experiments have not reached good reproducibility" (1979, 312). These later results were also presented at the 1979 conference, at which Dydak reviewed the situation.

According to Pickering, "The details of the Soviet experiment were not known to Western physicists, making a considered evaluation of its result problematic" (1984b, 299). This is simply not correct. During September 1979, an international workshop devoted to neutral-current interactions in atoms was held in Cargese (Williams 1980) and was attended by representatives of virtually all the groups actively working in the field, including Oxford, Washington, and Novosibirsk. At this workshop, not only did the Novosibirsk group present a very detailed account of their experiment (Barkov and Zolotorev 1980b), but, as C. Bouchiat remarked in his workshop summary paper, "Professor Barkov, in his talk, gave a very detailed account of the Novosibirsk experiment and answered many questions concerning possible systematic errors" (Bouchiat 1980, 364).

In early 1979, a Berkeley group reported an atomic physics result for thallium that agreed with the predictions of the W–S model (Conti et al. 1979). They investigated the polarization of light passing through thallium vapor and found a circular dichroism $\delta = (+5.2 \pm 2.4) \times 10^{-3}$, in comparison with the theoretical prediction of $(+2.3 \pm$

$0.9) \times 10^{-3}$. Although these were not definitive results—they were only two standard deviations from zero—they did agree with the W–S theory in both sign and magnitude.

It seems fair to say that in mid-1979, the various atomic physics experiments concerning the Weinberg–Salam theory were inconclusive. The Oxford and Washington groups had originally reported a discrepancy, but their more recent results, although preliminary, showed the presence of the predicted parity-nonconserving effects.In addition, both the Soviet and Berkeley results agreed with the model. Dydak (1979) summarized the situation in a talk at a 1979 conference. "It is difficult to choose between the conflicting results in order to determine the *eq* [electron-quark] coupling constants. Tentatively, we go along with the positive results from Novosibirsk and Berkeley groups and hope that future development will justify this step (it cannot be justified at present, on clear-cut experimental grounds)" (35). However, Pickering reports that "having decided not to take into account the Washington–Oxford results, Dydak concluded that parity violation in atomic physics was as predicted in the standard model" (1984b, 300).

I find no justification for Pickering's interpretation. Dydak was attempting to determine the best values for the parameters describing neutral-current electron scattering. Dydak had tentatively adopted the results in agreement with the W–S model, admitting that experiment did not, at the time, justify this. He concluded nothing about the validity of the standard model. Bouchiat, in his summary paper discussed earlier, was more positive. After reviewing the Novosibirsk experiment as well as the conflict between the earlier and later Washington and Oxford results he observed, *"As a conclusion on this Bismuth session, one can say that parity violation has been observed roughly with the magnitude predicted by the Weinberg–Salam theory"* C. Bouchiat 1980, 365),[13] but even this more positive statement does not conclude that the results agree precisely with the predictions of the theory, stating only that the experimental results were of the correct order of magnitude.

The situation was made even more complex when a group at the Stanford Linear Accelerator Center (SLAC) reported a result on the scattering of polarized electrons from deuterium that agreed with the W–S model (Prescott et al. 1978, 1979). This was the E122 experiment, discussed by Pickering. They not only found the predicted scattering asymmetry but also obtained a value for $\sin^2\theta_W = 0.20 \pm 0.03$ (1978) and 0.224 ± 0.020 (1979) in agreement with other measurements ($\sin^2\theta_W$ is an important parameter in the W–S theory). "We conclude that within experimental error our results are consistent with the W–S model, and furthermore our best value of $\sin^2\theta_W$ is in good agreement with the weighted average for the parameter obtained from neutrino experiments" (Prescott et al. 1979, 528).

Pickering concluded his story as follows:

In retrospect, it is easy to gloss the triumph of the standard model in the idiom of the "scientist's account": the Weinberg–Salam model, with an appropriate complement of quarks and leptons, made predictions which were verified by the facts. But missing from this gloss, as usual, is the element of choice. In assenting to the validity of the standard model, particle physicists chose to accept certain experimental reports and to reject others. The element of choice was most conspicuous in the communal change of heart over the Washington–Oxford atomic physics experiments and I will focus on that episode here. We saw in the preceding section that in 1977 many physicists were prepared to accept the null-results of the Washington and Oxford experiments and to construct new electroweak models to explain them. We also saw that by 1979 attitudes had hardened. In the wake of experiment E122, the Washington–Oxford results had come to be regarded as unreliable. In analysing this sequence, it is important to recognise that between 1977 and 1979 there had been no *intrinsic* change in the status of the Washington–Oxford experiments. No data were withdrawn, and no fatal flaws in the experimental practice of either group had been proposed. What had changed was the context within which the data were assessed. Crucial to this change of context were the results of experiment E122 at SLAC. In its own way E122 was just as innovatory as the Washington–Oxford experiments and its findings were, in principle, just as open to challenge. But particle physicists *chose* to accept the results of the SLAC experiment, *chose* to interpret them in terms of the standard model (rather than some alternative which might reconcile them with the atomic physics results) and therefore *chose* to regard the Washington–Oxford experiments as somehow defective in performance or interpretation. (1984b, 301)

Although I do not dispute Pickering's contention that choice was involved in the decision to accept the Weinberg–Salam model, I do disagree with him about the reasons for that choice. In Pickering's view, social interests of competing scientific traditions were paramount: "The standard electroweak model unified not only the weak and electromagnetic interactions: it served also to unify practice within otherwise diverse traditions of HEP [high-energy physics] theory and experiment. . . . Matched against the mighty traditions of HEP, the handful of atomic physicists at Washington and Oxford stood little chance" (1984b, 301–2). In my view, the choice was a reasonable one based on convincing, if not overwhelming, experimental evidence. As we shall see, this was also the judgment of atomic physicists.

The issue seems to turn on the relative evidential weight one assigns to the original Oxford and Washington atomic physics results and to the SLAC E122 experiment on the scattering of polarized electrons. Pickering seems to regard them as having equal weight. I do not. I argued earlier that both the original experimental results on bismuth and the comparison between experiment and theory were quite uncertain. When one adds to this original uncertainty the later contradictory results of both the Washington and Oxford groups, the parity-nonconserving measurement of the Novosibirsk group, and the Berkeley result on thallium, the original results were, at the very least, very uncertain.[14]

By 1981, when they published their latest result "that agrees in sign and approximate magnitude with recent calculations based upon the Weinberg–Salam theory" (Hollister et al. 1981, 643), the Washington group conceded that their earlier results were unreliable. Although theoretical calculations had reduced the size of the expected effect by a factor of approximately three or four (depending on which method of calculation one chose), the theoretical results still did not agree with the 1977 Washington–Oxford measurements. Furthermore, the 1981 Washington paper stated, "Our experiment and the bismuth optical-rotation experiments by three other groups [Oxford, Moscow, and Novosibirsk] have yielded results with significant mutual discrepancies far larger than the quoted errors" (Hollister et al. 1981, 643). They also pointed out, as I mentioned earlier, that their earlier measurements "were not mutually consistent" (643).

The moral of the story is clear. These attempts to detect atomic parity violation were extremely difficult experiments, beset with systematic errors of approximately the same size as the predicted effects. There is no reason to give priority to the earliest measurements, as Pickering does. It seems much more likely that these earlier results were perhaps less reliable because not all the systematic errors were known.

Strong Support for the Weinberg–Salam Theory: The SLAC E122 Experiment

I will now examine the arguments presented by the SLAC group in favor of the validity and reliability of their measurement. I agree with Pickering that "in its own way E122 was just as innovatory as the Washington–Oxford experiments and its findings were, in principle, just as open to challenge" (1984b, 301). But Pickering neglects to emphasize that for this reason, the SLAC group presented a very detailed discussion of their experimental apparatus and result and performed many checks on their experiment.[15]

The experiment depended in large part on a new high-intensity source of longitudinally polarized electrons. The polarization of the electron beam could be varied by changing the voltage on a Pockels cell. "This reversal was done randomly on a pulse to pulse basis. The rapid reversals minimized the effects of drifts in the experiment, and the randomization avoided changing the helicity synchronously with periodic changes in experimental parameters" (Prescott et al. 1978, 348). It had been demonstrated in an earlier experiment that polarized electrons could be accelerated with negligible depolarization. In addition, both the sign and magnitude of the beam polarization were measured periodically by observing the known asymmetry in elastic electron–electron scattering from a magnetized iron foil.

The experimenters also checked whether or not the apparatus produced spurious asymmetries. They measured the scattering using the unpolarized beam from the regular SLAC electron gun, for which the asymmetry should be zero. They assigned polarizations to the beam using the same random number generator that determined the sign of the voltage on the Pockels cell. They obtained a value for the experimental asymmetry that was consistent with zero and that also demonstrated that the apparatus could measure asymmetries of the order of 10^{-5}.

They also varied the polarization of the beam by changing the angle of a calcite prism, thereby changing the polarization of the light striking the Pockels cell. The results are shown in figure 10.1. Not only do the data fit the theoretically expected curve, but the fact that the results at 45° are consistent with zero indicates that other sources of error in the asymmetry are small. The graph shows the results for two different detectors, a nitrogen-filled Cerenkov counter and a lead glass shower counter. The consistency of the results increases the belief in the validity of the measurements. "Although these two separate counters are not statistically independent, they were analyzed with independent electronics and respond quite differently to potential backgrounds. The consistency between these counters serves as a check that such backgrounds are small" (Prescott et al. 1978, 350).

The electron beam helicity also depended on E_0, the beam energy, because of the g-2 precession of the spin as the electrons passed through the beam transport mag-

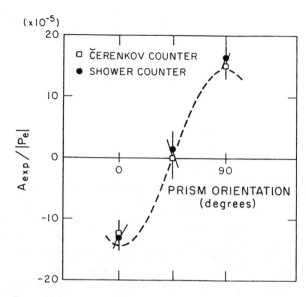

Figure 10.1 Experimental asymmetry as a function of prism angle for both the Cerenkov counter and the shower counter. The dashed line is the predicted behavior. (Prescott et al 1978).

nets. The expected distribution and the experimental data are shown in figure 10.2. The data quite clearly follow the g-2 modulation of the helicity, and the fact that the value at 17.8 GeV is close to zero demonstrated that any transverse spin effects were small.

A potential source of serious error came from small systematic differences in the beam parameters for the two helicities. Small changes in beam position, angle, current, or energy could influence the measured yield and, if correlated with reversals of beam helicity, could cause apparent, but spurious, parity-violating asymmetries. These quantities were carefully monitored, and a feedback system was used to stabilize them.

> Using the measured pulse to pulse beam information together with the measured sensitivities of the yield to each of the beam parameters, we made corrections to the asymmetries for helicity dependent differences in beam parameters. For these corrections, we have assigned a systematic error equal to the correction itself. The most significant imbalance was less than one part per million in Eo [the beam energy] which contributed -0.26×10^{-5} to A/Q² [the experimental asymmetry]. (Prescott et al. 1978, 351)

This is to be compared with their final result of $A/Q^2 = (-9.5 \pm 1.6) \times 10^{-5}$ (GeV/c)⁻², which is obviously much larger. This was regarded by the physics community as a reliable and convincing result.

Contrary to Pickering's claim, hybrid model alternatives to the W–S theory were both considered and tested by E122. In their first paper, they pointed out that the hybrid model was consistent with their data only for values of $\sin^2\theta_W < 0.1$, which was inconsistent with the measured value of approximately 0.23. In the second paper (1979), they plotted their data as a function of $y = (E_o - E')/E_o$, where E' is the energy

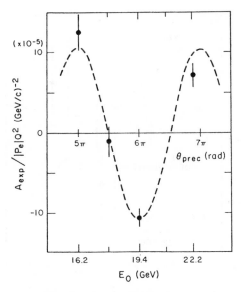

Figure 10.2 Experimental asymmetry as a function of beam energy. The expected behavior is the dashed line (Prescott et al 1978).

of the scattered electron. Both models, W–S and the hybrid, made definite predictions for this graph. The results are shown in figure 10.3, and the superiority of the W–S model is obvious. For W–S, they obtained a value of $\sin^2\theta_W = 0.224 \pm 0.020$, with a χ^2 probability of 40%. The hybrid model gave a value of 0.015 and a χ^2 probability of 6×10^{-4}, "which appears to rule out this model."

My interpretation of this episode differs dramatically from Pickering's. The physics community chose to accept an experimental result that confirmed the Weinberg–Salam theory because it had been extremely carefully done and carefully checked. This view is supported by Bouchiat's 1979 summary. After hearing a detailed account of the SLAC experiment by Prescott, he stated,

> To our opinion, this experiment gave the first truly convincing evidence for parity violation in neutral current processes. . . . I would like to say that I have been very much impressed by the care with which systematic errors have been treated in the experiment. It is certainly an example to be followed by all people working in this very difficult field. (1980, 358, 359–60)[16]

The physics community chose to await further developments in the atomic parity–violating experiments, which, as I have shown, were uncertain.

Pickering is correct that in stating that there had been no intrinsic change in the early Washington and Oxford results, but he fails to realize that they were never very convincing. They began as uncertain, although worrisome, and remained uncertain. The subsequent history (see Franklin 1990, chap. 8) shows that although other reliable atomic physics experiments confirm the W–S theory, the bismuth results, although generally in agreement with the predictions, are still somewhat uncertain. In addi-

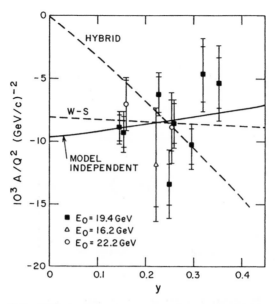

Figure 10.3 Asymmetries measured at three different energies plotted as a function of $y = (E_o - E')/E_o$. The predictions of the Hybrid model, the Weinberg–Salam theory, and a model independent calculation are shown. "The Weinberg–Salam model is an acceptable fit to the data; the hybrid model appears to be rules out" (Prescott et al. 1979).

tion, the most plausible alternative to the W–S model, which could reconcile the original atomic physics results with the electron-scattering data, was tested and found wanting. A choice certainly was made, but as the evidence model suggests, it was made on the basis of experimental evidence. The mutants died of natural causes, not from scientific opportunism.

Discussion

Several points are worth making about this episode of atomic parity violation experiments. Perhaps most important is that the comparison between experiment and theory can often be extremely difficult. This is particularly true in this episode, when one is at the limit of what one can calculate confidently and what one can measure reliably. In 1977, the atomic physics calculations of the expected parity-violating effects had a large uncertainty. Even today, estimates of the accuracy of the calculations are about 25%. The experimental results are also uncertain.[17] These experiments contain systematic errors, which may mimic or mask the expected result. Unlike statistical errors, which can be calculated precisely, systematic errors are extremely difficult both to detect and to estimate. The experimenters devoted considerable effort to finding such systematic errors and to eliminating them. Some of these errors have now been measured and corrected for. Others are still unknown. But that does not prevent us from having good reason to prefer the results from later, more sophisticated experiments.

Pickering remarks that by 1979 (and presumably to this day), there had been no intrinsic change in the early Washington and Oxford results. In the sense that no one knows with certainty why those early results were wrong, he is correct. Nevertheless, since those early experiments, physicists have found new sources of systematic error that were not dealt with in the early experiments. The redesign of the apparatus has, in many cases, precluded testing whether or not these effects were significant in the older apparatus. Although one cannot claim with certainty that these effects account for the earlier, presumably incorrect, results, one does have reasonable grounds for believing that the later results are more accurate. The internal consistency of the later measurements, especially those done by different groups, enhances that belief.

It is clear the relationship between theory and experiment is more complex than "Man proposes, Nature disposes." Our history has shown that it is not at all clear at times just what Man is proposing. The theoretical predictions in this episode had considerable uncertainty. The uncertain and discordant experimental results show that one may not be able to see just how Nature is disposing. What does seem clear, however, is that the evidence model fits the history of this episode better than Pickering's model does. Scientists chose, on the basis of reliable experimental evidence provided by the SLAC E122 experiment, to accept atomic parity violation as a strong confirmation of the Weinberg–Salam theory. They chose to leave an apparent, but also quite uncertain, anomaly in the atomic parity violation experiments for future investigation. Future work resolved the anomaly.

This episode also demonstrates that scientists make judgments about the reliability of experimental results that coincide with what one would decide on epistemological grounds. The E122 group argued for the validity of their experimental result using strategies that coincide with an epistemology of experiment (see n. 15). As we have seen, the scientific community accepted their arguments.

Notes

1. The constructivist view that I oppose is known as the "Strong Programme" and was summarized by Trevor Pinch: "In providing an explanation of the development of scientific knowledge, the sociologist should attempt to explain adherence to all beliefs about the natural world, whether perceived to be true or false, in a similar way. . . . What is being claimed is that many pictures can be painted, and furthermore, that the sociologist of science cannot say that any picture is a better representation of Nature than any other" (1986, 3, 8).

2. The detailed history is far more complex. Other experimental groups in Novosibirsk and Berkeley found atomic physics results in agreement with the W–S theory, and Moscow a group confirmed the anomaly.

3. The standard model includes quarks and the Weinberg–Salam theory.

4. To be fair, Pickering, as discussed later, couches his discussion in terms of the adjustment of theoretical and experimental research practice. I believe, however, that in this case, his meaning is essentially agreement between experiment and theory. In any event, I think that he made an incorrect judgment on the relative evidential weight of the two different experiments.

5. Pickering denies this allegation. For further discussion, see Ackermann 1991, Franklin 1993, Lynch 1991, and Pickering 1991.

6. Parity conservation is the same as space reflection or left–right symmetry. Parity nonconservation is a violation of that symmetry. For details, see Franklin 1990, chap. 4.

7. If parity is violated, but time reversal invariance holds, as experiment suggests, there will be an electric dipole transition element between two states of the same parity, such as two $S_{1/2}$ states of cesium. All experiments thus far have been based on measurements of an electroweak interference between such a weak interaction amplitude, E_{1PNC}, which does not conserve parity, and a parity-conserving electromagnetic amplitude. In the optical rotation experiments on bismuth and lead, one observes an interference between the allowed M_1 (magnetic dipole) amplitude and the E_{1PNC} amplitude. In highly forbidden M_1 transitions in heavy elements such as thallium or cesium, one observes an interference between the parity-nonconserving amplitude and the Stark-induced (by an electric field) amplitude. For details, see M. Bouchiat and Pottier 1984.

8. Pickering reported that the papers also "described two `hybrid' unified electroweak models, which used neutral heavy leptons to accommodate the divergence with the findings of high energy neutrino scattering" (1984b, 294). Although such models were discussed in the literature at the time, there is no mention of such speculation in these two experimental papers.

9. Carl Wieman, who made important contributions to the later atomic physics experiments, recalls such discussions (private communication).

10. The experiment with neutrinos (υ) and antineutrinos ($\bar{\upsilon}$) that Close referred to had concluded, "Measurements of R_υ and $R_{\bar{\upsilon}}$, the ratios of neutral current to charged current rates for υ and $\bar{\upsilon}$ cross sections, yield neutral current rates for υ and $\bar{\upsilon}$ that are consistent with a pure V-A interaction but 3 standard deviations from pure V or pure A, indicating the presence of parity nonconservation in the weak neutral current" (Benvenuti et al. 1976, 1039).

11. Pickering also discusses two other anomalies for the Weinberg–Salam model, the high-y anomaly and the trimuon events. However, these two anomalies were also soon shown experimentally to be incorrect.

12. My view is supported by the recollections of Carl Wieman, who was working on atomic parity violation at the time, and of James Scott, a physicist visiting at Oxford at the time of these experiments.

13. For another summary of the experimental situation at the time, see Commins and Bucksbaum 1980. They regarded the situation with respect to the bismuth results as unresolved.

14. For summaries of the situation by atomic physicists, not members of the high-energy physics community, see Commins and Bucksbaum 1980 and C. Bouchiat 1980.

15. The experimenters used several strategies to establish the validity of their result, which I have discussed elsewhere as parts of an epistemology of experiment (Franklin 1986, chap. 6). The experimenters intervened and observed the predicted effects when they changed the angle of the calcite prism and when they varied the beam energy. They checked and calibrated their apparatus by using the unpolarized SLAC beam and observed no instrumental asymmetries and found that their apparatus could measure asymmetries of the expected size. They also used different counters, the lead glass shower counter and the gas Cerenkov counter, and obtained independent confirmation of the validity of their measurement.

16. Bouchiat was a theoretical atomic physicist working on parity violation, and not a member of the high-energy physics community. As one can see, it was not the mighty traditions of high-energy physics that convinced him but, rather, the experimental evidence.

17. The smallest uncertainty in any experimental result is approximately 1%.

References

Ackermann, R. 1991. Allan Franklin, right or wrong. In *PSA 1990*, vol. 2, ed. A. Fine, M. Forbes, and L. Wessels, 451–57. East Lansing, Mich.: Philosophy of Science Association.

Baird, P. E. G., M. W. S. Brimicombe, R. G. Hunt, et al. 1977. Search for parity-nonconserving optical rotation in atomic bismuth. *Physical Review Letters* 39: 798–801.

Baird, P. E. G., M. W. S. Brimicombe, G. J. Roberts, et al. 1976. Search for parity non-conserving optical rotation in atomic bismuth. *Nature* 264: 528–29.

Barkov, L. M., and M. S. Zolotorev. 1978a. Measurement of optical activity of bismuth vapor. *JETP Letters* 28: 503–6.

———. 1978b. Observations of parity nonconservation in atomic transitions. *JETP Letters* 27: 357–61.

———. 1979. Parity violation in atomic bismuth. *Physics Letters* 85B: 308–13.

———. 1980a. Parity nonconservation in bismuth atoms and neutral weak-interaction currents. *JETP* 52: 360–69.

———. 1980b. Parity violation in bismuth: experiment. In *International Workshop on Neutral Current Interactions In Atoms (Cargese)*, ed. W. L. Williams, 52–76. Washington, D.C.: National Science Foundation.

Benvenuti, A., D. Cline, F. Messing, et al. 1976. Evidence for parity nonconservation in the weak neutral current. *Physical Review Letters* 37: 1039–42.

Bouchiat, C. 1980. Neutral current interactions in atoms. In *International Workshop on Neutral Current Interactions In Atoms*, ed. W. L. Williams, 357–69. Washington, D.C.: National Science Foundation.

Bouchiat, M. A., and L. Pottier. 1984. Atomic parity violation experiments. In *Atomic Physics 9*, ed. R. Van Dyck and E. Fortson, 246–71. Singapore: World Scientific.

Close, F. E. 1976. Parity violation in atoms? *Nature* 264: 505–6.

Commins, E., and P. Bucksbaum. 1980. The parity nonconserving electron–nucleon interaction. *Annual Reviews of Nuclear and Particle Science* 30: 1–52.

Conti, R., P. Bucksbaum, S. Chu, et al. 1979. Preliminary observation of parity nonconservation in atomic thallium. *Physical Review Letters* 42: 343–46.

Dydak, F. 1979. Neutral currents. *International Conference on High Energy Physics*, 25–49. Geneva: CERN.

Franklin, A. 1986. *The Neglect of Experiment*. Cambridge: Cambridge University Press.

———. 1990. *Experiment, Right or Wrong*. Cambridge: Cambridge University Press.

———. 1993. Discovery, pursuit, and justification. *Perspectives on Science* 1: 252–84.

Hollister, J. H., G. R. Apperson, L. L. Lewis, et al. 1981. Measurement of parity nonconservation in atomic bismuth. *Physical Review Letters* 46: 643–46.

Lewis, L. L., J. H. Hollister, D. C. Soreide, et al. 1977. Upper limit on parity-nonconserving optical rotation in atomic bismuth. *Physical Review Letters* 39: 795–98.

Lynch, M. 1991. Allan Franklin's transcendental physics. In *PSA 1990*, vol. 2, ed. A. Fine, M. Forbes, and L. Wessels, 471–85. East Lansing, Mich.: Philosophy of Science Association.

Miller, D. J. 1977. Elementary particles—a rich harvest. *Nature* 269: 286–88.

Pickering, A. 1984a. Against putting the phenomena first: the discovery of the weak neutral current. *Studies in the History and Philosophy of Science* 15: 85–117.

———. 1984b. *Constructing Quarks*. Chicago: University of Chicago Press.

———. 1991. Reason enough? More on parity violation experiments and electroweak gauge theory. In *PSA 1990*, vol. 2, ed. A. Fine, F. M. Forbes, and L. Wessels, 459–69. East Lansing, Mich.: Philosophy of Science Association.

Pinch, T. 1986. *Confronting Nature*. Dordrecht: Reidel.

Prescott, C. Y., W. B. Atwood, R. L. A. Cottrell, et al. 1978. Parity non-conservation in inelastic electron scattering. *Physics Letters* 77B: 347–52.

————. 1979. Further measurements of parity non-conservation in inelastic electron scattering. *Physics Letters* 84B: 524–28.

Williams, W. L., ed. 1980. *Proceedings, International Workshop on Neutral Current Interactions in Atoms (Cargese)*. Washington, D.C.: National Science Foundation.

11

Latour's Relativity

John Huth

In "A Relativistic Account of Einstein's Relativity,"[1] Bruno Latour attempts to demonstrate that even the "theoretical" sciences are "social through and through," thereby trying to buttress an important claim of the "Strong Programme" in the social studies of science. The second issue that he attempts to address in this article is what he sees as limitations in the applications of the "usual concepts" of the social sciences to the natural sciences:

> Instead of extending the social sciences' usual concepts to the natural sciences, I want to redefine these very social concepts in order to make them able to explain the more formal sciences. The task at hand is to keep the same strong programme, but to doubt what social sciences have to say about society. It is, in effect, a two-pronged enterprise, one that treats the natural and the social sciences symmetrically. (Latour 1988, 4)

This article is centered on Latour's fairly narrow and confused reading of a text on relativity, written by Einstein for laypersons, entitled "Relativity: The Special and General Theory."[2] I am not qualified to say whether the theory of relativity can provide inspiration to the social sciences, but it does appear that at least one "prong" of Latour's approach, namely, the demonstration that relativity is "social through and through" is based on several misinterpretations of Einstein's text, a failure to grasp how relativity is practiced by physicists as whole, a lack of research into the background of the origins of relativity, and a complete twisting of the meaning of the terms *society* and *abstraction* to fit the central thesis of his article. The description of Einstein's text and relativity appear to have very little relation to the issues that would be important to a practicing physicist or to one who teaches the subject to freshmen. Perhaps this issue is not germane, but the natural question arises that if he cannot get many of the details right about the seminal point of the article, can any of the conclusions drawn be valid?

Latour has voiced disdain for the notion that scientists could criticize the work of the practitioners of science studies:

> First, the opinions of scientists about science studies are not of much importance. Scientists are the informants for our investigations of science, not our judges. The vision

we develop of science does not have to resemble what scientists think about science. . . . With apologies to the narcissistic egos of scientists, the inner workings of their trade do not resemble commencement addresses and honorific speeches.[3]

Yet Latour's article very neatly demonstrates the role that scientists can play in enforcing a kind of reality check against severe misinterpretations of the content of scientific theories by practitioners in Latour's field.

"A Relativistic Account" has, in fact, become a poster child for what can go wrong in science studies. It has been quoted in several venues by scientists as a notorious example of how misinterpretations of the details of a theory can be used to draw bizarre conclusions about the social nature of the physical sciences. Among other places, one finds references to this article in Alan Sokal's famous *Social Text* hoax,[4] in an essay by Irving Klotz in *The Scientist*,[5] and by Mario Bunge.[6] These references use Latour's misinterpretation as a basis for criticism of what has been broadly characterized as the social constructivist school in science studies.

Has "A Relativistic Account of Einstein's Relativity" had a significant impact on the field of science studies? A quick consultation of various citation indices show that less than 1% of the 2,170 citations of Latour's work after 1988 refers to this article. Of this 1%, roughly half are from scientists using this as an example of science studies run amok. A colleague of mine working in the field of science studies referred to the article as "silly" and perhaps not worth my time reviewing.

Given this reception, one might properly ask why one should examine the "Relativistic Account," since it appears to have had little influence in the field of science studies and several authors have already identified this work as poor scholarship. The answer is simple. First, although the article itself may have had limited influence in the field, it is a classic example of how a misinterpreted theory ends up as grist for the social constructivist program. Other areas of physics that I'm aware of as being very poorly understood by many science studies practitioners include quantum mechanics (the uncertainty principle, the "many worlds" interpretation, wave-particle duality), particle physics, and even classical physics (chaotic systems). The article is worth dissecting to show how very limited knowledge of a field can propagate into bold and unjustified claims about the structure of a theory. Second, most of the previous critiques by scientists of this particular article have been rather brief and have always assumed that the reader is familiar with the theory of relativity. But if we are to reach our colleagues in the field of science studies, we must first explain the theory under scrutiny before critiquing some of the more notorious examples.

To this end, I briefly describe the important points of the special theory of relativity that are germane to issues raised in Latour's article.

Relativity as Seen by a Physicist

Relativity was not a new concept at the time of Einstein's work but was an important principle of physics dating, at least from Newton and Galileo. Two critical concepts that are cornerstones of classical mechanics are (1) the existence of "inertial frames of reference," that is, systems that are moving, one relative to the other, with uniform velocities; and (2) the notion that the fundamental laws of mechanics would be the same, regardless of the inertial frame from which one performs an experiment.

In Newtonian physics, some of the "invariant" quantities—ones that would appear the same in different frames of reference—would include forces, mass, and conservation of energy. In the era before Einstein, there were prescriptions that related how measurements would appear to observers in different frames of reference. These prescriptions, termed *transformation laws*, were based on what appeared to be sensible assumptions about space and time from the more limited experience of that period. These transformation laws, termed *Galilean transformations,* would seem to agree with some commonsense notions about how velocities might add. A classic example is to imagine what the velocity of a bullet would be when viewed, on the one hand, by a person on a train who fired the gun and, on the other hand, from the perspective of a person on a station platform who sees an addition of the velocity of the train and the velocity of the bullet relative to the train.

Toward the end of the nineteenth century, with the development of the theory of electromagnetism, it became apparent that a new fundamental, and hence invariant, quantity was appearing: the speed of light. The speed of light appears in the equations (the Maxwell equations) describing the behavior of electric and magnetic fields. Since the speed of light appears as a fundamental constant of the theory, it followed that to be consistent with the construction of classical mechanics, this, too, must be an invariant quantity.

The major problem was that the speed of light is not an invariant quantity under the classical Galilean transformations that include what one might call "intuitive" concepts of how velocities should add. The way to "fix" the problem was to modify the transformation laws expressing how measurements would appear in different frames of reference so that the speed of light remained an invariant for all frames of reference. In addition, for velocities much slower than the speed of light, one would retain the Galilean transformation laws as a practical "limiting" case.

Einstein's major contribution to the resolution of this conflict was to realize that one had to abandon the concept of a "universal" time—that is, a time that was independent of the frame of reference—in favor of a time that varied from one frame of reference to another. The implication of this was that certain ideas, such as the simultaneity of events, also were dependent on one's reference frame. The new transformation laws, the Einstein–Lorentz transformations, have this frame dependence of time built into them, whereas the Galilean transformations did not.

With the Einstein–Lorentz transformations, one obtains the speed of light as an invariant quantity for all frames of reference. In addition, for inertial frames traveling with relative velocities approaching the speed of light, new effects arise, such as the apparent contraction of objects in the direction of motion, the slowing of the passage of time in other frames, and a change in the form of the laws for energy and momentum conservation. These effects, however, must be precisely defined in terms of how they correspond to "events" measured in different frames. All these effects have been verified experimentally, and to date, nothing has challenged the validity of the Einstein–Lorentz transformation laws.

One important point is that the description of the new transformation laws requires a bit more care in how one defines measurements in different inertial frames. This is because of the lack of simultaneity in different frames. For the practicing physicist, an inertial frame of reference can be represented as a rigid grid of points. At each point in that frame are a precise location and a local time that can be defined consistently. Measurements of the same "event" can be described by the local time and position in any one of an infinite number of frames of reference. The measurements of the same event (e.g., an "event" may be the decay of an atom) or relation among multiple events can be related from one frame to others via the Einstein–Lorentz transformation laws.

Einstein published his account of relativity in 1905 in the journal *Annalen der Physik*. For a physicist, this remains an amazing paper for its lucidity and direct exposition. About a decade later, in response to much pressure for a popular account of relativity, Einstein published a layperson's text entitled "Relativity: The Special and General Theory." In "Relativity," Einstein tries to communicate the concepts of relativity to a general audience. In fact, the original title of the book was *Über die spezielle und die allgemeine Relativitätstheorie (Gemeinverständlich)*. Here, *Germeinverständlich* translates roughly as "generally understandable" or "popular version." The method of exposition in "Relativity" is quite different from a scientific paper, as should be the case if one does not want to lose the lay reader. Einstein spends much of his time establishing how one measures events in different inertial frames via an anthropomorphic example of observers on a fast-moving train and on an embankment.

"Observers"—either humans or other beings—are not a requirement for the theory. However, the use of humans holding clocks and rulers in the text of "Relativity," does help explain the theory to a person who is not used to dealing on a daily basis with transformation laws between different inertial frames; this is the basis for Einstein's lay text on the subject. I have no doubt that most readers can more readily assimilate these concepts if they are related to devices common to everyday experience, but these are not prerequisites to understanding the theory. Often the consequences of relativity to subatomic particles or galaxies are indirect and never employ the idea of an "observer."

The critical point of Einstein's relativity is neither the role of the observer, which is not necessary, nor the specification of inertial frames, which was already part of Galilean–Newtonian relativity, but, rather, the new form of the transformation laws and the abandonment of the idea that time would be universal for all frames of reference. The result was that the speed of light is now satisfactorily described as an invariant for all inertial frames of reference.

Latour's "Relativistic Account"

Now let us turn to Latour's "A Relativistic Account of Special Relativity." As stated in his abstract, Latour's main purpose is to show why "Einstein's text is read as a contribution to the sociology of delegation" (1988, 3). Latour complains about the so-called Strong Programme in the social studies of science: "Its achievements are far from impressive in the mathematical sciences. The more formalized a field of science, the less field studies there exist and the less convincing they are" (3).

One might suppose that a very direct and straightforward interpretation of the failure to demonstrate the "Strong Programme" in the natural sciences is that it lacks sufficient explanatory power and hence must be discarded. This, however, would not lead to articles such as "A Relativistic Account," so Latour explains, that rather than abandon the Strong Programme with its concept that science is "social through and through" (1988, 3), he seeks a redefinition of the term *social* to demonstrate the relevance of the Strong Programme: "The second way of interpreting this failure is to consider that the definition of 'society' brought into play in order to explain the sciences is unfit for the task" (1988, 3). How does Latour propose to achieve this? He wants "to redefine these very social concepts in order to make them able to explain the more formal sciences" (1988, 4) and so sets out the main premise of his critique of relativity: "In what ways can we, by reformulating the concept of society, see Einstein's work [the text of "Relativity"] as explicitly social?" (1988, 5).

Here Latour is engaging in a kind of semantic subterfuge; that is, if one cannot prove a statement within the confines of normal scholarly discourse, why not redefine the terms until one can? In this article, the meanings of *society* and *abstraction* are so stretched to fit his interpretation of relativity that they lose any semblance of the common meanings and shed no new light on the theory itself.

A focus of Latour's analysis are the semiotic concepts of narration referred to as "shifting in" and "shifting out." "Shifting in and out" is the narrative operation in which the author of a work in effect changes the personae of a narrative. For example, an author may say, "I want to tell you the story of something that happened to a friend of mine . . ." and so uses the first person pronoun to refer to the experiences of the friend. This is an example of "shifting in." In "shifting out," the process is reversed, so that the writer reverts to the original persona in the narrative.

In his text, Einstein tries to explain relativity to the lay reader by describing events as viewed by observers on a train moving near the velocity of light relative to an embankment where, also, observers are standing. The train represents one frame of reference, and the embankment represents another. As the narrative develops, Einstein explains the peculiar results of relativity in terms of what each observer sees in his or her frame of reference relative to the other. The use of "humans" as observers in the narrative is not necessary; in fact, the consequences of relativity can be inferred from very indirect measurements. Nonetheless, the anthropomorphization of the observers in the two frames of reference makes much sense for a lay text so that the reader may better visualize the transformations from one frame of reference to another.

In examining Einstein's text, Latour notes that when the discussion turns from the imaginary observer in one frame of reference (the train) to an imaginary observer in another (the embankment), this represents a "shifting in" of the narrative. I quote Latour on the issue of shifting in and out:

> The peculiarity of Einstein's narration is not that it puts to use shifting in and out, since every narration does the same, but that it focuses the reader's attention upon these very operations.
>
> He [Einstein] is interested only in the way in which we send any actor to any other frame of reference. Instead of describing laws of nature, he sets out to describe how any description is possible. . . .
>
> Technically, his book is about delegations . . . and is a book of meta-linguistics or of semiotics, one which tries to understand how any narration is constructed.

Playing the idiot, the author-in-the-text redefines what an event is, what a space and a time are, by the practical activity of a little character holding firmly a rigid little rod (no cheap psychoanalysis intended) who superimposes the readings of the hands of watches and the notches of rulers. (1988, 9)

This is a critical issue. Either Latour's description is completely trivial in the sense that it is addressing how a scientist might produce a narrative for a lay audience by making an anthropomorphic description, or it demonstrates a complete misunderstanding of the meaning of the concept *frames of reference*. The existence of a frame of reference does not depend on the presence of a human or the comparison of readings of watches, rulers, and the like. In fact, the vast majority, if not the entirety, of the experimental consequences of special relativity involve humans only indirectly, and not as observers "holding rigid little rods" (Latour's attempt to belittle the narrative notwithstanding).

For example, when protons move toward each other at high energies, they undergo a contraction in their direction of motion, one relative to the other, so that they evolve with higher velocities from spheres to pancake-shaped objects. When they collide, the fact that they have pancake (rather than spheroidal) shapes has definite consequences for the distribution of particles that results from the collision. These consequences, predicted by Richard Feynman, have been observed in high-energy proton–proton collisions in many experiments. Rather than saying that little observers are sitting on the protons, the experimental consequences of special relativity are to produce a definite prediction for the angles and energies of particles emerging from the collision.

To say that Einstein "is interested only in the way in which we send any actor to any other frame of reference" does a gross injustice to the words and meaning of the text, particularly when one reads (and realizes) that Einstein is interested mainly in the way one can obtain a set of transformation laws consistent with the speed of light being an invariant quantity. The critical issue is what one takes to be fundamental—the Galilean transformation laws or the speed of light. Hence, to say that "instead of describing laws of nature, he sets out to describe how any description is possible" is really to assert something quite contrary to the text and to what one might construe as even a remotely reasonable interpretation of the text and the theory.

Here Latour appears to be confusing the term *observer* for *inertial frame*. Einstein does take some pains to explain that the train/embankment description is merely an example of what happens in frames of reference, but in any case, there can be an infinite number of potential observers in any given reference frame, since there are an infinite number of points in that frame where one can locate an observer.

The Mysterious "Third Observer"

For a person who routinely deals with particles traveling near the speed of light, Latour's description is perplexing. To add even more to the bewilderment, Latour conjures up the existence of a "third" observer. Here it appears that Latour has not grasped the idea that the postulation of the frames of reference is really a way of organizing the calculations, in which observers (as humans) don't need to exist, either explicitly or implicitly:

> Then to the hard and lowly work of building a rigid scaffolding to frame any event is added the practical management of at least three delegates shifted in other frames of reference. (1988, 10)

> The characters may be shifted out, but not shifted back in, running the risk of falling into relativism. Einstein's solution is to consider three actors; one in the train, on the embankment and a third one, the author or one of its representants, who tries to super-impose the coded observations sent back by the two others. (1988, 11)

> [In a footnote, Latour adds:] Most of the difficulties related to the ancient history of the inertia principle are related to the existence of two frames only; the solution is always to add a third frame that collects the information sent by the two others. (1988, 43)

One can search Einstein's text in vain for any mention of a "third" observer. There is certainly no explicit reference of a "third" observer, and even in a careful reading, there is nothing that can be remotely construed as the implicit existence of a third observer. It is simply a figment of Latour's imagination. Certainly, in classical mechanics (the "ancient history of the inertia principle") there is no need for a third frame. To assert this, as Latour does, without support from Einstein's text or supplementary references, and to declare that this also is true for all what I understand to be Galilean transformations is truly poor scholarship.

Furthermore, as I shall now show, to posit the existence of a third privileged/executive observer is inconsistent with Einstein's theory. In one sense, there are two possible answers to the question of how many "observers" could exist in the theory. One might properly claim that there are either two or an infinite number of observers implicit in the theory. The mathematical prescription for moving from one frame of reference to another, the Einstein–Lorentz transformations, deals with two frames of reference. Thus one might claim that there are "two" observers. Unfortunately, this misses a major point, as in any frame of reference, there are an infinite number of potential observers at every point in space. Second, since the relative velocity of the two frames is a variable quantity, an infinite number of possible frames can be compared, one to the other. Finally, multiple frames of reference—two, three, five, whatever—are possible. One simply has to perform the transformation multiple times. So, in almost any context, the theory really has an infinite number of potential observers in an infinite number of potential frames of reference. Only in a very general sense do the transformation prescriptions refer to the relation between two frames.

In either case, the notion of a third observer is a complete fabrication of Latour's. The most charitable interpretation is that the scientist thinking about a calculation in special relativity is a "third" observer, but this is so trivial that it does not add anything useful to a discussion of special relativity and is certainly not a very deep insight. Even if one were to claim that the "third observer" is a device only relevant to Latour's discussion of social relativism, it is nonetheless a tremendous error to attribute this to Einstein's text or to the "principle of inertia."

Do Clocks "Generate" Time?

To advance his discussion, Latour comments on the "meaning" of space and time with respect to the measuring instruments: "Instead of considering instruments (rulers and clocks) as ways of representing abstract notions like space and time, Einstein takes the instruments to be what generates space and time" (1988, 11). Latour's (sole) basis for this assertion appears to be his complete misinterpretation of the following quotation from Einstein's text: "We understand by the 'time' of an event the reading (position of the hands) of that one of these clocks which is in the immediate vicinity (in space) of the event. In this manner a time-value is associated with every event which is essentially capable of observation" (1931, 24).

Here there is a huge rift between the text and Latour's interpretation. The idea that clocks and rulers "generate" space and time has no basis in Einstein's quotation (above) cited by Latour. It would seem that the use of the word *generate* is only to reinforce Latour's assertion that the theory is a social construct. All of this is even clearer if one takes the trouble of going back to the original German text, where the statement was:

> Dann versteht man unter der "Zeit" eines Ereignisses die Zeitangabe (Zeigerstellung) derjenigen dieser Uhren, welche dem Ereignis (räumlich) unmittelbar benachbart ist.

The critical word *Zeitangabe* means "statement of time," and *Zeigerstellung* translates to "placement of the clock hands." This is what Einstein says one "understands" by the "time."

And this fact again illustrates a major problem in Latour's formulation. If his article was meant to be a formal, or semiotic, reading of Einstein's text, it is not relevant to arbitrarily substitute words whose meanings are not justified by the text. Furthermore, the very premise of his "semiotic" reading of a text in translation begs the question of whether he is imputing meaning to the author or the translator.

Beyond this more narrow question of the text itself, there is the larger issue of whether Einstein subscribed to this instrumentalist, Machian viewpoint. In an extreme form of the instrumentalist position, one might assert that only measurements with devices have any physical meaning, and that concepts such as "time" are only extensions of these tools. While I do not intend to venture into the philosophical issues associated with this position, it is worth saying that, at the time that Einstein wrote *Relativity*, he did not subscribe to this position. This has been discussed at length elsewhere (reference [7] for example). To quote Einstein at roughly the time wrote the popular text.

> I am anxious to draw attention to the fact that this theory is not speculative in origin; it owes its invention entirely to the desire to make physical theory fit observed fact as well as possible. We have here no revolutionary act, but the natural continuation of a line that can be traced through centuries. The abandonment of certain notions connected with space, time, and motion, hitherto treated as fundamentals must not be regarded as arbitrary, but only as conditioned by observed facts. (Quoted in Holton 1988, 251)

Latour does say in a footnote [1, p42] that he is "perfectly well aware that this paper depends on a Machian interpretation by Einstein of his own work, an interpretation that he later recanted" and cites reference 7 below. But he goes on to claim that "semiotics is concerned with what the text does, not with what the enunciator thinks."

Despite what appears as a disclaimer to allay criticism along the lines above, it does not come close to justifying the tremendous extrapolation Latour is attempting to make to *justify his* claim that the formal sciences are "social through and through." His "awareness" does not excuse a misreading of the text.

Ultimately, the question of this "Machian interpretation" does not require Einstein or special relativity at all. In physics, as other sciences, one does rely on concepts that ultimately find their basis on measurable quantities. The philosophical stance that the measurements by means of devices provide the operational meaning of these concepts is not uniquely associated with Einstein. After all, one could claim that positions and times in pre-Einsteinian physics were simply synonymous with measurements on rulers and clocks.

Let us return to Latour's central thesis. He now builds on his two major misconceptions about special relativity, the concept of shifting between imaginary delegates and the existence of a third observer and veers off into an invented drama:

> It is the enunciator [third observer] that has the privilege of accumulating all the descriptions of all the scenes he has delegated observers to. The above dilemma boils down to a struggle for the control of privileges, for the disciplining of docile bodies, as Foucault would have said. . . . If, on the contrary, there are no privileged points of view, there is nothing to prevent one of them getting an edge over the other. (1988, 15)

All this discussion about the power of one observer over another is quite alien to the whole body of relativity, in which practitioners in the field rarely, if ever, actually employ imaginary observers, let alone discuss power relations. Nor does Einstein's text.

For the practitioner, however, the Einstein–Lorentz transformation prescriptions for moving from one frame of reference to another are the critical issue. For Latour, by contrast, these are almost a footnote in his whole discourse. In this regard, the prescriptions are viewed by Latour as a kind of trivial paperwork that is somehow secondary to this issue of the "relativism":

> Given the importance of the gain, the paperwork imposed by the retranscription of each document appears to be quite light. Given any set of coordinates sent by any one of the delegated observers, it is possible for the enunciator to shift them back in his own frame of reference by substituting each with another, through the set of equations known as a "Lorentz Transformation." (1988, 18)

And in a caption showing the Einstein–Lorentz transformation, "The Lorentz transformation defines the paperwork necessary to move documents from one frame to the other and still maintain superimposition of the traces at the end" (1988, 18). Here the "gain" has something to do with how one regains absolute relativism between the observers. The "superimposition of the traces" is simply a fancy way of saying that they compare observations.

Rather than being a simple way of comparing "paperwork," the Einstein–Lorentz transformation is the entire crux of the theory. The issues of erecting an imaginary scaffold for a frame of reference is important and puts the discussion in concrete terms, but the transformation laws themselves are what actually guarantee the invariance of the speed of light for all reference frames. It is the form of these transformation laws that is critical (and testable). Indeed, nothing antecedent to the form of the transformation laws says anything specific about nature, but the actual structure of the transformations does. Nonetheless, Latour seems to shrug off the transformation laws as being of secondary importance, when in fact they are the entire point.

After describing the "first prong" of his approach—whether Einstein's text is "social through and through," Latour turns to the "second prong," namely, the question of whether relativity has relevance for relativism in the social sciences. Perhaps relativity can provide inspiration for Latour's version of relativism. I really don't know whether it can or cannot. One thing is clear, however; it is not "Einstein's" relativity that is being applied to issues of relativism in the social sciences but, rather, Latour's own particular brand of relativism, one that has no solid basis in Einstein's text or the theory of relativity. It would also be a mistake to conclude that Latour's primary objective is a treatise on social relativism. As the preceding discussion illustrates, at least half of the article is devoted primarily to the application of social constructs to relativity. I will say nothing more about this but leave it to practitioners in the social sciences to decide whether Latour's concepts are useful in this realm.

At the end of his article, Latour returns to relativity and seems intent on resolving some problems. I find this part of his article fascinating, as it seems that someone may have read a proof of Latour's manuscript and pointed out to him the obvious glaring problems—one of which is the simple fact that the "observers" need not be real humans. Einstein is very clear that the descriptions of his observers are abstracted and generalized in the physicists theoretical practice. Since Latour entirely missed this point in his discussion, he must now address the concept of abstraction:

> Einstein replaces coordinates by train and embankment or walking men by beams of light, or trains and embankment by earth and sun. The process by which abstract notions are replaced by characters is usually called, in semiotics, figurativity, or figuration. . . . At face value, it seems that Einstein is writing a popular book. . . . One could claim that Einstein, like the Lord, masters the abstract structure but knowing the weakness of his readers, feeds them figures, stories and parables instead. (1988, 32)

Well, not only at face value is Einstein writing a popular book; at all levels, the book was intended for a popular audience. It is difficult to buy into the notion that Einstein should be faulted for trying to make his work accessible to an audience beyond physicists. The absence of references to any other treatise on relativity by physicists, including Einstein himself, is indicative of the level (or lack thereof) of Latour's scholarship.

Latour must somehow preserve the idea of "social" relations when the objects under discussion are nonhuman abstractions in the actual practical applications of relativity. He thus tries to redefine "abstraction," referring to the so-called three observers: "After these three transformations, the figures of the train and of the embankment have become geometric coordinates. Are these coordinates less figurative than the train and embankment? Are they more abstract?" (1988, 33).

In the commonsense meaning of the word *abstract*, in the commonsense reading of Einstein's text, the answer from most physicists would be, "Yes, the geometric coordinates are abstractions of Einstein's description of the train and embankment," but here Latour answers his own question: "No they simply have different details and keep only some of the elements of the train" (1988, 33).

For Latour, the meaning of "abstraction" is taken way beyond its common meaning to something that is quite unintelligible: "Abstraction does not designate a higher level of figuration but a fast circulation from one repertoire to another" (1988, 35). I confess that I really don't know what this means. Perhaps Latour is claiming that if one can give a lay description of the basic ideas of a scientific theory using social analogies, then the theory becomes "social," even though its canonical representation makes no reference whatsoever to members of a society.

Latour's new meaning of the term *abstraction* is extended far beyond its common usage and that the sole utility of this reinterpretation is to explain away Einstein's move from the train/embankment construction to the generalities so that the social interpretation can remain. If indeed this is the c??e, then it is an example of a distortion of terms in a way that does not shed any new light but, rather, serves only to satisfy the author's original premise.

It has been asserted by David Mermin (1997) that much of Latour's article has been misunderstood by scientists, partly because it was intended as "fun."[8] As one can see in many of the preceding quotes, Latour indeed makes attempts at wordplay. One can make one's own judgment as to whether they are amusing. Notwithstanding these attempts at humor, it is reasonably clear where puns are being attempted ("rigid little rods") and where Latour is trying to formulate a serious argument. This question of humor appears to be irrelevant to any critique of "a relativistic account."

Summary

One issue that I brought up at the beginning of this essay is whether scientists have the right to comment on the social studies of science. Readers may simply judge for themselves on which cases the commentary of scientists is appropriate; I submit that Latour's article is one such case. By now, we have seen in some detail that Latour greatly distorted the meaning of the theory of relativity to suit his ideological needs. He did this through a distorted reading of Einstein's text and also a very sloppy or deliberate misinterpretation of the text (the "third observer") and a redefinition of terms whose only purpose appears to suit Latour's own ideological needs. The fact that his only direct reference to Einstein is his lay text "Relativity" is a good indication of how little research Latour did on the subject. This kind of misinterpretation of the theory of relativity is nothing new and has appeared almost continuously since the inception of the theory. Given the vast body of work accumulated on Einstein, his background, his philosophical leanings, his influence (rightly or wrongly) on some philosophers, the practical issues of how physicists ωeal with relativity, and so forth, it is astonishing that Latour's article could pass even a very modest threshold for scholarly discourse in a journal devoted to social issues in science.

Notes

1. Bruno Latour, "A Relativistic Account of Einstein's Relativity," *Social Studies of Science* 18 (1988): 3–44.

2. A. Einstein, *Über die spezielle und die allgemeine Relativitätstheorie (Gemeinverständlich)* (Braunschweig: Friedr. Vieweg und Sohn, 1917). Latour quotes the translation: *Relativity: The Special and General Theory* (London: Methuen, 1920, paperback ed., 1960).

3. Bruno Latour, "Who Speaks for Science?" *The Sciences*, March–April 1995, 6–7.

4. Alan D. Sokal, "Transgressing the Boundaries: Toward a Transformative Hermeneutics of Quantum Gravity," *Social Text* 46–47 (Spring–Summer 1996): 217–52.

5. Irving Klotz, *The Scientist*, July 22, 1996.

6. Mario Bunge, "A Critical Examination of the New Sociology of Science," *Philosophy of the Social Sciences* 21 (1991): 524.

7. See, for example, Gerald Holton, *Thematic Origins of Scientific Thought* (Cambridge, Mass.: Harvard University Press, 1988).

8. David Mermin, "Reference Frame," *Physics Today* 50 (October 1997): 11.

Part IV

Art, Nature, and the Rise of Experimental Method

In traditional accounts of the Enlightenment, the Scientific Revolution plays an important causal and symbolic role. Many historians of science have questioned and revised the more romantic accounts of the rise of modern science. The rise of early modern science is an exciting locus of scholarship these days—Floris Cohen is already at work on a companion volume to his monumental 1994 historiographic survey. Not surprisingly, STSers have tried their hand at deconstructing and deflating the achievements of the seventeenth century. Their program is to show that science was conceived in sin, as it were—namely, that from the beginning, science was compromised by sexism, power struggles, political influence, and delusions of epistemic grandeur.

The essays in this section deal with two important strands in that discussion: the feminist allegation that the experimental method, as described by Bacon and others, is intrinsically misogynist and the claim, rising out of Shapin and Schaffer's influential *Leviathan and the Air Pump*, that the choice of experimental methodology and the adoption of a Newtonian worldview had more to do with Restoration politics than with the quest for scientific understanding.

Alan Soble analyzes and answers the myriad feminist commentaries on Bacon, especially those of Merchant, Harding, and Keller. He defends Bacon's approach to nature against the charges of rape and torture by deploying the traditional tools of responsible scholarship. That is, quotations must be accurate and read in context; if one is to draw inferences from the metaphors used to describe nature, one must look at positive images as well as negative ones; words and expressions must be interpreted as they would have been understood at the time by the audience addressed by the author. Soble concludes that Bacon's methodology was not inimical to women and nature.

William Newman addresses the feminist corollary to their claim about scientific method—the proposition that the alchemical worldview, because it was more holistic, organic, and androgynous, incorporated a less hostile and domineering attitude toward women and nature. Newman suggests that Merchant and Keller must have been thinking about a Jungian reconstruction of alchemy. His portrayal of the sadistic sexual imagery rampant in alchemical writings serves as a chilling reminder that the opponents of mechanistic explanations are not necessarily caring, warmhearted people.

Cassandra Pinnick looks at Shapin and Schaffer's account of the Hobbes–Boyle dispute, which they present as an example of the fertility of the Strong Programme as a guide for doing history. Focusing on their descriptions of the methodological differences between the two natural philosophers, Pinnick argues that the dispute over the proper role of experiment in science in general and the reliability of the air pump in particular was much more nuanced and less polarized than this influential case study reveals. Given these historiographic flaws, *Leviathan and the Air Pump* cannot be used as historical evidence to support the Strong Programme.

Margaret Jacob looks at Latour's postmodernist glosses on the Shapin and Schaffer narrative. Intent on projecting his relativistic standpoint onto the rise of science in the seventeenth century, Latour fails to recognize either the scientific benefits of Boylean experimental methodology or the political progressiveness of the new natural religion that Jacob calls "Newtonian physicotheology." Whereas Latour believes that Hobbes basically got it right when he said that what counts is never truth but power, Jacob argues that the new science, with its insistence on experimentation that can be verified by an independent audience, assisted in the evolution of a democratic society. If we are to continue the process of democratization, we need to have an accurate understanding of its history.

12

In Defense of Bacon

Alan Soble

> What a man had rather were true he more readily believes. There-
> fore he rejects difficult things from impatience of research.
> Francis Bacon, *Novum organum*, book I, 49

Feminist science critics, in particular Sandra Harding, Carolyn Merchant, and Evelyn Fox Keller, claim that sexist sexual metaphors played an important role in the rise of modern science, and they single out the writings of Francis Bacon as an especially egregious instance. I defend Bacon.

Science and Rape

In an article printed in the *New York Times*, Sandra Harding introduced to the paper's readers one of the more shocking ideas to emerge from feminist science studies:

> Carolyn Merchant, who wrote a book called "Death of Nature," and Evelyn Keller's collection of papers called "Reflections on Gender & Science" talk about the important role that sexual metaphors played in the development of modern science. They see these notions of dominating mother nature by the good husband scientist. If we put it in the most blatant feminist terms used today, we'd talk about marital rape, the husband as scientist forcing nature to his wishes.[1]

Harding asserts elsewhere, too, that sexist metaphors played an important role in the development of science.[2] But here she understates the point by referring to "marital rape" and so does not convey it in the "most blatant feminist terms," because her usual way of making the point is to talk about rape and torture in the same breath, not mentioning marriage. For example, Harding refers to "the rape and torture metaphors in the writings of Sir Francis Bacon and others (e.g., Machiavelli) enthusiastic about the new scientific method" (*SQIF*, 113). By associating rape metaphors with science, Harding expects the unsavoriness of rape to spill over onto science:

> Understanding nature as a woman indifferent to or even welcoming rape was . . . fundamental to the interpretations of these new conceptions of nature and inquiry. . . . There does . . . appear to be reason to be concerned about the intellectual, moral, and political structures of modern science when we think about how, from its very beginning, misogynous and defensive gender politics and . . . scientific method have provided resources for each other. (*SQIF*, 113, 116)

195

I dare not hazard a guess as to how many people read Harding's article in the *Times*, how many clipped out that scandalous bit of bad publicity for science and put it on the refrigerator, or how many still have some vague idea tying science to rape. But the belief that vicious sexual metaphors were important to science has gained some currency in the academy.[3]

Contemporary Sexual Metaphors

In *Whose Science? Whose Knowledge?* Harding proposes that the "sexist and misogynistic metaphors" that have thus far "infused" science be replaced by "positive images of strong, independent women," metaphors based on "womanliness" and "female eroticism woman-designed for women" (*WSWK*, 267, 301). Harding defends her proposal by claiming that "the prevalence of such alternative metaphors" would lead to "less partial and distorted descriptions and explanations" and would "foster the growth of knowledge": "If they were to excite people's imaginations in the way that rape, torture, and other misogynistic metaphors have apparently energized generations of male science enthusiasts, there is no doubt that thought would move in new and fruitful directions" (267). What are the misogynistic metaphors that have already "energized" science and that must be replaced? In a footnote, Harding sends us to chapter 2 of *WSWK*. There we find a section entitled "The Sexual Meanings of Nature and Inquiry" (42–46), which contains merely four examples of metaphors in the writings of two philosophers (Francis Bacon and Paul Feyerabend), one scientist (Richard Feynman), and the unnamed preparers of a National Academy of Sciences (NAS) booklet, "On Being a Scientist."

In the passage from Feynman's Nobel lecture quoted by Harding, the physicist reminisces about a theory in physics as if it were a woman with whom he fell in love, a woman who has become old yet has been a good mother and left many children (*WSWK*, 43–44; *SQIF*, 120). Harding incredibly interprets this passage as "thinking of mature women as good for nothing but mothering" (*SQIF*, 112). From the NAS booklet, Harding quotes: "The laws of nature are not . . . waiting to be plucked like fruit from a tree. They are hidden and unyielding, and the difficulties of grasping them add greatly to the satisfaction of success" (*WSWK*, 44). Here, says Harding, one can hear "restrained but clear echoes" of sexuality. Perhaps the metaphors used by Feynman and the NAS are sexual, but they are hardly misogynistic or vicious, and I wonder why Harding put them on display.[4] These were supposed to be examples of how viciously sexist metaphors "energized" science, but they seem feeble. In fact, about her four examples Harding claims only that in Bacon's writings is there a rape metaphor.[5] But let us examine her treatment of Feyerabend first, for there are significant connections between them.

Harding quotes the closing lines of a critique of Kuhn and Lakatos by Feyerabend, who ends his technical paper with the joke that his view "changes science from a stern and demanding mistress into an attractive and yielding courtesan who tries to anticipate every wish of her lover. Of course, it is up to us to choose either a dragon or a pussy cat for our company. I do not think I need to explain my own preferences" (Feyerabend, 229).[6] Harding's complaint is not that Feyerabend employed a sexual metaphor, for in *WSWK* Harding condones "alternative" sexual images reflecting "female eroticism woman-designed for women" (267). Rather, Feyerabend's metaphor—science is either a selfish shrew who exploits men or a prostitute who waits on men hand and foot—is the wrong kind of sexual metaphor. Harding quotes the same passage in her earlier *SQIF*, giving it as an example of the attribution of gender to scientific inquiry (*SQIF*, 120).

In her view, this passage conveys, as does Feynman's, a cultural image of "manliness." Whereas Feynman's notion of manliness is "the good husband and father," Feyerabend's notion is "the sexually competitive, locker-room jock" (*SQIF*, 120). Thus science, in Feyerabend's metaphor, is a sexually passive, accommodating woman, and the scientist and the philosopher of science are the jocks she sexually pleases.[7] I do not see how portraying science as a courtesan implies that the men who visit her, scientists and philosophers, are locker-room jocks. The fancy word *courtesan*, if it implies anything at all, vaguely alludes to a debonair Hugh Hefner puffing on his pipe, not to a Terry Bradshaw swatting bare male butt with a wet towel. (Should we homogenize men, or think of the philosopher of science as a locker-room jock wannabe?)

Harding concludes her brief discussion of Feyerabend in *SQIF* by claiming that his metaphor, coming strategically at the end of his paper, serves a pernicious purpose. He depicts "science and its theories" as "exploitable women" and the scientist as a masculine, manly man to imply to his (male) audience that his philosophical "proposal should be appreciated *because* it replicates gender politics" of a sort they find congenial (*SQIF*, 121). In *WSWK*, Harding similarly asserts that this metaphor was the way Feyerabend "recommended" his view (43).

This line of thought is not very promising. Some men readers prefer strict to submissive women; would Feyerabend's contrary preference for kittens tend to undermine for them his critique of Kuhn, because it does not match their tastes? I agree that some women reading his paper would probably not empathize with the metaphor, even if they concurred with the critique of Kuhn that preceded it. But they could, if they wished, ignore it as irrelevant to Feyerabend's arguments—at least because the metaphor comes at the end, tacked onto the arguments already made and digested. Had Bacon employed rape metaphors, Harding would be right that "it is . . . difficult to imagine women as an enthusiastic audience" (*SQIF*, 116). Still, had there been any women in Bacon's audience, they could have disregarded his metaphors and accepted (or rejected) the rest on its merits. For Harding to assert that the men in Feyerabend's audience would be in part persuaded by this appeal and that Feyerabend thought that he could seduce them with his "conscientious efforts at gender symbolism" (120) insults men and exaggerates the chicanery of philosophy.

Harding on Bacon

According to Harding, vicious sexual metaphors were infused into modern science at its very beginning, were instrumental in its ascent, and eventually became "a substantive part of science" (*WSWK*, 44). Harding thinks Francis Bacon (1561–1626) was crucial to this process.[8] What she says about Feyerabend, that he hoped his view would "be appreciated *because* it replicates gender politics" (*SQIF*, 121), is what she claims about Bacon, although in more extreme terms: "Francis Bacon appealed to rape meta-phors to persuade his audience that experimental method is a good thing" (*WSWK*, 43; see *SQIF*, 237).[9]

This is a damning criticism. Bacon is not depicted as a negligible Feyerabend mak-ing silly jokes about science the prostitute. Rather, Harding is claiming that Bacon drew an analogy between the experimental method and rape and tried to gain advan-tage from it (see *SQIF*, 116). Imagine the scene that Harding implies. Bacon wants to persuade fellow scholars to study nature systematically by using experimental meth-ods that elicit changes in nature, rather than to study nature by accumulating speci-mens and observing phenomena passively. So, thinking that his audience found rape desirable, attractive, permissible, or at least that it would be fun, even if despicable, Bacon champions experimentalism by drawing an analogy between it and rape. Bacon says to them: Think of doing science my way as forcing apart with your knees the slender thighs of an unwilling woman, pinning her under the weight of your body as she kicks and screams in your ears, grabbing her poor little jaw roughly with your fist to shut her mouth, and trying to thrust your penis into her dry vagina; that, boys, is what the experimental method is all about.

What did Bacon do or say to deserve such an abusive accusation? Is there any evidence that Bacon's writings contain a rape metaphor? Here is the entire text that Harding offers to support her charge:

> For you have but to hound nature in her wanderings, and you will be able when you like to lead and drive her afterwards to the same place again. Neither ought a man to make scruple of entering and penetrating into those holes and corners when the inqui-sition of truth is his whole object. (*WSWK*, 43)[10]

I suppose that a man who made no scruple of penetrating holes (and corners?) might be a rapist, but he also might be a foxhunter, a proctologist, or a billiard player. And I suppose that to "hound" nature could be seen as raping her. But the spirited stu-dent who storms my office and too often sits down next to me in the cafeteria, hoping for some words of wisdom—no more than that—also is hounding me.

Perhaps Harding, in reading "hound" as about rape, has in mind Robin Morgan's definition of rape:

> *Rape exists any time sexual intercourse occurs when it has not been initiated by the woman, out of her own genuine affection and desire.* . . . How many millions of times have women had sex "willingly" with men they didn't want to have sex with? . . . How many times have women wished just to sleep instead or read or watch the Late Show? . . . Most of the decently married bedrooms across America are settings for nightly rape. (Morgan, 165–66)

The implication of Morgan's definition of rape for doing science seems to be that the search for knowledge must proceed passively, letting Nature "initiate" our congress with her; if science pressures Nature, it is raping her. In Morgan's view, a man who pesters ("hounds") his wife for sex when she prefers to watch television has commit-ted rape if she caves in. But Bacon's audience would never have recognized this pro-saic sexual phenomenon as rape. Nor would most reasonable men and women today judge it to be rape, even if we lament the fact that men's economic power sometimes gives them an advantage in marital bargaining. Morgan's definition of rape trivializes that phenomenon, and even Harding thinks the rape metaphor in Bacon is "violent," not mere pushiness (*SQIF*, 116).

Keller's work might help us discern the reasoning behind Harding's reading of Bacon's *De augmentis*. In her essay "Baconian Science," Keller repeats the first of the two sentences quoted by Harding, to illustrate her claim that for Bacon, even though "Nature may be coy," she can still "be conquered": "For you have but to follow and as it were hound nature in her wanderings, and you will be able, when you like, to lead and drive her afterwards to the same place again."[11] What "leads to [this] conquest," in Keller's view, is "not simple violation, or rape, but forceful and aggressive seduction." Just because Keller interprets this passage from *De augmentis* as a rape-free zone does not mean there is no rape image there, for Keller warns us that "the distinction between rape and conquest sometimes seems too subtle" (*REF*, 37).

Harding might have concluded, therefore, that the conquest of nature implied by "hound" is more accurately described as rape. Her rape interpretation of Bacon would then depend on eradicating the difference between rape and seduction. But Bacon recognizes this difference; he advises that science would be more successful by patiently wooing nature than by raping her: "Art . . . when it endeavours by much vexing of bodies to force Nature to its will and conquer and subdue her . . . rarely attains the particular end it aims at. . . . Men being too intent upon their end . . . struggle with Nature than woo her embraces with due observation and attention" ("Erichthonius," myth 20, *Wisdom of the Ancients,* in Robertson, 843). The seduction, in this passage at least, seems considerate and delicate, not, as Keller says, "forceful and aggressive."[12]

Something else can be gleaned from Keller. Compare the sentence Keller quoted from Bacon's *De augmentis* with the first sentence Harding attributes to Bacon. The sentence as quoted by Keller correctly includes the phrase "follow and as it were" (*Works* 4, 296), which are missing from Harding's quotation. I think there is some difference between Bacon's nuanced "follow and as it were hound" nature and Harding's rendition, the crude and unqualified "hound" her. Harding's misquotation of Bacon (really a misquotation of Merchant's quotation of Bacon) is serious, since students of Harding who read only *WSWK* will be misguided. Five years earlier, in *SQIF*, Harding quoted this *De augmentis* passage twice and almost got it right.[13] Here is one instance:

> To say "nature is rapable"—or, in Bacon's words: "For you have but to follow and as it were hound nature in her wanderings, and you will be able when you like to lead and drive her afterward to the same place again. . . . Neither ought a man to make scruple of entering and penetrating into those holes and corners when the inquisition of truth is his whole object"—is to *recommend* that similar benefits can be gained from nature if it is conceptualized and treated like a woman resisting sexual advances. (*SQIF*, 237; ellipsis and italics are Harding's)

In repeating her accusation against Bacon, Harding does not sense that "follow and as it were" is a substantial qualification of "hound."

Furthermore, the passage's "penetrating" need not be taken as having any sexual implications.[14] And even if "penetration" is sexual (was it for Bacon and his audience?), "penetration" does not entail or suggest rape. Perhaps Harding construes the *unscru-*

pulous penetration of holes to be an allusion to rape.[15] But this reading makes sense only by wrenching "scruple" out of context. Bacon's point, which he repeats elsewhere, is that any scientist determined to find the truth about nature should be prepared to get his hands dirty; when truth is the goal, everything must be investigated, even if to prissy minds the methods employed and the objects studied are foul. (Think about survey researchers justifying the study of human sexual behavior.) Thus, a few lines after "scruple" in *De augmentis*, Bacon complains that "it is esteemed a kind of dishonour . . . for learned men to descend to inquiry or meditation upon matters mechanical" (*Works* 4, 296; see *Advancement, Works* 3, 332).

Parts of *Novum organum* and *De augmentis* (whose title begins *De dignitate*) were intended to establish the dignity, despite the dirty hands, of engaging in science to improve the human condition (see *Parasceve, Works* 4, 257–59). *Novum organum* is especially clear on this. In one aphorism, Bacon condemns "an opinion . . . vain and hurtful; namely, that the dignity of the human mind is impaired by long and close intercourse with experiments."[16] Bacon returns to this theme later in *Novum organum*:

> And for things that are mean or even filthy,—things which (as Pliny says) must be introduced with an apology,—such things, no less than the most splendid and costly, must be admitted into natural history. Nor is natural history polluted thereby; for the sun enters the sewer no less than the palace, yet takes no pollution. . . . For whatever deserves to exist deserves also to be known, for knowledge is the image of existence; and things mean and splendid exist alike. Moreover as from certain putrid substances— musk, for instance, and civet—the sweetest odours are sometimes generated, so too from mean and sordid instances there sometimes emanates excellent light and information. But enough and more than enough of this; such fastidiousness being merely childish and effeminate. (120; see 121)

Bacon is, like Calvin, a rascal. He prefers to dissect bugs and chase snakes than play house or have an afternoon tea with Susie.

Bacon's two sentences from *De augmentis* appear once again in Harding's *SQIF*:

> Bacon uses bold sexual imagery to explain key features of the experimental method as the inquisition of nature: "For you have but to follow and as it were hound nature in her wanderings, and you will be able when you like to lead and drive her afterward to the same place again. . . . Neither ought a man to make scruple of entering and penetrating into those holes and corners, when the inquisition of truth is his whole object— as your majesty has shown in your own example." . . . It might not be immediately obvious to the modern reader that this is Bacon's way of explaining the necessity of aggressive and controlled experiments in order to make the results of research replicable! (*SQIF*, 116; the first ellipsis is Harding's)

Harding is right about one thing: contrary to her interpretation, it is not obvious that Bacon, with the phrase "to the same place," is referring to experimental replicability.[17] Nor is Keller's reading plausible, that Bacon here asserts that nature can be "conquered" (*REF*, 36–37). William Leiss earlier suggested this reading: "Having discovered the course leading to the end result, we are able to duplicate the process at will" (59).[18] This makes sense, because Bacon's immediately preceding sentence is "From the wonders of nature is the most clear and open passage to the wonders of art," and then by way of explaining or defending this idea, that nature herself teaches us how to fabricate artificial devices, Bacon now writes "For you have but to follow."

That is, to learn how to achieve one of nature's effects (to use an anachronistic example, the overcoming [conquering] of bacterial infection), we must study how nature accomplishes it (we learn how to follow nature by pestering her in the lab). Once we discover Nature's Way (the various mechanisms of the immune system), we can then copy, modify, and rearrange its main ingredients (develop "artificial" devices, vaccinations, that elicit antibodies) to "lead" nature "to the same place again." Bacon is modifying a point that appears elsewhere in his writings and that he made just a page before in De augmentis:

> The artificial does not differ from the natural in form or essence, but only in the efficient. . . . Nor matters it, provided things are put in the way to produce an effect, whether it be done by human means or otherwise. Gold is sometimes refined in the fire and sometimes found pure in the sands, nature having done the work for herself. So also the rainbow is made in the sky out of a dripping cloud; it is also made here below with a jet of water. Still therefore it is nature which governs everything.[19]

In Cogitata et visa, Bacon goes so far as to say that phenomena found in nature (he praises silk spun by a worm) and from which we can learn "are such as to elude and mock the imagination and thought of men" (Farrington, 96). Bacon's example in De augmentis of obtaining a rainbow from a spray of water is serene, even lovely, and makes it improbable that he viewed experimental manipulations as nothing but mere acts of aggression. Bacon's affirmation that nature "governs everything"—the ways of nature are responsible even for the artificial rainbow we make with a spray of water—is reason to doubt that he conceived of the relationship between science and nature principally as that between man the master and woman the dominated. At the very beginning of Novum organum, as well as in "The Plan" of The Great Instauration (Works 4, 32)—that is, often and in prominent places—Bacon writes that science is "the servant and interpreter of Nature" and "Nature to be commanded must be obeyed" (1, 3).[20]

In SQIF, Harding introduces the De augmentis passage by saying that it contains "bold sexual imagery,"[21] but after quoting the passage, she escalates the charge: experimentalism, the "testing of hypotheses," is "here formulated by the father of scientific method in clearly sexist metaphors" (SQIF, 116). Harding immediately takes the next step: "Both nature and inquiry appear conceptualized in ways modeled on rape and torture—on men's most violent and misogynous relationships to women—and this modeling is advanced as a reason to value science." Of course, if the passage contains no rape metaphor, Harding's thesis that Bacon employed a rape metaphor to recommend experimentalism falls apart. But even if the passage does contain a rape image, why think that Bacon deliberately used it to promote experimentalism? Or that the metaphor really did "energize" the new science?

Examine the two sentences Harding quotes from the 1623 De augmentis as they appeared in its 1605 English-language predecessor of this text, The Advancement of Learning: "For it is no more but by following and as it were hounding Nature in her wanderings, to be able to lead her afterwards to the same place again. . . . Neither ought a man to make scruple of entering into these things for inquisition of truth" (*Works*, 3, 331). Three candidate offenders, "drive," "penetrate," and "holes," are missing.[22] Which of these two texts was more momentous for Bacon's program? In the *Advancement*, written soon after the accession of James I, Bacon was surely trying to win over his king; when writing *De augmentis*, the older Bacon had been stripped of his official positions and was writing for posterity. In which situation should we expect Bacon to use harsh (or soft) language to do the persuading? This is treacherous terrain; the contrast between *Advancement* and *De augmentis* should give us pause.

In addition, in *Novum organum* and elsewhere, Bacon argues on behalf of science in terms more likely to convince his audience and "energize" science. Science will improve the human condition (81, 129; see *De augmentis*, *Works* 4, 297); in this way, the works of science are, for Bacon, works of love.[23] And in a different vein, at the end of book 1 of *Novum organum* (129), Bacon reminds his audience that science fulfills the biblical command (Genesis 1:26) that humans should rule the world (see Leiss, 53). Both these themes in Bacon are typical and familiar.[24] Why conjecture that Bacon also appealed to a rape metaphor, as if that were the icing on the cake of his vindication of science? Harding seems to assume that from the (purported) fact that the text contains a rape metaphor, it follows that Bacon used the metaphor to persuade his audience to embrace his philosophy. But it is dangerous to draw that conclusion solely from the presence of a rape or any weaker sexual metaphor. This mistake is similar to one made about erotica, namely, arguing solely from the presence in a text of a photographic or linguistic depiction of a certain sexual act that the text recommends or endorses the depicted act (Soble 1985, 73–74).

My mentioning erotica here is pertinent. Harding entitles the section of *SQIF* in which she first quotes Bacon "Should the History and Philosophy of Science Be X-Rated?" "This question is only slightly antic" because (previewing her comments on Bacon and Feyerabend) "we will see assumptions that . . . the best scientific activity and philosophical thinking about science are to be modeled on men's most misogynous relationships to women" (*SQIF*, 112). Harding thinks that science, its history and philosophy, should be rated "x" because it contains, in her view, explicit and nasty sex. But I cannot perceive, as she does, the sexually aggressive locker-room jock in Feyerabend's metaphor, nor can I perceive, as she does, the rape metaphor in Bacon.

Perhaps Harding's philosophy itself should be rated "x": she injects sex or rape where there is none to begin with. Consider the NAS metaphor: the laws of nature are "not waiting to be plucked like fruit from a tree [but] are hidden and unyielding." Harding finds here a "clear echo" of sexuality. But these few words can be read, without effort, as innocent and nonsexual; so it is Harding, like the person who feels squeamish at the sight of uncovered piano legs, who has infused the sex into them. Furthermore, if Harding is uncommonly sensitive to the nuances of language and this enables her to extract a rape metaphor out of "hound" and "holes," or if the metaphor is one that mostly or only women could sense, this would undermine Harding's claim that Bacon used a rape metaphor to persuade his (male) audience.[25]

One more example: Bacon's portrayal of inquiry as a "disclosing of the secrets of nature" (*Works* 4, 296; see *Novum organum*, 43, 89). We could construe the language of discovering, or uncovering, the hidden secrets of nature as alluding to a quest for carnal knowledge of a deeply concealed female sexuality that is not keen on being exposed (for whatever reason, be it prudish modesty, girlish self-doubt, or lazy re-luctance). We might interpret such language in this way to suggest that this was the latent meaning of the philosophical claim that underneath the appearances of events are unobservable structures and forces about which we have no direct knowledge and about which we will remain ignorant unless we investigate, experimentally, their phenomenal manifestations. But the sexual metaphor, if we insist on digging it out, is tame; there is no rape and no need to twist the metaphor against its will (as Harding does) to *be* rape.

Merchant on Bacon

According to Harding, rape is Bacon's metaphor for and model of the experimental method. For the historian Carolyn Merchant, "the interrogation of witches" by torture (*DON*, 172)[26] is, instead, Bacon's "symbol for the interrogation of nature":

> Much of the imagery he used in delineating his new scientific objectives and methods . . . treats nature as a female to be tortured through mechanical interventions [and] . . . strongly suggests the interrogations of the witch trials and the mechanical devices used to torture witches. In a relevant passage, Bacon stated that the method by which nature's secrets might be discovered consisted in investigating the secrets of witchcraft by inquisition, referring to the example of James I. (*DON*, 168)

In which "relevant" passage does Bacon state such a thing? Merchant calls on our passage from *De augmentis* (*Works* 4, 296; italics and ellipses are Merchant's) to substantiate her assertion:

> *For you have but to follow and as it were hound nature in her wanderings, and you will be able when you like to lead and drive her afterward to the same place again.* Neither am I of opinion in this history of marvels, that superstitious narratives of *sorceries, witchcrafts, charms,* dreams, divinations, and the like, where there is an assurance and clear evidence of the fact, should be altogether excluded. . . . howsoever the use and practice of such arts is to be condemned, yet from the speculation and consideration of them . . . a useful light may be gained, not only for a true judgment of the offenses of persons charged with such practices, *but likewise for the further disclosing of the secrets of nature. Neither ought a man to make scruple of entering and penetrating into these holes and corners, when the inquisition of truth is his whole object—as your Majesty has shown in your own example.* (*DON*, 168)[27]

Bacon does not state that nature should be tortured in the way witches are interrogated. I do not perceive any torture metaphor here. The two sentences Merchant italicizes (roughly the two in which Harding finds rape) do not deserve this interpretation, nor do the sentences that fall between.

Merchant misreads why Bacon mentions James I. In referring to his king in this passage, Bacon is not alluding to cruel methods of inquisition but is pointing out that James I was willing to get his hands dirty by studying witchcraft. What James I "show[ed] in [his] own example," says Bacon, is that everything in nature is an appropriate object for scientific study—one of Bacon's principles—not that science should torture nature as if it were a witch.[28] Thus the text provides no reason to think that "Bacon's recommendation that experimental method should characterize the new science was couched in terms of the method James I had successfully used to 'expose' witches" (Nelson, 353, n. 137).

Moreover, by 1622, when *De augmentis* was being written, James I had already changed his mind about witches and had intervened to save some of the accused from execution (Robbins, 278–79).[29] Thus Bacon could not have been appealing here to a beloved pastime of James I in devising a metaphor for experimentalism. Bacon might have had better luck appealing to a torture-the-witch metaphor in the 1605 *Advancement*, right after James's Statute of 1604, but as I have pointed out, the language of that early version of *De augmentis* is not very provocative and even less amenable to a Harding–Merchant type of reading.

If we reinsert into this passage from *De augmentis* the words that Merchant deleted (indicated in the passage by italics), we can better understand Bacon:

> Neither am I of opinion in this history of marvels, that superstitious narratives of sorceries, witchcrafts, charms, dreams, divinations, and the like, where there is an assurance and clear evidence of the fact, should be altogether excluded. *For it is not yet known in what cases, and how far, effects attributed to superstition participate of natural causes; and therefore* howsoever the use and practice of such arts is to be condemned, yet from the speculation and consideration of them (*if they be diligently unravelled*) a useful light may be gained, not only for a true judgment of the offenses of persons charged with such practices, but likewise for the further disclosing of the secrets of nature.

Bacon is not recommending that witches or nature be tortured. Rather, he is telling his audience to pay attention to the distinction between the context of discovery and the context of justification: regardless of the source of certain claims ("narratives"), their content might be true, and this can be known only by investigating them scientifically. This is what Harding calls the "desirable legacy" of modern philosophy of science, the notion that "we should be able to decide the validity of a knowledge claim apart from who speaks it" (*WSWK*, 269). Bacon, a true Wiccan, suggests that scientists might learn something from witches.

The passage from *De augmentis* was the first and longest quotation that Merchant used in making her case, but her best shot missed the mark. She quotes other passages, mostly scattered words and partial sentences, to round out her argument. Although I cannot deal with them all, her frequent reference to Proteus in Bacon's works does deserve discussion. Introducing one of them, Merchant claims that Bacon drew an analogy between the "inquisition" of nature and "the torture chamber" (*DON*, 169). Here are his words: "For like as a man's disposition is never well known or proved till he be crossed, nor Proteus ever changed shapes till he was *straitened* and *held fast*; so nature exhibits herself more clearly under the *trials* and *vexations* of art than when left to herself" (*De augmentis, Works* 4, 298, Merchant's italics. See also *Advancement, Works* 3, 333).

Here, force is applied to Proteus in particular. But if we take Bacon's analogy between nature and Proteus literally, the implication is that we must be smart enough to outfox nature to get a hearing for our questions. Then as we try to hold fast to nature, nature will almost always escape by changing unpredictably and uncontrollably, slithering or leaping away or disappearing as a gas, and we will not get an answer. Our attempts to bind her will be largely fruitless. The Proteus image, then, is a tribute to the sagacity and subtlety of nature (see *Novum organum*, 10, 24).

If we do not take the analogy literally, we can be content with the core of Bacon's idea, which has nothing to do with torture and everything to do with the advantages to be gained from experimental manipulations. For example, a person "left to herself" might have little opportunity to reveal the greed that lies buried in her heart. But if she is tempted by a stuffed wallet dropped on an empty street—the deliberate "vexations" arranged by a conniving social scientist—she might respond by keeping the money and not returning the wallet to its owner, thereby revealing the greedy part of her character.

Merchant quotes another Proteus passage that perhaps more strongly supports her reading: "The vexations of art are certainly as the bonds and handcuffs of Proteus, which betray the ultimate struggles and efforts of matter" (*Parasceve, Works* 4, 257; *DON,* 171). "Bonds and handcuffs" look damning when equated with "vexations of art" (experimental techniques). But these devices are being used on Proteus, the supreme Houdini, in which case we need not be so anxious about his safety (besides, Proteus is a guy). What does Bacon mean by the less than transparent "betray the ultimate struggles and efforts of matter"? In the next sentence (not quoted by Merchant), Bacon explains: "For bodies will not be destroyed or annihilated; rather than that they will turn themselves into various forms." Here Proteus stands for matter; so, no matter how much we bind matter, says Bacon, it is indestructible. Bacon is not issuing a normative claim, as if urging us to bind matter to prevent it from behaving perversely (Merchant's reading, *DON,* 171; see Bordo, 109); he is making the ontological point that no amount of binding will allow us to annihilate matter.

Bacon states the idea nicely in *Wisdom of the Ancients*:

> if any skilful Servant of Nature shall bring force to bear on matter, and shall vex it and drive it to extremities as if with the purpose of reducing it to nothing, then will matter (since annihilation or true destruction is not possible except by the omnipotence of God) finding itself in these straits, turn and transform itself into strange shapes, passing from one change to another till it has gone through the whole circle ("Proteus," myth 13 [Robertson, 838]).

Likewise, Proteus in the *Odyssey* (book 4), at the hands of Menelaus and his crew, goes through the whole cycle and is never destroyed. Note that Bacon speaks here again of the scientist as a "servant of nature."

Let us also look at Bacon's frequent use of "vex" and its congeners. Merchant italicized "vexations" in *De augmentis*, and both Leiss (59) and Keller (*REF,* 36) think "vex" conveys sexual aggression. Even though Bacon's use of "vex" is occasionally strong (e.g., in relation to imperishable matter; *Thoughts on the Nature of Things, Works* 5, 427–28), "vex" does not always or usually carry a pernicious connotation but is meant, innocuously, along the lines of his "hound" and my "pester." For example, in *Novum organum* Bacon writes:

> For even as in the business of life a man's disposition and the secret workings of his mind and affections are better discovered when he is in trouble than at other times; so likewise the secrets of nature reveal themselves more readily under the vexations of art than when they go their own way. (98)[30]

This mature and polished statement of Bacon's philosophy contains no rape, no torture, no bondage, just the thought that to know nature it is not enough to watch it; nature must be provoked into showing us its inner workings. I find nothing to complain about in this thought, especially when I consider how much of my knowledge of human nature I never would have acquired had not my family, friends, and colleagues, let alone myself, been "crossed" (*De augmentis, Works* 4, 298) and thereby goaded into exposing features of our personalities we do not ordinarily broadcast.

Keller on Bacon

Going beyond Harding and Merchant, Evelyn Fox Keller argues that Bacon's sexual images involve a more "complex sexual dialectic" (Keller 1980, 302; *REF*, 35). In Bacon's language, the scientist both aggressively dominates and is "subservient" and "responsive" to nature (*REF*, 36–37); science is both master of and obedient to nature. Keller's essay is in effect a reply to Leiss, who says about Bacon's "famous formula"—we command nature by obeying her—that "some commentators have claimed that it sounds a note of humility in man's attitude toward nature. But this interpretation . . . invents inconsistencies which do not really exist in Bacon's work" (58–59). I think Keller is right to find inconsistency in Bacon's metaphors but wrong in what she makes of it.

In arguing for one side of the inconsistency, that Baconian science is "aggressive" in its "conquest" of nature, Keller assembles seven passages (*REF*, 36). The first is from Bacon's *Refutation of Philosophies*: "Let us establish a chaste and lawful marriage between Mind and Nature" (Keller 1978, 413; 1980, 301; *REF*, 36).[31] The metaphor seems benign, as does that of a "chaste, holy and legal wedlock" (Keller's second passage, from *The Masculine Birth of Time*; Farrington, 72), but Keller thinks of the marital imagery in Bacon as aggressive (*REF*, 19).[32] There is no need to. Bacon uses marriage imagery promiscuously in his writings, with no hint of aggression: see the Preface to *The Great Instauration*, in which Bacon speaks about the "true and lawful marriage" he is attempting to effect "between the empirical and the rational faculty" (*Works* 4, 19) and *Novum organum*, in which Bacon criticizes those who would "deduce" Christianity from the principles of philosophy, thereby "pompously solemnising this union of sense and faith as a lawful marriage" (89). And in the two marriage passages quoted by Keller, there is no intimation of aggression at all; Bacon immediately proceeds to say (typical for him; see "The Plan" of *The Great Instauration, Works* 4, 27) that the marriage will issue in "wholesome and useful inventions . . . to bring relief" from "human necessities" (*Refutation*, Farrington, 131) and "will overcome the immeasurable . . . poverty of the human race" (*Masculine Birth*, Farrington, 72).

Thus it is difficult to agree with Keller that the marital image in Bacon "constitutes an invitation to the 'domination of nature'" (Keller 1978, 429, n. 5; *REF*, 91, n. 6) because it "sets the scientific project squarely in the midst of our unmistakably patriarchal tradition" (Keller 1978, 423). When Bacon writes in *Masculine Birth* (Keller's third passage), "I am come in very truth leading to you Nature with all her children to bind her to your service and make her your slave" (*REF*, 36, see 48; Farrington, 62), Keller sees a wedding announcement in which nature is fingered as the bride.

She is a slave, since that will be her married lot. The image of binding nature as a slave is surely an ugly one, but it is not necessarily about marriage.

Keller cites four more passages that she believes exhibit the aggression toward nature in Bacon's images. In one, from the Preface to *Novum organum*, Bacon writes of trying to "penetrate further" and "find a way at length into her inner chambers" (*Works* 4, 42), which is hardly pushy. Keller then turns to the *De augmentis* "hound . . . and drive" passage that I have already examined, which (contra Keller) contains no obvious conquest, sexual or otherwise. Keller also refers to Bacon's thought that more can be learned about nature by vexing her than by observing her in freedom, "left to herself."

The final passage cited by Keller is from *Cogitata et visa* (1607, published in 1653; Farrington, 57). According to Keller, Bacon says about "the discipline of scientific knowledge, and the mechanical inventions it leads to" (*REF*, 36) that they do not "merely exert a gentle guidance over nature's course; they have the power to conquer and subdue her, to shake her to her foundations."[33] This is, apparently, crude aggression. But our judgment of Bacon will be improved by realizing three things. First, Bacon is in part referring here not to mechanical devices in general but specifically to "Printing, Gunpowder, and the Nautical Needle," which he thinks had more effect on "human affairs" than any "empire," "school," or "star" (Farrington, 93). Bacon makes the point hyperbolically—that these exemplary inventions are things that can conquer and shake nature to her foundations—but given his point, the sentence is a prime candidate for being read generously. Second, *Cogitata et visa* is a polished work, even though unpublished during Bacon's life, and, except for this line, is tranquil throughout. Much of *Cogitata* went right into *Novum organum*; it contains one questionable sentence that was not destined to join its sisters there and on which Keller pounces.[34] Third, in the later 1612 *Description of the Intellectual Globe*, Bacon repeats the line but softens it: he warns us against making the "subtle error," one that causes "despair," of thinking that science has "no power to make radical changes, and shake her in the foundations" (*Works* 5, 506). Both *conquer* and *subdue* are gone. In an analogous passage in the late *De augmentis*, Bacon redeems himself linguistically by advising against the "subtle error" of thinking, with "premature despair," that science has "by no means [the power] to change, transmute, or fundamentally alter nature" (*Works* 4, 294). Now the rest, *shaking* the foundations, is gone.

Guided into Bacon by Harding, Merchant, and Keller, one expects to find his work cluttered with scandalous metaphors. But Keller unveils only one clearly ugly line out of thousands of pages of Bacon's lifework, and this, "make her your slave," occurs in a tiny fragment of a manuscript written at the dawn of his philosophical career (the 1603 *Masculine Birth*, published posthumously in 1653; Farrington, 57). Furthermore, "shake her to her foundations" (*Cogitata*) is either expunged from or revised in Bacon's later writings. In "Feminism and Science," a paper widely reprinted (e.g., Boyd, Gasper, and Trout, 279–88; Harding and O'Barr, 233–46), Keller bizarrely reproduces these two lines, and only these two, to discredit the new science (1982, 598).[35] To poke through these essays and parade their meanest two lines as the truth about Bacon and his scientific philosophy, without any methodological reluctance, is uncharitable if not hostile.[36]

In arguing on behalf of the other side of the inconsistency, that Bacon's images sometimes express a different attitude, Keller mentions the well-known passages I cited in which Bacon speaks about the scientist as the servant of nature who obeys her. So Keller finds a "puzzle" in which the "ambiguities" of Bacon's images "become contradictions." Science is "aggressive yet responsive, powerful yet benign, masterful yet subservient, shrewd yet innocent" (*REF*, 37). Keller solves the problem to her own satisfaction by sensing that Bacon viewed, in *Masculine Birth*, the human mind as hermaphroditic or as sequentially bisexual. As she interprets Bacon:

> To receive God's truth, the mind must be pure and clean, submissive and open. Only then can it give birth to a masculine and virile science. That is, if the mind is pure, receptive, and submissive in its relation to God, it can be transformed by God into a forceful, potent, and virile agent in its relation to nature. Cleansed of contamination, the mind can be impregnated by God and, in that act, virilized: made potent and capable of generating virile offspring in its union with Nature. (*REF*, 38; see Keller 1980, 304, 307)

Keller's proposal, that Bacon's model of the mind is hermaphroditic or bisexual, since the mind begins as a passive female and is transformed into a virile male, does not obviously represent progress in understanding Bacon's philosophy. There is as little in the extremely brief *Masculine Birth* to support Keller's elaborate reading of Bacon as there is in *De augmentis* to support Harding's perception of rape. The fragments of *Masculine Birth* are themselves a puzzle and not to be entrusted with the task of clarifying vexatious passages in Bacon's mature works. The abundance of contrary images in Bacon's writings suggests that gender images in Bacon are *less* interesting than Keller's esoteric reading makes them out to be.[37]

Methodology

To make the case that Bacon deliberately used rape, sexual, or torture metaphors to convince his audience of the virtues of experimental science, the following (only necessary) conditions must be satisfied. The metaphors (1) should clearly be in his texts (which they are not), (2) should be located in vital places (not merely in posthumously published manuscripts), (3a) should appear in several passages (not just once) yet also (3b) not indiscriminately (see Koertge, 353–54), and (4) should not have their thrust diluted by other, contrasting images (see Landau, 49–50). If Feyerabend had likened not only science but also such disparate things as rivers, music, and champagne to a kitten, or if at one point he called science a kitten but elsewhere a stern mistress, we would not be inclined to take his metaphors seriously. Similarly, to the extent that Bacon applied his images to a wide assortment of things or used diverse, even contradictory, metaphors, it would be implausible to claim that he relied on them to vindicate science.

On the one hand, Bacon sees nature as wise and subtle, so discovering Nature's Ways requires not just diligence but shrewd intelligence. Similarly, in Bacon's world—he and his peers are gentleman, not barroom bruisers—wary women are wooed with poetry or bribed or promised a love that will not be forthcoming. Comprehending the secrets of nature might be like uncovering the secrets of a woman, but brain, not brawn, yields the joys of science and sex. On the other hand, nature is also, for Bacon, one tough cookie, whose floods and hurricanes and famines and pestilences kill us and destroy our property. But neither image in this dualism—nature is smart but, with luck and skill, can be seduced; nature is cruel but, with luck and skill, we can avoid the worst of it—is obnoxious. Bacon's mistake was similar to Feyerabend's, who conceived of women only as kittens or stern mistresses, imagery that excessively narrows women's modes of existence. Maybe both Baconians and Feyerabendians can be prodded into avoiding these mildly sexist polarities and to think, instead or in addition, in terms of an equal relationship with an independent and capable woman of substance. But Harding has other metaphors in mind. In calling for the "prevalence" of metaphors reflecting "female eroticism woman-designed for women" and "woman-to-woman relations" (WSWK, 267), she leaves little or no room for rehabilitated heterosexual metaphors.

The incoherence of Bacon's images goes beyond the mastery/obedience contrast Keller exhibits. Consider, for example, how indiscriminately, in *Novum organum* alone, Bacon uses bondage imagery. In one passage, he complains that "men's powers" have been "bound up" by the "enchantments of antiquity." In another, he criticizes logical demonstrations on the grounds that they "make the world the bond-slave of human thought, and human thought the bond-slave of words." And lest we get the impression that nature is the only object of bondage, Bacon recommends that the human mind should be bound, for it will do us some good. For the sake of the improvement of knowledge, scientists should "bind themselves to two rules" and "the understanding must not . . . be supplied with wings, but rather hung with weights, to keep it from leaping and flying" (84, 69, 130, 104, respectively; see 20). This is good advice for those reading Bacon today, as it might prevent them from claiming too quickly that the mind is, for Bacon, an appropriate object of bondage just because the mind is sometimes or partially female and from claiming, too ideologically, that the metaphor, as patriarchal and sexually oppressive, "energized" or provided a "resource" for the new science.

Also consider that the effect of matter on the course of nature is, almost incomprehensibly, more dramatic and graphic, in Bacon's language, than what science does, or can do, to nature ("hound" and "drive"). Bacon writes in *De augmentis* that nature

is either free, and follows her ordinary course of development; . . . or she is driven out of her ordinary course by the perverseness, insolence, and frowardness [sic] of matter, and violence of impediments; as in the case of monsters; or lastly, she is put in constraint, moulded, and made as it were new by art and the hand of man; as in things artificial. (*Works* 4, 294; see *Parasceve*, *Works* 4, 253)

The contrast is sharper in Description of the Intellectual Globe:

For nature is either free . . . or again she is forced and driven quite out of her course by the perversities and insubordination of wayward and rebellious matter, and by the violence of impediments; . . . or lastly, she is constrained, moulded, translated, and made as it were new by art and the hand of man. (*Works* 5, 505–6)

Constraint of nature by the human hand need not be vicious, violent, a bit of torture, or perverse. A bush—as natural a piece of nature as we can imagine—can be gently pruned, thereby constrained, "made as it were new by art," watered and fed and, as a result, it will both thrive and bring us pleasure.

How should we understand Bacon's using the bearded Pan to represent nature in myth 6 of *Wisdom of the Ancients* and his using Proteus to represent matter in myth 13 and nature in *De augmentis*? (Robertson, 828–32, 838; *De augmentis, Works* 4, 298.) Similarly, what are we to make of "Atalanta," myth 25? (Robertson, 847–48.) There Bacon says, "Art remains subject to Nature, as the wife to the husband." He has reversed his purportedly favorite sexist image. Instead of the husband science dominating his woman nature in a patriarchal marriage, science is the unfortunate wife dominated by nature the man. Bacon's point, which has nothing to do with gender, is that science (Atalanta) will not win the race with nature (Hippomenes) if she allows herself to be distracted by baubles, that is, impatiently, quickly gained research results that eventually prove worthless.[38] More good advice.

I suggest, given the wide variety of Bacon's metaphors, that we not take them seriously as attempts at deliberate manipulation of his audience or as the smoke signals of his seething unconscious. And it is unbelievable that "the increasing empirical success of the scientific world view *depended* on the continued presence in it—either explicitly or tacitly—of the gender politics metaphors" (Harding 1981, 316; italics added; see also Koertge, 353–54). Bacon's metaphors are more plausibly understood as "literary embellishments" than as a "substantive part of science" (contra Harding, *WSWK*, 44; see Landau, 49). As Bacon says in *Description of the Intellectual Globe*, the metaphors are irrelevant:

> If any one dislike that arts should be called the bonds of nature, thinking that they should rather be counted as her deliverers and champions, because in some cases they enable her to fulfil her own intention by reducing obstacles to order; for my part I do not care about these refinements and elegancies of speech; all I mean is, that nature, like Proteus, is forced by art to do that which without art would not be done; call it what you will,—force and bonds, or help and perfection. (*Works* 5, 506)

As we are rightly rereading the canon through feminist lenses, let us take care lest we succumb to the "impatience of research." Otherwise, in our investigations, be they philosophical or historical, we will discover precisely what we hoped to discover, and we will project into the canon the horrors that frighten us, horrors that are not there.

> It is not the pleasure of curiosity, nor the quiet of resolution, nor the raising of the spirit, nor victory of wit, nor faculty of speech, nor lucre of profession, nor ambition of honour or fame, nor inablement for business, that are the true ends of knowledge; . . . but it is a restitution and reinvesting . . . of man to the sovereignty and power . . . which he had in his first state of creation. . . . It is a discovery of all operations and possibilities of operations from immortality (if it were possible) to the meanest mechanical practice. And therefore knowledge that tendeth but to satisfaction is but as a courtesan, which is for pleasure and not for fruit or generation. (*Valerius terminus, Works* 3, 222)[39]

Notes

Assistance was provided by the University of New Orleans and its College of Liberal Arts, through the release time of a research professor appointment and by the Research Support

Scheme of the Open Society Institute (grant 1520/706/94). This chapter was published in *Philosophy of the Social Sciences* 25 (1995): 192–215. This version contains additions and corrections, some of which were kindly suggested by Noretta Koertge.

1. "Value-Free Research Is a Delusion," *New York Times*, October 22, 1989, E24. Copyright ©1995 by The New York Times Co. Reprinted by permission.

2. *The Science Question in Feminism* (hereafter cited as *SQIF*), 112, 113, 116; *Whose Science? Whose Knowledge?* (hereafter cited as *WSWK*), 43, 267.

3. (1) In her review of Sandra Harding and Jean O'Barr's anthology *Sex and Scientific Inquiry*, Nancy Tuana writes: "Evelyn Fox Keller and Susan Bordo argue that the methods and epistemology of modern science are not in fact neutral, but are male-biased, grounded in metaphors of dominance and rape" (62). (2) In her review of Keller's *Reflections on Gender and Science*, Helen Longino writes: "Keller notes that Francis Bacon . . . first made explicit the connections between knowledge and power and . . . described experimentation in language appropriate to rape and seduction" (1988, 563). But as we will see, it is not Keller but Harding who thinks that Bacon conceives of the experimental method as rape. In *Science as Social Knowledge* (205), while laying out Keller's views, Longino writes that the new science's "conception of inquiry . . . envisioned the seeker after knowledge as male and the object of knowledge (nature) as female, and which described the activity of inquiry in language used to describe the male pursuit of females: rape and courtship" (3). "The relationship of scientists and science to nature has been described as 'rape,' 'forced penetration,' and 'domination,'" reports Lynn Hankinson Nelson, citing Harding, Keller, and Merchant (*Who Knows: From Quine to a Feminist Empiricism*, 213, 353, n. 136). (4) Joseph Agassi, too, perpetuates this noxious idea: "Now Bacon did oppose instrumentalism, yet he did not denounce it as dogmatism—as the putting of nature in chains and the rape of nature, to use his metaphors" (92).

4. The second edition of *On Being a Scientist* does not contain the first-edition passage that Harding found objectionable. Indeed, the introduction begins in politically correct fashion by quoting Barbara McClintock.

5. Don Ihde read Harding too quickly: "The rise of Modern Science [according to Harding in *SQIF*] was . . . a movement into the Baconian, masculinist context of an aggression upon nature betrayed in the metaphors of science 'twisting the tail of nature' or even the use of rape metaphors which proceeded from Bacon on into very contemporary speeches by Nobel Prize acceptees" (70–71). Harding, as uncharitable as her reading of Feynman is, did not accuse him, too, of trading in rape metaphors.

6. I quote Feyerabend because Harding (*WSWK*, 43) provides the wrong "I think I do not have to explain my own preferences." In *SQIF*, Harding quotes it correctly (120).

7. Although Harding interprets Feyerabend's metaphor as being about "science and its theories" (*SQIF*, 121) and he says as much (229), it makes more sense to read it as being about Nature. We should not view Nature as a stern and demanding mistress, which it is for Popper and Lakatos: their Nature yells "False!" or "Incompatible!" when it does not like our scientific theories. Instead, Nature is an indulgent courtesan, one who lets us do whatever we want—in theory construction. It is Nature that whispers, "Anything goes, big guy," not Science.

8. For an overview of Bacon's life and philosophy, see Thomas Macaulay's 1837 essay "Francis Bacon" and John Robertson's 1905 critical reply (vii–xvi).

9. As far as I know, Harding first accused Bacon of relying on rape metaphors in her 1980 *PSA* paper "The Norms of Social Inquiry and Masculine Experience," ?18. Noretta Koertge critically discussed Harding's claim in the same colloquium (353–54). In her later works, *SQIF* and *WSWK*, Harding repeats the accusation against Bacon but neither rebuts nor acknowledges Koertge's doubts. Is this a case of political fervor (or mere personal stubbornness) trouncing scholarship?

10. The two sentences are from Bacon's 1623 *Of the Dignity and Advancement of Learning, Works* 4 (hereafter cited as *De augmentis*), 296. Harding took the passage not from Bacon but from Merchant, *The Death of Nature* (hereafter cited as *DON*), 168, who did take it from *Works* 4.

11. *Reflections on Gender and Science* (hereafter cited as *REF*), 36.

12. Paolo Rossi suggests that in "Erichthonius," Bacon expressed his view that to be successful with nature, science has to "humbly beg her assistance" (101; see "humble respect," 105). Keller (*REF*, 37) quotes Rossi, but not this phrase, thereby creating the impression that he agrees with her "forceful and aggressive seduction" reading.

13. Harding quotes these two sentences from *De augmentis* three times in her two books, always informing us that her source is Merchant's *DON*, 168. Merchant includes the five missing words. In addition to failing to mark an ellipsis in the first of Bacon's two sentences, Harding makes a second mistake in *WSWK*: ellipsis points belong between the two sentences, since Harding omitted a large chunk. Any hint of rape created by the juxtaposition of these sentences in *WSWK* is therefore artificial. There are other errors (see *Works* 4, 296). In Merchant, we correctly find "these holes," but in Harding "those." Merchant and Harding write "whole object," but both are wrong; "sole object" is correct. Merchant, and Harding in *SQIF*, gives "drive her afterward," but both are wrong; in *WSWK*, Harding got "afterwards" right. Iddo Landau continues this comedy of errors. In "How Androcentric Is Western Philosophy?" (49), he accurately reproduces Harding's misquotations of Bacon and Feyerabend from *WSWK*, thereby misquoting Bacon and Feyerabend. If you want to check what you have read in today's newspaper, buy another copy (Wittgenstein).

14. Let alone, as in Merchant, "strong sexual implications" (*DON*, 168).

15. See Bordo's remark on this sentence from *De augmentis*, apparently provoked by Keller and Merchant: it illustrates "the famous Baconian imagery of sexual assault" (Bordo, 107–8).

16. Book 1, 83 (in *Works* 4); see *Cogitata et visa* (Farrington, 82). Henceforth I supply for *Novum organum* only the book 1 aphorism number.

17. Landau too quickly grants to Harding that the passage is about replicability (49).

18. Harding never mentions Leiss's reading of a sentence from Bacon that she quotes twice in *SQIF* and once again in *WSWK*. Leiss's *Domination of Nature* is included in Harding's bibliography (*SQIF*, 258), but Leiss is not listed in the index (*SQIF*, 267), nor is his book mentioned in any of *SQIF*'s innumerable footnotes.

19. *Works* 4, 294–95; see *A Description of the Intellectual Globe, Works* 5, 507.

20. See 129 and *Cogitata et visa*: "No force avails to break the chain of natural causation. Nature cannot be conquered but by obeying her" (Farrington, 93). Koertge ("Methodology," 353) responded to Harding's 1980 *PSA* paper ("The Norms of Social Inquiry") by quoting five passages in Bacon that indicate "humility and respect" for nature.

21. This phrase is more than reminiscent of Merchant's description of Bacon's experimentalism: "Here, in bold sexual imagery, is the key feature of the modern experimental method" (*DON*, 171).

22. Susan Griffin is wrong to paraphrase Bacon as having written in 1609, "nature must be hounded in her wanderings before one can lead her and drive her" (16). Griffin says in her bibliographical notes (234, 258, 259) that the lines she attributes to Bacon were taken from an unpublished manuscript by Carolyn Iltis and from *The Domination of Nature* by "Leis" [*sic*]. But Leiss quotes "lead and drive" from the 1623 *De augmentis* (59).

23. *Valerius terminus, Works* 3, 217, 221–22; Preface to *The Great Instauration, Works* 4, 21; and see Farrington, 28–29.

24. But Bacon was no blind optimist: he recognized that science (and philosophy and history) done poorly would go wrong. See "Daedalus," myth 19, *Wisdom of the Ancients* (Robertson, 842–43).

25. Similarly, it will not help Harding to claim that her experiences and social location as a feminist and/or woman grant her an epistemic advantage ("standpoint epistemology")—in this case, make her an especially perceptive reader of early-seventeenth-century texts (*WSWK*, 121–33, 150–51; see my review of *WSWK*). Harding's reading of Bacon is a politically inspired reading that goes wrong; hence it subverts her claim that feminist scholarship is better because it is political. See n. 9, above.

26. Harding might have believed that Merchant found a rape metaphor in Bacon's works. After all, much later, according to Daphne Patai and Noretta Koertge, the claim that "modern science . . . has viewed nature as a woman who exists only to be raped" was made by Merchant (*Professing Feminism*, 123). But the only occurrence of the word *rape* in Merchant's chapter on Bacon is in quotation marks (*DON*, 171). And when Merchant talks about the "rape" of nature, she is not voicing Harding's thesis that for Bacon, the experimental method is rape, but the different thesis that the new science justified the "commercial exploration" or exploitation of the earth (e.g., mining; *DON*, 41).

27. Merchant's "a" true judgment should be "the." See n. 13, above.

28. Recall from *Novum organum*: "The sun enters the sewer no less than the palace, yet takes no pollution" (120).

29. Robbins suggests, sounding a contemporary note, that James's realization that children had been "falsely charging people as witches" was crucial to his change of mind.

30. See a variant of these lines in "The Plan" of *The Great Instauration, Works* 4, 29, which is the "vex" passage to which Keller calls attention.

31. Keller took this sentence from Leiss (25), who took it from Farrington's translation (131). *Redargutio philosophiarum* is not translated from the Latin in *Works* 3 and was not published until 1734 (Farrington, 57).

32. Londa Schiebinger thinks that the "masculine" in *Masculine Birth of Time* "was at most a tangential attack upon women" (137). Instead, it was meant to reject an Aristotelian, "passive, speculative, and effeminate philosophy." Similarly, Patai and Koertge understand "the violence of [Bacon's] images [as] directed at Aristotelians, not women" (124).

33. In her 1980 "Baconian Science" (308, n. 11) and its revision in *REF* (36), Keller claims to have taken this passage from Spedding's *Description of the Intellectual Globe, Works* 5, 506; it is not there. In her 1982 "Feminism and Science" (598, n. 22), she again says that it is from *Description of the Intellectual Globe*, but now cites p. 506 of Robertson's collection instead of Spedding's *Works*. On that page in Robertson, however, is *De augmentis*, book 5, chap. 2. (The passage is actually in Farrington's translation of *Cogitata et visa*, 93.) A clue to solving this mystery can be found in Leiss, 58, 216, n. 18; and n. 31, above.

34. When revising "Baconian Science" for inclusion in *REF*, Keller made three changes. She changed the subtitle from "A Hermaphroditic Birth" to "The Arts of Mastery and Obedience." She deleted a passage she had quoted from Bacon's *Valerius terminus (Works* 3, 222) and her comments on it (compare the first, 303, with the second, 37). And she changed (on the same pages) "when nature becomes divine [for Bacon] . . . the scientific mind becomes 'he'" to the *opposite* "when nature becomes divine [for Bacon] . . . the scientific mind becomes more nearly female." Does Keller want to be held to the earlier (presumably mistaken) version? Why hold Bacon to a different standard? Note that Keller says in a later chapter in *REF* (48), "As Bacon's metaphoric ideal was the virile superman, the alchemist's ideal was the hermaphrodite. . . . Whereas Bacon sought domination, the alchemists asserted the necessity of allegorical, if not actual, cooperation between male and female."

35. In another essay, Keller, right after quoting the "slave" passage from Bacon, writes: "This conjunction between scientific and masculine norms has been historically functional in guaranteeing a sexual division of emotional and intellectual labor that effectively excludes most women from scientific professions" (1987, 79). Incomprehensibly, when discussing this

essay, Christina Sommers (491) and Joseph Adelson (110) call the author "Elizabeth Fox-Keller." No one can read these days. (Her name is "Evelyn.")

36. Merchant mentions the "slave" passage twice in the space of two pages (169, 170); and Ruth Bleier (204–5) condemns Bacon by reproducing only the "slave" passage, which she took from Keller's "Feminism and Science."

37. Graham Hammill goes one step beyond Keller. Whereas Keller relies on the contradictory sexual images in Bacon to argue that Bacon's model of the mind is hermaphroditic or bisexual, not simply a masculine domination of female nature, Hammill relies on Bacon's purported homosexual activities with his men servants and his anally erotic penchant for enemas— as well as a few pieces of his texts—to argue that "for Bacon the anus becomes more properly the seat of knowledge" and the penetration of nature is "not an odyssey into the vagina but . . . wanderlusts into the anus." The masculine birth is accomplished by "anal purging" (244, 251, 247). Bacon's imagery, we might say, expresses not an effeminate homosexuality or a woman-hating homosexuality but homosexuality nonetheless, or at least a woman-envying heterosexuality. Keller might appreciate Hammill's reading of Bacon; see her *Secrets of Life. Secrets of Death* (45–52), in which she speaks about male science as anal "production." (For a discussion of *Secrets*, see my "Gender, Objectivity, and Realism.")

38. Bacon uses the Atalanta story often, without any obfuscating marital or gender images; see *Novum organum* (70, 117) and "The Plan" of *The Great Instauration (Works* 4, 29).

39. The last sentence of this beautiful passage is the quotation Keller deleted from "Baconian Science" when revising that essay for *REF*; see n. 34, above.

References

Adelson, J. 1988. An academy of one's own. *The Public Interest* no. 91 (Spring): 107–10.

Agassi, J. 1989. The lark and the tortoise. *Philosophy of the Social Sciences* 19: 89–94.

Bacon, F. 1962–63. *The Works of Francis Bacon.* 14 vols. Ed. J. Spedding, R. L. Ellis, and D. D. Heath. 1857–74. Reprint, Stuttgart: Verlag.

Bleier, R. 1984. *Science and Gender.* New York: Pergamon Press.

Bordo, S. 1987. *The Flight to Objectivity.* Albany: State University of New York Press.

Boyd, R., P. Gasper, and J. D. Trout, eds. 1991. *The Philosophy of Science.* Cambridge, Mass.: MIT Press.

Farrington, B. 1966. *The Philosophy of Francis Bacon.* 1964. Reprint, Chicago: University of Chicago Press.

Feyerabend, 1970. Consolations for the specialist. In *Criticism and the Growth of Knowledge,* ed. I. Lakatos and A. Musgrave, 197–230. Cambridge: Cambridge University Press.

Griffin, S. 1978. *Woman and Nature.* New York: Harper & Row.

Hammill, G. 1994. The epistemology of expurgation: Bacon and *The Masculine Birth of Time.* In *Queering the Renaissance,* ed. J. Goldberg, 236–52. Durham, N.C.: Duke University Press.

Harding, S. 1981. The norms of social inquiry and masculine experience. *PSA 1980* 2: 305–24.

———. 1986. *The Science Question in Feminism.* Ithaca, N.Y.: Cornell University Press.

———. 1989. Value-free research is a delusion. *New York Times,* October 22, E24.

———. 1991. *Whose Science? Whose Knowledge?* Ithaca, N.Y.: Cornell University Press.

Harding, S., and J. F. O'Barr, eds. 1987. *Sex and Scientific Inquiry.* Chicago: University of Chicago Press.

Ihde, D. 1993. *Philosophy of Technology: An Introduction.* New York: Paragon House.

Keller, E. F. 1978. Gender and science. *Psychoanalysis and Contemporary Thought* 1: 409–33.

———. 1980. Baconian science: a hermaphroditic birth. *Philosophical Forum* 11: 299–308.

———. 1982. Feminism and science. *Signs* 7: 589–602.

———. 1985. *Reflections on Gender and Science.* New Haven, Conn.: Yale University Press.

———. 1987. Women scientists and feminist critics of science. *Daedalus* 116(4): 77–91.

———. 1992. *Secrets of Life. Secrets of Death: Essays on Language, Gender and Science.* New York: Routledge.

Koertge, N. 1981. Methodology, ideology and feminist critiques of science. *PSA 1980* 2: 346–59.

Landau, I. 1996. How androcentric is Western philosophy? *Philosophical Quarterly* 46(182): 48–59.

Leiss, W. 1972. *The Domination of Nature.* New York: Braziller.

Longino, H. 1988. Review essay: science, objectivity, and feminist values. *Feminist Studies* 14: 561–74.

———. 1990. *Science as Social Knowledge.* Princeton, N.J.: Princeton University Press.

Macaulay, T. 1967. Francis Bacon. In *Critical and Historical Essays.* Vol. 2, 290–398. New York: Dutton.

MacKinnon, C. 1989. *Toward a Feminist Theory of the State.* Cambridge, Mass.: Harvard University Press.

Merchant, C. 1980. *The Death of Nature.* New York: Harper & Row.

Morgan, R. 1977. *Going Too Far.* New York: Random House.

Nelson, L. H. 1990. *Who Knows: From Quine to a Feminist Empiricism.* Philadelphia: Temple University Press.

On Being a Scientist: Responsible Conduct in Research. 1995. Washington, D.C.: National Academy [of Sciences] Press.

Patai, D., and N. Koertge. 1994. *Professing Feminism.* New York: Basic Books.

Robbins, R. H. 1959. *The Encyclopedia of Witchcraft and Demonology.* New York: Crown.

Robertson, J. M., ed. 1905. *The Philosophical Works of Francis Bacon.* London: Routledge and Sons.

Rossi, P. 1968. *Francis Bacon: From Magic to Science.* London: Routledge & Kegan Paul.

Schiebinger, L. 1989. *The Mind Has No Sex? Women in the Origins of Modern Science.* Cambridge, Mass.: Harvard University Press.

Soble, A. 1985. Pornography: defamation and the endorsement of degradation. *Social Theory and Practice* 11(1): 61–87.

———. 1992. *Review of Whose Science? Whose Knowledge?* by S. Harding. *International Studies in the Philosophy of Science* 69: 159–62.

———. 1994. Gender, objectivity, and realism. *Monist* 77: 509–30.

Sommers, C. 1993. The feminist revelation. In *Morality and Moral Controversies.* 3rd ed., ed. J. Arthur, 479–93. Englewood Cliffs, N.J.: Prentice Hall.

Tuana, N. 1990. Review of *Sex and Scientific Inquiry,* ed. S. Harding and J. F. O'Barr. *American Philosophical Association Newsletter on Feminism and Philosophy* 89(2): 61–62.

13

Alchemy, Domination, and Gender

William R. Newman

Among influential feminist historians such as Evelyn Fox Keller and Caroline Merchant, the weltanschauung of premodern alchemy has come to represent an attractive alternative to the "repressive" and "patriarchal" world of contemporary science (Keller 1985, Merchant 1983; see also Merchant 1995). Both Keller and Merchant contrast the emphasis on mechanism and domination of nature that they view as characteristic of post–sixteenth-century science with an earlier, organic approach to nature supposedly embodied in alchemy and an amorphous "hermetic tradition." In this essay, I argue that this rose-tinted view of alchemy is unrealistic when one examines the alchemical literature at first hand. Indeed, early modern science may well have derived from the alchemical tradition itself some of the very themes that Merchant and Keller decry.[1] The domination of nature and the attempt to force it to serve human goals are prominent themes in alchemy. But Keller and Merchant ignore these alchemical themes primarily because they have adopted a distorted picture of alchemy from twentieth-century sources.

The tendency to contrast the mysterious imagery of alchemy and its frequent use of sexual metaphor with the mechanical cosmos of post-Cartesian science was in existence long before the current dissatisfaction with science. Carl Jung, the father of "analytical psychology," developed an elaborate theory in the 1920s and 1930s that alchemy was really a matter of "psychic processes" expressed in "pseudochemical language."[2] The goal of alchemy was not the transmutation of metals but the transcendent reintegration of the feminine unconscious with the male consciousness, a healing rendered necessary by the excessive rationalism of the scientific worldview. Jung's follower Mircea Eliade, by training a scholar of comparative religion, furthered this viewpoint in his popular exposition of alchemical belief, *The Forge and the Crucible* (published in French in 1956, in English in 1962). Eliade argued that alchemy was above all a soteriological discipline, teaching the alchemist how to perfect his own inner nature. In Eliade's view, the purpose of alchemy was a reintegration with the sacred, and in expressing this aim, the alchemist saw himself surrounded by a living, holy nature. Eliade went so far as to contrast the alchemists' vitalistic "worldview" with the emphasis on mechanism that characterizes modern science. According to him, the alchemists of the Renaissance were among the last Europeans to experience nature as a living being: after them came the "death" of the cosmos with the advent of mechanism.[3]

A romantically colored rendition of this view has resurfaced to form a central thesis of Carolyn Merchant's popular *The Death of Nature* (1980), whose very title could be a restatement of Eliade. Merchant uses Eliade and Jung to argue that, considered as a homogeneous whole, the alchemists held a sacred view of nature in which the earth was revered as female (Merchant 1980, 17–20, 25–27).[4] She sees the triumph of the mechanical philosophy over alchemy in the seventeenth century as a central example of "the transition from the organism to the machine" (xxii). These themes were elaborated still further by Evelyn Fox Keller in her 1985 *Reflections on Gender and Science* (43–65). According to Keller, "the hermetic tradition" (equated with alchemy) and "mechanism" provided the two poles available to natural scientists in mid-seventeenth-century Britain (1985, 44). Unlike the upholders of a mechanical worldview, the alchemists employed a highly gendered language whose "basic images" were "the hermaphrodite and the marital couple" (1985, 48). Hence, although she admits that the alchemists were not outright feminists, Keller sees them as having championed the "view of nature and woman as Godly," a position that she claims was defeated by the mechanical philosophers (1985, 53–54).

Another element in Merchant's and Keller's approaches is their denigration of Francis Bacon, whom both view as having been a key figure in the birth of mechanistic science. They attack Bacon repeatedly for his "patriarchal" goal of dominating nature and for his use of images involving the "hounding," "torture," and subordination of nature, portrayed anthropomorphically as female (Keller 1985, 33–42; Merchant 1980, 164–91). My goal here is not to defend Bacon, which Alan Soble has already done, but to point out that many of the same themes can be found in the very alchemical writers whom Keller and Merchant hold up as the antithesis of Baconian science (Soble 1995, 192–215). Alchemical writers frequently express their view that males are superior to females and often use the imagery of torture in describing their processes.[5] Surprisingly, the very alchemists whom Merchant and Keller examined are among those who carry these tendencies the furthest. Accordingly, I shall focus in the following on writings attributed to the Swiss magus and alchemist Paracelsus (1493–1541), the late antique author Zosimos of Panopolis, and the medieval physician Arnald of Villanova (ca. 1240–1311), all of whom appear in Keller's and Merchant's works (Keller 1985, 43, 45, 49–53; Merchant 1980, 18, 22, 26–27, 102, 104, 117–21, 161, 197, 201, 243, 254, 283).

Art Triumphing over Nature: The Paracelsian Homunculus

Keller argues that unlike Bacon, the alchemists emphasized "coition, the conjunction of mind and matter, the merging of male and female" (Keller 1985, 48). Whereas Bacon idealized "the virile superman," the alchemists promoted the hermaphrodite:

> In depicting hermaphroditic union, sexual union, or simply the collaborative effort of man and woman, their graphic images represent the conjunction, or marriage, of male

and female principles that was central to the hermetic philosophy. . . . Alchemical writings suggest a principle of symmetry (one might almost say of equality) between male and female principles. (Keller 1985, 48–49)

In support of this purported alchemical equality between male and female, Keller quotes two long passages from Paracelsus. One describes his belief that both man and woman are endowed with seed and that human generation is brought about by the fruitful concourse of these two principles. It is unfortunate that Keller did not explore the Paracelsian corpus a bit more deeply before claiming that the cooperative effort of male and female seed implied that man and woman must cooperate. One of the more widely cited treatises ascribed to Paracelsus, *De natura rerum*, contains a recipe for making the homunculus, or artificial human. The Paracelsian author views this as the crowning piece of man's creative power, making its artificer a sort of demiurge on the level of a lesser god. But the artificial human created by alchemical means would not be merely an image of man on a smaller scale. As I shall show, the author of the principal homunculus tract planned to improve on human nature by producing a purely masculine being devoid of the usual defects inherited from a female parent. Instead of "merging" male and female, as Keller tells us the alchemists were wont to do, *De natura rerum* tells us precisely how to segregate them.

De natura rerum opens with a discussion of the powers of art (i.e., technology) and nature. In a way that is remarkably reminiscent of Francis Bacon's promotion of human technical prowess, the Paracelsian author compares the unconstrained generation of natural things with what the alchemist can induce in a flask. Although the author is not overly concerned with philosophical niceties, he at once assimilates natural and artificial generations by saying that both come from "the earth" by means of warm, moist putrefaction. Thus the putrefaction that occurs in the stomach is not essentially different from what occurs in a glass vessel.

After a few words on the wonders of putrefaction, which allows one thing to be transmuted into another, *De natura rerum* extends the foregoing logic to a discussion of eggs. In incubating her egg, the hen merely supplies the necessary heat for the "mucilaginous phlegm" within to rot and, in so doing, to become the living matter that will develop into a chick.[6] The key agent, once again, is putrefaction. But this incubation and ensuing putrefaction can be implemented artificially by means of warm ashes and without the brooding hen. More than this, if a living bird is burned to powder and ashes in a sealed vessel and its remains be left to rot into mucilaginous phlegm in "a horse's womb" (*venter equinus*—a technical term for hot, decaying dung), the same phlegm may again be incubated to produce "a renovated and restored bird." In this fashion, all birds may be killed and reborn, so that the alchemist becomes a sort of little god who brings about a miniature conflagration complete with a "rebirth and clarification" of matter like that which will accompany the Final Judgment. This clarification of matter by the fire of the Day of Judgment is one of Paracelsus' habitual themes, which he expounds at length in his late *Astronomia magna*, the definitive statement of his late philosophy.[7] *De natura rerum* goes on to announce that the death and rebirth of birds forms "the highest and greatest *magnale* and mystery of God, the highest secret and wonder-work."[8]

Despite this categorical statement, *De natura rerum* has even greater marvels to offer, as the author then observes:

> You must also know that men too may be born without natural fathers and mothers. That is, they are not born from the female body in natural fashion as other children are born, but a man may be born and raised by means of art and by the skill of an experienced spagyrist, as is shown hereafter.[9]

Having introduced the homunculus, the text then digresses to discuss the unnatural union of man with animals, which can also produce offspring, though "not without heresy." Still, one should not automatically treat a woman who gives birth to an animal as a heretic, "as if she has acted against nature," for the monstrous offspring may only be a product of her disordered imagination.

Animals, too, can produce monsters, as when their offspring do not belong to the same race as the parents. But the author of *De natura rerum* is more interested in the case of monsters that "are brought to pass by art, in a glass" (315). A good example of such artificial monsters is the basilisk, which is made from menstrual blood sealed up in a flask and subjected to the heat of the "horse's womb."[10] The basilisk is "a monster above all monsters," for it can kill by its glance alone. Being made from menstrual blood, it is like a menstruating woman, "who also has a hidden poison in her eyes" (316) and can ruin mirrors with her glance and make wounds impossible to heal or spoil wine with her breath. But the poison of the basilisk is much stronger than that of the woman per se, because it is the living and undiluted embodiment of her poisonous excrescence:

> Now I return to my subject, to explain why and for what reason the basilisk has the poison in its glance and eyes. It must be known, then, that it has such a characteristic and origin from impure [i.e., menstruating] women, as was said above. For the basilisk grows and is born out of and from the greatest impurity of women, from the menses and the blood of the sperm.[11]

One could therefore say that for the author of *De natura rerum*, the basilisk is the epitome of the female itself, a valuation that does not seem to contradict the undisputed corpus of Paracelsus.[12]

Soon after this memorable account, *De natura rerum* arrives at a lengthy description of the homunculus and its mode of generation. Coming on the heels of the basilisk, which was made by a sort of artificial parthenogenesis, the homunculus is its masculine twin. Just as the basilisk embodied the quintessence of feminine impurity, so the homunculus, created without any feminine matter, serves as a magnification of the intellectual and heroic virtues of masculinity. But first let us relate its mode of production:

> We must now by no means forget the generation of homunculi. For there is something to it, although it has been kept in great secrecy and kept hidden up to now, and there was not a little doubt and question among the old philosophers, whether it even be possible to nature and art that a man can be born outside the female body and [without] a natural mother. I give this answer—that it is by no means opposed to the spagyric art and to nature but that it is indeed possible. But how this should happen and proceed—its process is thus—that the sperm of a man be putrefied by itself in a cucurbit

for forty days with the highest degree of putrefaction in a horse's womb or at least so long that it comes to life and moves itself, and stirs, which is easily observed. After this time, it will look somewhat like a man, but transparent, without a body. If after this, it is fed wisely with the arcanum of human blood and is nourished for up to forty weeks and is kept in the even heat of the horse's womb, a living human child will grow therefrom, with all its members like another child, which is born of a woman but much smaller.[13]

As we can see, the Paracelsian author of *De natura rerum* introduces his homunculus within the framework of the traditional question of the limits of human art. Unlike the timid philosophers of old, the author says, he is willing to affirm the powers of human art in making a test-tube baby. And doubly marvelous will this creature be, having grown out of sperm alone, unpolluted by the poisonous matrix from which the basilisk originated. Because of its freedom from the gross materiality of the female, the homunculus is translucent and, as it were, bodiless. Like the "clarified" birds produced by alchemical techniques, the homunculus is almost incorporeal. Hence the author can use the homunculus as yet another excuse to vaunt the powers of human art, which he immediately sets out to do. *De natura rerum* announces that if they reach adulthood, such homunculi can give rise to further marvelous beings such as giants and dwarves. These creatures have wonderful strength and powers, such as the ability to defeat their enemies with "great, forceful victory" (317) and to know "all hidden and secret things" (317). Why are they so gifted? Because "they receive their life from art, through art they receive their body, flesh, bone, and blood. Through art they are born, and therefore art is embodied and inborn in them, and they need learn it from no one" (317).[14]

The reasoning here is straightforward. Because the homunculus is a product of art, in its mature state it has an automatic and intimate acquaintance with the arts and consequently knows "all secret and hidden things." Thus the homunculus is not merely an artificial marvel in itself but a key to further marvels. It is the final expression of man's power over nature, as the author says, "a miracle . . . and a secret above all secrets" (317).[15]

One can see, then, that the Paracelsian dream of making a homunculus embodies supposedly "Baconian" themes despite having originated in a "hermetic" context. On the one hand, *De natura rerum* is clearly intent on dominating nature by subverting natural processes. The author sets his homuncular ruminations in the broader context of the traditional debate between nature's powers and those of human art, and he evidently thinks that art has gotten the upper hand. But how has it done so? Precisely by engaging in a sort of "eugenic" experiment, whose goal is the elimination of female characteristics in the offspring. The making of the homunculus is not merely a usurpation of the female role in generation, as Sally G. Allen, Joanna Hubbs, and Nancy Tuana have argued (Allen and Hubbs 1980, 213; Tuana 1993, 145–47). Rather, it is the explicit exclusion of "femaleness" from the experimental progeny.

The Theme of Torture in Alchemy

The theme of human domination over nature, so obvious in the Paracelsian *De natura rerum*, is frequently expressed by alchemists in the language of torture. The so-called hermetic tradition was not composed of gentle nature worshipers, as Merchant and Keller would have us believe, but of active interventionists intent on turning nature to their own purpose.

Let us briefly consider the work of Zosimos of Panopolis (4th c. c.e.), who appears as one of the heroes in Merchant's work (1980, 18). One of the most famous works attributed to Zosimos is his *On Virtue*, which consists primarily of a succession of dreams linked together by interpretation. In the first dream, Zosimos sees a "sacrificer," a priest standing on an altar shaped like a piece of alchemical glassware.[16] The priest announces that his name is Ion and that he is enduring "an intolerable violence." What follows is quoted in its entirety:

> For someone has come at dawn, running, and has made himself my master, cutting me apart with a knife, tearing me asunder according to the constitution of harmony, and skinning my entire head with the sword that he clasped. He intertwined the bones with the flesh and burned me up with the scorching fire from his hand until I had learned to become a spirit by metamorphosing my body. (Mertens 1995, 36)

The priest Ion, standing on his alchemical altar, is describing the passage from base materiality to volatile pneuma that must be carried out in the work of alchemy. But Zosimos has chosen to express this laboratory process in the unforgettable terms of torture and human sacrifice. Nor is this all. After hearing Ion's soliloquy, Zosimos then sees the priest begin to vomit out his own entrails while his eyes turn to blood. Ion then turns into a sort of midget, biting and ripping himself with his own teeth. Upon seeing this grisly spectacle, Zosimos wakes up and interprets the dream to refer to the production of alchemical "waters."

What follows is hardly less picturesque. Zosimos has a second dream, in which the alchemical altar is filled with a boiling water. Writhing and moaning within the vessel is an "innumerable crowd" of men being boiled alive. They, too, must undergo a transmutation into pneuma, which requires that they undergo this "punishment" (*kolasis*). Zosimos decides that this is an alchemical allegory also and concludes with some general remarks on the method of the "art." Here he passes from a discussion of individual reagents, anthropomorphized as men undergoing punishment, to the torture of nature as a whole. He argues that for the alchemist to succeed, Nature (*physis*) must be "forced to the investigation" (*ekthlibomenē pros tēn zētēsin*), whereupon she, suffering (*talaina*), assumes successive forms until her punishment renders her spiritual (Mertens 1995, 40–41). As Michelle Mertens points out in her excellent commentary to the text, the personified Nature is pushed into a state of "confusion," which makes her want to change her state. The alchemist then induces her to assume various intermediate states until she is on the verge of death: only by this means can she become spiritual and therefore useful for the alchemical work (Mertens 1995, 224–25).

One could hardly ask for a clearer statement of the supposedly Baconian principle that nature must be constrained, even tortured, than is found in the work of Zosimos. But unlike Bacon, Zosimos gives a graphic account of how such torture is to occur— the horrific descriptions of self-devouring dwarves and boiling men clearly refer to the "torture" inflicted on metals and minerals by processes carried out at high heat with corrosives. Alchemy itself is seen as a means of torturing Nature in order to "spiritualize" her and, ultimately, to acquire the means of transmuting metals.

The depiction of alchemical processes as a form of torture is not restricted to late antiquity but is widespread in medieval alchemy as well. Let us consider the Catalan physician Arnald of Villanova, from whom Merchant borrows an illustration of a copulating king and queen, who represent to her "the sacred marriage of male and female in alchemy" (1980, 22). Although it is unlikely that the genuine Arnald of Villanova wrote an alchemical treatise, a large corpus of alchemical works is attributed to him (Newman 1991, 194–204). Some of these contain highly developed figurative treatments of alchemy expressed in terms of Christianity, possibly related to the fact that the genuine Arnald became a follower of the prophet Joachim of Fiore late in his own life. Several Arnaldian treatises, such as *On the Secrets of Nature*, compare the "great work" of the alchemists with the life and death of Jesus. Here, the author says that like Jesus, mercury is to "be taken and beaten and scourged lest by reason of pride he perish." Elaborating on this theme, he continues,

> Therefore take the son [i.e., mercury] after he has been beaten and put him to bed to enjoy himself for a while, and when you feel that he is enjoying himself, then take him pure and extinguish in cold water. And when you have repeated this process, hand him over to the Jews to be crucified. (Thorndike 1934, 76)

Note that the Latin term for torture in general was "crucifixion"; *cruciatus* meant anything from flogging to strangulation. The "crucifixion" of mercury, then, could mean any severe treatment that caused it to change its form. This striking command to subject mercury to the same ill treatment as Jesus receives even further elaboration in another Arnaldian treatise, the *Exempla* or *Parabolae*. Here a detailed comparison is made between the transformation of mercury and the passions or torments of Jesus. Just as Jesus was first scourged until he bled, then made to wear a crown of thorns, next nailed to the cross, and finally treated to gall and vinegar, so mercury must be tortured in four stages. Although these are described rather unclearly in the text, one can get a good idea of the author's intent by considering the third passion: "The third passion was the cross of Christ where he was hanged and his soul received torment. In mercury, too, when it reddens by means of cooking, this redness denotes the body of Christ" (214).[17]

The Arnaldian author of the *Parabolae*, like Zosimos of Panopolis, believed that in its present fallen condition, the natural world must be purged by suffering. Although he chooses Jesus rather than a personified nature to exemplify this belief, the intention is the same. Only the torture of alchemical purification by fire can return natural substances to their prelapsarian state. Like Jesus, the mercury of the *Parabolae* must be killed and laid in a sepulcher. "There it will stay for three days and nights just as Christ, and then it will be reborn brilliant and ruddy."[18] Viewed in the light of this particular tradition, the alchemist has as his peculiar function the "torture" and "murder" of natural substances.

Conclusion

It is a strange irony that Merchant and Keller should have chosen alchemy as the premodern contrast to Baconian science. As I have argued here and elsewhere, the alchemical tradition was deeply imbued with the desire to improve on nature, and this quest often adopted the language of dismemberment and torture.[19] In some instances, such as the work of Zosimos, the alchemist is even advised to constrain and punish a personified "nature" directly. If one wishes to see misogyny in this decree, he or she is free to do so, but there are much more obvious targets, such as the homunculus recipe of the Paracelsian *De natura rerum*. Here misogyny is elevated to undreamed-of heights, for the very summum bonum of alchemy—indeed of the arts in general—is announced to be the exclusion of femininity from the artificial human.

But the real irony of Merchant's and Keller's choice lies in the fact that Francis Bacon himself, the patriarchal founder of the "mechanist" tradition, has long been known to have borrowed from the work of alchemists. Graham Rees, for example, showed in a series of well-known articles beginning in 1975 that Bacon employed a "semi-Paracelsian" cosmology modeled in important respects on the cosmological views of Paracelsus and his followers (Rees 1975, 83–101; also see Rees and Edwards 1996). More significant for our purposes, however, is the fact the Italian scholar Paolo Rossi had already pointed out in 1957 that Bacon's goal of "scientific domination of nature" owed a heavy debt to alchemical and magical sources (16). Although Keller and Merchant are aware of Rossi's work, they both minimized the degree of Bacon's debt to alchemical writers for the theme of domination (Keller 1985, 37, refers to Rossi, as does Merchant 1980, 316–17). How can Merchant and Keller have missed such an obvious feature of the alchemical tradition as the desire to outdo nature? The answer to this riddle probably stems from the fact that neither Keller nor Merchant has made a concerted effort to come to grips with the history of alchemy on its own terms.

Despite having done original work on the iatrochemist Francis Mercurius van Helmont, Merchant is quite content to rely on such occult compendia as Kurt Seligmann's *Magic, Supernaturalism, and Religion* (1948) for her knowledge of Greek alchemy and the Jungian Jolande Jacobi's collection of excerpts from Paracelsus (1951) for her knowledge of the latter, not to mention Eliade and Jung themselves.[20] Keller draws most of what she knows of alchemy from Merchant and from Merchant's secondary sources. In the absence of any sustained attempt to deal with the alchemical sources themselves, it seems clear that Merchant and Keller have merely repackaged the romantic image of alchemy propounded long ago by Jung and Eliade, who were themselves anything but historians of science. Jung's and Eliade's view of alchemy as the representative of an archaic, holistic, premechanistic past belongs to a rigorously unhistorical literature that mixes together modern African smiths, Renaissance German metallurgists, and ancient Babylonian rituals. This is not the place to subject Jung and Eliade to the scrutiny that they deserve, but the alchemical examples that I have adduced against the arguments of Keller and Merchant should cast serious doubt on the romantic interpretation of alchemy as an inherently philogynous enterprise.[21]

Notes

1. See Newman 1989, 423–45, for the prominent defense of the technological alteration of nature by alchemists.

2. Jung's earliest writing on alchemy is found in *The Secret of the Golden Flower*, published with Richard Wilhelm in 1929: see Martin 1992, 16. Sustained treatments are found in the

following of Jung's texts: *Aion, Collected Works,* vol. 9, part 2, 1959; *Psychology and Alchemy, Collected Works,* vol. 12, 1953; *Alchemical Studies, Collected Works,* vol. 13, 1964; and *Mysterium conjunctionis, Collected Works,* vol. 14, 1963. For the quotation, see Jung, "The Idea of Redemption in Alchemy," in Dell 1939, 210.

3. "After the mental revolution accomplished by the Renaissance, the physico-chemical operations and cosmic events achieve their autonomy from the laws of universal *life,* enclosing themselves, though, into a system of 'dead' mechanical laws" (Eliade, "Metallurgy, Magic and Alchemy," 1938, 23). See also Eliade, *Forgerons et alchimistes,* 1956. Jung's views are enthusiastically embraced on pp. 161–74 and 201–4.

4. Merchant's use of Eliade is documented on p. 296, n. 1; her use of Jung is on p. 298, n. 21.

5. I do not, of course, wish to deny that Bacon's language of *natura vexata* may also have had legal or juridical sources but merely to point out that alchemy is full of similar images and that these may have been among Bacon's sources. For a recent treatment of the English legal system and its influence on Bacon, see Martin 1992, 164–66.

6. [pseudo?] Paracelsus, *De natura rerum,* in Sudhoff 1928, vol. 11, 313. Sudhoff rejected the authenticity of *De natura rerum* in its present form, though he suggested that it might contain "Hohenheimische Ausarbeitungen oder Entwuerfe" (vol. 11, xxxiii), but Will-Erich Peuckert questions this rejection: see Peuckert, 1968, vol. 5, ix. Kurt Goldammer also accepts the authenticity of *De natura rerum,* though with reservations: "Der Gedanke der Substanzenseparierung hat dann auch die paracelsische Todesanschauung in jenen beruehmt gewordenen Ausfuehrungen der umstrittenen Schrift 'De natura rerum' geliefert, von der ich annehme, das sie in ihrer Grundidee echt ist, wenn auch eine Ueberarbeitung durch Schuelerhaende sich nicht ausschliessen laesst," from Goldammer, "Paracelsische Eschatologie, Zum Verstaendnis der Anthropologie und Kosmologie Hohenheims I," *Nova acta Paracelsica* 5 (1948): 52. It is clear, then, that the substance of *De natura rerum* is genuine, even if the document in its present form is suspect.

7. Paracelsus, *Astronomia magna,* in Sudhoff 1929a, vol. 12, 322: According to Paracelsus, after the world has been consumed by fire in the final conflagration, everything will be as "ein eidotter ligt im clar." This will be a *perspicuum,* and this will be both a *chaos* and also "das wasser, von dem die geschrift sagt, auf welchem der geist gottes getragen wird."

8. [pseudo] Paracelsus, *De natura rerum,* in Sudhoff 1928, vol. 11, 313: "Das ist auch das hoechst und groessest magnale und mysterium dei, das hoechst geheimnus und wunderwerk."

9. [pseudo] Paracelsus, *De natura rerum,* in Sudhoff 1928, vol. 11, 313: "Es ist auch zu wissen, das also menschen moegen geboren werden one natuerliche veter und muetter. Das ist sie werden nit von weiblichem leib auf natuerliche weis wie andere kinder geboren, sonder durch kunst und eines erfarnen spagirici geschicklikeit mag ein mensch wachsen und geboren werden, wie hernach wird angezeigt &c."

10. [pseudo] Paracelsus, *De natura rerum,* in Sudhoff 1928, vol. 11, 315–16: "Dan der basiliscus wechst und wird geboren aus und von der groessten unreinikeit der weiber, aus den menstruis und aus dem blut spermatis, so dasselbig in ein glas und cucurbit geton und in ventre equino putreficirt, in solcher putrefaction der basiliscus geboren wird."

11. [pseudo] Paracelsus, *De natura rerum,* in Sudhoff 1928, vol. 11, 315: "Nun aber damit ich widerumb auf mein fuernemen kom, von dem basilisco zuschreiben, warum und was ursach er doch das gift in seinem gesicht und augen habe. Da ist nun zu wissen, das er solche eigenschaft und herkomen von den unreinen weibern hat, wie oben ist gemelt worden. Dan der basiliscus wechst und wird geboren aus und von der groessten unreinikeit der weiber, aus den menstruis und aus dem blut spermatis."

12. Paracelsus, *De generatione hominis,* in Sudhoff 1929c, vol. 1, 305, where the female is viewed as the principle of all evil: "Das aber ein mensch vil lieber stilet als der ander, ist die

ursach also, das alles erbars in Adam gewesen ist und das widerwertige der erbarkeit, uner-barkeit in Eva. Solches ist auch also durch die wage herab gestigen in die samen nach dem ein ietlichs sein teil davon gebracht hat, nach dem ist er in seiner natur. Denn etwan hat die diebisch art uberwunden, etwan die hurisch, etwan die spilerisch &c." Cf. the parallel locus in Paracelsus, *Das Buch von der Geberung der Empfindlichen Dinge in der Vernunft*, in Sudhoff 1929b, vol. 1, 278–81.

13. [pseudo] Paracelsus, *De natura rerum*, in Sudhoff 1928, vol. 11, 316–17: "Nun ist aber auch die generation der homunculi in keinen weg zu vergessen. Dan etwas ist daran, wiewol solches bisher in grosser heimlichkeit und gar verborgen ist gehalten worden und nit ein kleiner zweifel und frag under etlichen der alten philosophis gewesen, ob auch der natur und kunst moeglich sei, dass ein mensch ausserthalben weiblichs leibs und einer natuerlichen muter moege geboren werden? Darauf gib ich die antwort das es der kunst spagirica und der natur in keinem weg zuwider, sonder gar wol moeglich sei. Wie aber solches zugang und geschehen moege, ist nun sein process also, nemlich das der sperma eines mans in verschlossnen cucurbiten per se mit der hoechsten putrefaction, ventre equino, putrificirt werde auf 40 tag oder so lang bis er lebendig werde und sich beweg und rege, welches leichtlich zu sehen ist. Nach solcher zeit wird es etlicher massen einem menschen gleich sehen, doch durchsichtig on ein corpus. So er nun nach disem teglich mit dem arcano sanguinis humani gar weislich gespeiset und erneret wird bis auf 40 wochen und in steter gleicher werme ventris equini erhalten, wird ein recht lebendig menschlich kint daraus mit allen glitmassen wie ein ander kint, das von einem weib geboren wird, doch viel kleiner."

14. [pseudo] Paracelsus, *De natura rerum*, in Sudhoff 1928, vol. 11, 317: "Dan durch kunst uberkomen sie ir leben, durch kunst uberkomen sie leib, fleisch, bein und blut, durch kunst werden sie geboren, darumb so wirt inen die kunst eingeleibt und angeboren und doerfen es von niemandts lernen."

15. [pseudo] Paracelsus, *De natura rerum*, in Sudhoff 1928, vol. 11, 317: "Dan es ist ein mirakel und magnale dei und ein geheimnis uber alle geheimnus."

16. Mertens 1995, 35. See her note 6 for the significance of the altar.

17. MS. Venice, San Marco 6, 214, f. 164v–168r: "Tertia passio fuit crux christi ubi pependit et anima penam recepit. Etiam in mercurio per decoctionem ut rubescat et denotat illa rubedo corpus christi."

18. MS. Venice, San Marco 6, 214, f. 168r: "Postea debet in sepulcro poni. Et ibi stet per tres dies et noctes sicut christus. et resuscitabitur candidus et rubicundus."

19. For the theme of improving on nature by means of alchemy, see Newman 1989, 423–45.

20. Merchant 1980, 298, nn. 16–19. For an interesting anecdote about Jacobi and Jung, see Noll 1994, 285–86.

21. For a thorough debunking of Jungian analytical psychology, see Noll 1994. For a critique of the Jungian interpretation of alchemy, see Newman 1994, 115–69. See also Robert Halleux 1979, 55–56; and Obrist 1982.

References

Allen, Sally G., and Joanna Hubbs. Outrunning Atalanta: Feminine Destiny in Alchemical Transmutation. *Signs* 6: 210–229, 1980.

Eliade, Mircea. Metallurgy, Magic and Alchemy. In *Cahiers de Zalmoxis*, publiés *par Mircea Eliade, vol 1*. Paris: Librairie Orientaliste Paul Geuthner, [1938].

———, *Forgerons et alchimistes*. Paris: Flammarion, 1956.

Halleux, Robert, *Les Textes alchimiques*. Turnhout: Brepols, 1979.

Jung, C. G. The Idea of Redemption in Alchemy. In Stanley Dell, ed. *The Integration of the Personality*. New York: Farrar & Rinehart, 1939.

———,.*Psychology and Alchemy: Collected Works*. Vol. XII. London: Routledge, 1953.

———, *Aion: Collected Works*. Vol. IX, Part II. London: Routledge, 1959.

———, *Mysterium Conjunctionis: Collected Works*. Vol. XIV. London: Routledge, 1963.

———, *Alchemical Studies: Collected Works*. Vol. XIII. London: Routledge, 1967.

Keller, Evelyn Fox, *Reflections on Gender and Science*. New Haven: Yale University Press, 1985.

Martin, Julian, *Francis Bacon, The State, and the Reform of Natural Philosophy*. Cambridge: Cambridge University Press, 1992.

Merchant, Carolyn, *The Death of Nature*. San Francisco: Harper and Row, 1983; first edition, 1980.

———, *Earthcare: Women and the Environment*. New York: Routledge, 1995.

Mertens, Michele, ed., *Zosime de Panopolis, Mémoires authentiques. Les Alchimistes grecs*. Paris: Belles lettres, 1995. Vol. 4.

MS. Venice, San Marco 6, 214, f. 164v–168r.

Newman, William R. Technology and Alchemical Debate in the Late Middle Ages. *Isis* 80: 423–445, 1989.

———, *The "Summa perfectionis" of pseudo-Geber*. Leiden: Brill, 1991.

———,*Gehennical Fire: the Lives of George Starkey, an American Alchemist*. Cambridge, Mass.: Harvard University Press, 1994.

Noll, Richard. *The Jung Cult*. Princeton: Princeton University Press, 1994.

Obrist, Barbara. *Les débuts de l'imagerie alchimique*. Paris: Le Sycomore, 1982.

Paracelsus, *Astronomia magna*, Karl Sudhoff, ed. *Theophrastus von Hohenheim, genannt Paracelsus, Saemtliche Werke, I. Abteilung*. Munich: Oldenbourg. 1929a, Vol. 12.

———, *Das Buch von der Geberung der Empfindlichen Dinge in der Vernunft*, Karl Sudhoff, ed. *Theophrastus von Hohenheim, genannt Paracelsus, Saemtliche Werke, I. Abteilung*. Munich: Oldenbourg, 1929b, Vol. 1.

———, *De generatione hominis*, Karl Sudhoff, ed. *Theophrastus von Hohenheim, genannt Paracelsus, Saemtliche Werke, I. Abteilung*. Munich: Oldenbourg, Vol. 1, 1929c.

[pseudo?] Paracelsus. *De natura rerum*, Karl Sudhoff, ed. *Theophrastus von Hohenheim, genannt Paracelsus, Saemtliche Werke, I. Abteilung*. Munich: Oldenbourg, 1928, Vol. 11.

Peuckert, Will-Erich, *Theophrastus Paracelsus: Werke*. Basel: Schwabe & Co., 1968, Vol. 5.

Rees, Graham. Francis Bacon's Semi-Paracelsian Cosmology. *Ambix* 22: 83–101, 1975.

Rees, Graham, and Michael Edwards, eds., *The Oxford Francis Bacon*. Oxford: Clarendon Press, 1996, Vol. 6.

Rossi, Paolo, *Francis Bacon: From Magic to Science*. London: Routledge and Kegan Paul, 1968.

Soble, Alan. In Defense of Bacon. *Philosophy of the Social Sciences* 25: 192–215, 1995.

Thorndike, Lynn, *History of Magic and Experimental Science*. New York: Columbia University Press, 1934, Vol. 3.

Tuana, Nancy. *The Less Noble Sex*. Bloomington: Indiana University Press, 1993.

14

What Is Wrong with the Strong Programme's Case Study of the "Hobbes–Boyle Dispute"?

Cassandra L. Pinnick

Is the Strong Programme "History"?

The title of this chapter asks a question: What is wrong with the Strong Programme's case study of the "Hobbes–Boyle dispute"? The answer to this question, perhaps surprisingly, is—the history. This answer may be unexpected because the usual complaints about the Strong Programme are philosophical; they indict the causal claims it makes or the incoherent nature of its project. Although I shall say something about these philosophical failures, I first want to confirm that an essay about the Strong Programme is properly considered a part of the history of philosophy of science.

Is the Strong Programme "history"? Would that it were! It is difficult to imagine another theory that has suffered so many definitive philosophical deaths (Laudan 1984, Roth and Barrett 1990), and yet it does not go away. Despite the Strong Programme's retreat from its interest model, it continues to play the philosophical elder to social constructivism. Despite philosophers' hopes that proponents of a demonstrably incoherent theory would go quietly into the night, this apparently is not to be. Instead, David Bloor blusters that philosophical critics "have not convinced me of the need to give ground on any matter of substance" (1991, ix, afterword)—as though he could promise to achieve the "science of science" but never show (as he specifically promised) the decisive role of sociological factors in the causal analysis of scientific belief.[1]

Philosophical criticism of the Strong Programme's principal texts provoked a bold response. Declaring that they were wasting time discoursing with the philosophers (and that, in any case, there was no need to vanquish the philosophy of science), Bloor and his supporters announced their intention to get on with the "science of science." True to the claimed empirical foundations of their program, they determined to construct their own historical underpinnings. A record of success, in the form of case studies built on and reflexively demonstrating the Strong Programme tenets, would ultimately justify the Strong Programme over philosophy of science. In Bloor's words:

All that can be done is to check the internal consistency of the different theories [the Strong Programme and philosophy of science] and then see what happens when prac-

227

tical research and theorising is based upon them. If their truth can be decided at all it will only be after they have been adopted and used, not before. So the sociology of knowledge is not bound to eliminate a rival standpoint. It only has to separate itself from it, and make sure that its own house is in logical order. (1976, 9)

Sympathetic philosophers of science supported the Strong Programme's move. For example, Mary Hesse wrote that the hegemony of philosophical accounts of science was too long-standing, a circumstance that cried out for counterbalance. She said that the prima facie plausibility of a sociological redefinition of scientific knowledge made it good practical sense to call a moratorium on the philosophical attacks:

But suppose we treat the [Strong Programme] thesis not as a demonstrable conclusion from acceptable premises but as a hypothesis in the light of which we decide to view knowledge, and consider whether its consequences are consistent with the rest of what we wish to affirm about knowledge, and whether it does in the end provide a more adequate and plausible account than the various rationalist positions we have found questionable. (1980, 42)

The already 25-year-long moratorium must finally end. A hands-off stance toward the Strong Programme's case studies is no longer justifiable.[2] The promised track record—the compelling empirical record of successes—has not materialized. The Strong Programme and its influence should be part of the (recent) history of philosophy of science. Yet, it prospers, and this brings me back to my point.

Why regard the Strong Programme as belonging to the history of philosophy of science? Because

- The Strong Programme is a philosophy of science.
- Its views have been around for about as long as Kuhn's views have been (and Kuhn counts as history).
- And in any case, a lot of philosophers of science wish it were history.

In a recent conversation, a philosopher/historian of science suggested to me what I'll call the "sandwich theory." It goes something like this:

Sandwich Theory: The Shapin and Schaffer book, Leviathan and the Air Pump: Hobbes, Boyle, and the Experimental Life (1985), like the other Strong Programme case studies, just needs to be read with charity. You take the first and last chapters with a large degree of intellectual tolerance (those are the parts to ignore, with all the bad argumentation and the trumpeting about sociology of knowledge); then, in between, there is a good historical account to read.

I contend, without qualification, that the sandwich theory is wrong because—as I hope to show in this essay—Shapin and Schaffer's historical account is wrong.[3] I will attempt to demonstrate that the sandwich theory (and Hesse's moratorium) have lulled philosophers of science for too long now into a sanguine attitude, specifically toward Shapin and Schaffer's Leviathan and the Air Pump. The failure of philosophers of science to engage the historical basis of the Shapin and Schaffer case study has allowed flawed history and historiographical technique to assume iconic proportions, in science studies and in philosophy. It's time for historians of science, philosophers of science, and historians of philosophy of science to have something to say—in the negative—about Shapin and Schaffer's book. It's not just the sociology that belongs in the dust bin. It's the historical goods, too.

Methodological Failures

Because my emphasis here is on the historical flaws in the Shapin–Schaffer argument, I will only briefly mention the philosophical problems that arise for their case study via their use of the Strong Programme theoretic.

Key to these problems are criticisms that the Strong Programme would make against philosophy of science in the name of the Underdetermination (or the "Duhem–Quine") Thesis but that come back to haunt Shapin and Schaffer's own analysis. It is simply wrong to believe that underdetermination provides any motivation, much less support, for a bid to replace philosophy of science. The very thesis that the Strong Programme proponents use so widely is transparently damaging to their own programmatic aims.[4] Here is why:

If undetermination opens a logical gap for philosophical explanations, then the gap extends equally to interest- (and socially) based explanations. That is, if the underdetermination thesis succeeds in showing that reasons and evidence are infinitely plastic, then sociopolitical forces must be infinitely plastic as well. This holds precisely for the forces Shapin and Schaffer cite to explain the "Hobbes–Boyle dispute."

Even worse, the same arguments that would show—for these theorists qua historians—that Boyle and Hobbes fixed their scientific beliefs because of political commitments and social alliances with Restoration society, would also show that Shapin and Schaffer's preferred analysis is merely a reflection of their (Shapin and Schaffer's) political commitments and social alliances—unless, of course, historians are immune, in ways that scientists are not, to social determination.

Moreover, even if we grant that underdetermination opens a huge, logical gap, all that follows is that some factor, other than logical ones, must come into play. But for all the Strong Programme argument shows, the precipitating causes may be astrological rather than sociopolitical (see Slezak 1991).

All these problems for the sociologists rest on their mistaken faith that underdetermination puts teeth into antiphilosophical arguments. In fact, the supposed challenge that underdetermination presents to the philosophy of science rests on a fallacious equation of deductive logic and the scope of what is rational (see Laudan 1990). For too long now (at least since Hume), deductive proof has not constituted the sole standard of rational belief. One wonders why any theory that seeks to trade on this fallacy for its initial motivation should deserve notice at all—unless, of course, it is acceptable to ignore the history of rationality theory from Hume onward.

Still, it is one thing (1) to equate wrongly the rational with the deductively provable, but it is quite another then to proceed, as the sociologists do, (2) to make fallacious inferential moves, concluding that the purported deductive gap must (and can only) be plugged with sociopolitical causes (Slezak 1991). The former (1) is a rather subtle mistake, but the latter (2) is sloppy reasoning. In the end, showing the existence of a logical gap is nothing like showing that there are no rational causes that might span, or close, the gap.

I wish I could report acknowledgment of these philosophical problems among those concerned with accounting for scientific belief. Instead, many people believe that the earlier incarnations of the Strong Programme failed for quite different reasons, namely its too-narrow focus on macrointerests or on economics or because it was blind to the privileged epistemic vantage point of certain "marginalized" groups in society (see Harding 1992; and, contra, Pinnick 1994). Never, to recall Bloor's promise, is there an owning up to the fact that this house is not in "logical order." My brief survey here should suffice to rehearse the deep methodological problems inherent in the Strong Programme.[5]

Bad History

I now turn to the historical account presented in Shapin and Schaffer's book *Leviathan and the Air Pump: Hobbes, Boyle and the Experimental Life*. Let me use their own words to describe their project: "It will not escape our readers' notice that this book is an exercise in the sociology of scientific knowledge. One can either debate the possibility of the sociology of knowledge, or one can get on with the job of doing the thing. We have chosen the latter option" (1985, p. 15).

This passage announces the undertaking of a typical and familiar methodology: testing an account of science against a historical case. So far, we find no problems, but beyond this, the situation immediately changes.

A major thesis of *Leviathan and the Air Pump* is that Hobbes has been incorrectly relegated to the ranks of political philosophers. Rather, Shapin and Schaffer say, Hobbes should be regarded as a noteworthy philosopher of science (or "natural philosopher," 7) because, as they promise to demonstrate, the Hobbesian analysis of science, though different from Boyle's analysis, was just as credible. "They [Hobbes's views] were not widely credited or believed—but they were *believable*" (13; italics in original). And so because it is *equi*credible, Hobbes's work has important things to say about the philosophy of science:

> Of course, our ambition is not to rewrite the clear judgment of history: Hobbes's views found little support in the English natural philosophical community. Yet we want to show that there was nothing self-evident or inevitable about the series of historical judgements in that context which yielded a natural philosophical consensus in favour of the experimental programme. Given other circumstances bearing upon that philosophical community, Hobbes's views might well have found a different reception. (13)

This claim places Shapin and Schaffer's work explicitly and precisely in the history of science and in the history of the philosophy of science. The integrity of their historiography hinges on the successful demonstration of two conclusions, one historical and the other philosophical. They must show first that Hobbes and Boyle advocate rival philosophies of science (the history). And derived from that rivalry, they must demonstrate the equicredibility they assert exists between Hobbes's and Boyle's philosophies of science (the philosophy). This equicredibility is the means to their programmatic end, namely, to link the history to the sociological explanation, providing a positive test of the Strong Programme "hypothesis" (*pace* Hesse 1980).

This claimed equicredibility requires that Shapin and Schaffer argue counterfactual claims, such as that Hobbes's views could have triumphed if the Restoration settlement had been different. But their strategy goes offtrack even before the difficulties of counterfactual analysis are encountered. As we will see, the "Hobbes–Boyle dispute" was not as clear or as sharp as Shapin and Schaffer make it sound and as the equicredibility thesis requires.[6]

In short, my point is that only by doing bad history can Shapin and Schaffer claim to discover and, in their book represent, significant differences between Hobbes and Boyle. I support this charge by examining the works of the historical figures involved.

Strategy

Shapin and Schaffer intend to investigate what they call the "Hobbes–Boyle dispute." They write that "Hobbes and Boyle proffered radically different solutions to the problem of knowledge" (147) and that the controversy between Hobbes and Boyle centered on the propriety of experimental trials performed with the Torricellian air pump. Shapin and Schaffer do not in any place argue that this controversy is pivotal in the rise of experimentalism; rather, they present its prominence as a simple given.[7] Against this, note that the present historical reading holds that no exchange between Hobbes and Boyle was at all central to the general debate in seventeenth-century natural philosophy.[8]

But saying this is not to object to Shapin and Schaffer's breaking new historical ground. Rather, it is to point out that the logic of their larger argument—to demonstrate Hobbes's importance to the philosophy of science—requires that they motivate their unorthodox centering of the debate and the persons. They cannot simply assume their new historical view as a given. This is especially so when we are faced with the "sandwich theory," and it is the history (more than the philosophy) that we are expected to admire. Shapin and Schaffer's new view of Hobbes (and Boyle) calls out for justification. Without it, suspicion is raised concerning the case study's overall strategic design. Since they provide no initial plausibility for their new perspective on Hobbes, the historical arm of their book's argument begins with the air of a petitio principii, for it presumes what it purports to prove: that a genuine and important debate—one that is held to be potentially instructive for philosophy of science—took place but has heretofore been unjustly ignored.

Shapin and Schaffer stake their case on a new reading of Hobbes's texts—especially what have been taken as political texts—as philosophy of science. But this new reading is deficient, as there are evident discrepancies between Hobbes-on-method and Shapin and Schaffer's representation of Hobbes-on-method. Such discrepancies undercut the contention that Hobbes and Boyle differed fundamentally over method and aims. And if the historical principals were not "radically different," we ought not heed Shapin and Schaffer's call to restructure the history and the philosophy of science.

History—Hobbes

According to Shapin and Schaffer, Hobbes was always committed to the virtues of a priori demonstration. This view has an initial plausibility. It is clear that Hobbes frequently stressed that the goal of inquiry should be true claims deduced from a fund of basic, indubitable truths. Knowledge, Hobbes maintained, arises from the examination of causes and effects: "Philosophy is such knowledge of effects or appearances, as we acquire by true ratiocination from the knowledge we have first of their causes or generation: And again, of such causes or generations as may be from knowing first their effects" (Hobbes, *Elementorum philosophiae sectio prima de corpore*, part 1, chaps. 1, 3; *English Works*, 1, 3.)[9] For Hobbes, "true ratiocination" is purely deductive. Thus, Hobbes's preference, expressed here, is for demonstrable knowledge.

If this were all that Hobbes ever said on method, then Shapin and Schaffer's claim that Hobbes was unwavering in his allegiance to a priori demonstration would be correct. But, unfortunately for Shapin and Schaffer, Hobbes's views were not fixed. In remarks published a year later, Hobbes modified his earlier unqualified view, denying that science is, or could be, based on necessarily, or demonstrably, true claims.

> Of arts some are demonstrable, others indemonstrable; and demonstrable are those constructions of the subject whereof is in the power of the artist himself, who, in his demonstration, does no more but deduce the consequences of his own operation. The reason whereof is this, that the science of every subject is derived from a precognition of the causes, generation, and construction of the same; and consequently where the causes are known, there is place for demonstration, but not where the causes are to seek for. Geometry therefore is demonstrable, for the lines and figures from which we reason are drawn and described by ourselves; and civil philosophy is demonstrable, because we make the commonwealth ourselves. But because of natural bodies we know not the construction, but seek it from the effects, there lies no demonstration of what the causes be we seek for, but only what they may be. (*Six Lessons*, 1656, Epistle Dedicatory, *EW* 7, 183–84)

Similar passages can be found in other of Hobbes's major works. For example, he writes of "sufficiently demonstrated" true causes as "the end of physical contemplation" (*EW* 1, 531). In another place, he says that although the natural philosopher "prove not that the thing was thus produced, yet he proves that thus it may be produced when the materials and the power of moving are in our hands: which is as useful as if the causes themselves were known" (*EW* 7, 4).

Despite clear evidence of the subtlety of Hobbes's views, Shapin and Schaffer insist that Hobbes argued doggedly that nothing short of demonstrable truth was properly deemed to be knowledge in any domain of inquiry, science included. In contrast to Shapin and Schaffer's claim that Hobbes desired to unite human inquiry under a single deductive model (see their chap. 3), there is good reason to believe that Hobbes allowed for an explanatory dualism: "Of arts some arts are demonstrable, others indemonstrable."

Thus, whereas Hobbes (and others) may have preferred an epistemology of science built ideally on deductively closed demonstration, in fact he probed for a new way to address epistemological ("knowledge") questions in natural philosophy. The new way reflected an awareness that scientific claims should be measured against a probabilistic scale, or, to repeat Hobbes's own words, "there lies no demonstration of what the causes be we seek for, but only of what they *may* be" (*EW* 7, 184; italics added).

History—Boyle

Let us now compare Boyle's views. According to Shapin and Schaffer, Boyle was always an experimentalist. They assert that unlike Hobbes, Boyle did not countenance the virtues of a priori demonstration. They are wrong. Consider, for example, a comparatively early statement of methodological principle, in which Boyle addresses the question of proving hypotheses.

> And on this occasion let me observe, that it is not always necessary, though it be always desirable, that he, that propounds an hypothesis in astronomy, chemistry, anatomy, or other parts of physics be able a priori to prove his hypothesis to be true, or demonstratively to show, that the other hypotheses proposed about the same subject must be false.[10]

It is clear that Boyle's preference here is for epistemic foundations in natural philosophy that are with the "always desirable" a priori demonstration. Boyle's views as stated here are in marked agreement with Hobbes when Hobbes's adherence is to "true ratiocination." Also note the striking agreement between Boyle and Hobbes that in natural philosophy, demonstration is not the correct epistemic standard. For Boyle, demonstration "is not always necessary." For Hobbes, in the case of "natural bodies," "there lies no demonstration."

In fact, Boyle's mature scientific writings continue to contemplate, as an integral epistemic feature, a role for a priori demonstration. In the next quoted passage from Boyle's discussion of "the Mechanical Hypothesis," we detect tokens of Boyle's deductivism. In this lengthy, but important, quotation, he argues that the mechanical hypothesis deserves preference over rivals to it. Boyle's reasons to favor a particular hypothesis are not because that hypothesis proves that "other hypotheses proposed about the same subject must be false, but because it has the methodologically desirable ability to unite sects." Rivals, by contrast, would "multiply them" (*Works* 4, 7).

> And now at length I come to consider that, which I observe the most to alienate other sects from the mechanical philosophy; namely, that they think it pretends to have principles so universal and so mathematical, that no other physical hypothesis can comport with it, or be tolerated by it. . . . But this I look upon as an easy, indeed, but an important mistake; because by this very thing, that the mechanical principles are so universal, and therefore applicable to so many things, they are rather fitted to include, than necessitate to exclude, any other hypothesis, that is founded in nature, as far as it is so. And such hypotheses, if prudently considered by a skillful and moderate person, who is rather disposed to unite sects than multiply them, will be found, as far they have truth in them, to be either legitimately (though perhaps not immediately) deducible from the mechanical principles, or fairly reconcilable to them. (*Works* 4, 7)

It is clear that Boyle considers the three mechanical hypotheses to be the axioms of a deductive system of such commanding universality that all other (true) systems—the chemists' system, for example—can be deduced from them. The preceding passage is typical of Boyle's view of the deductive virtues of the mechanical hypothesis. In the next passage, Boyle describes the methodological grounds for preferring the three mechanical hypotheses. The basis for preference is something other than deductive proof, but deductive means are still featured in Boyle's argument for the explanatory richness of the mechanical hypotheses.

> I know it may be objected, in favour of the chemists, that as their hypostatical principles, salt, sulphur, and mercury, are but three, so the corpuscularian principles are but very few; and the chief of them bulk, size, and motion, are three neither, so that it appears not, why the chemical principles should be more barren than the mechanical. [The three principles apply to] the almost infinitely diversifiable contextures of the small parts [as opposed to the chemists' principles, which lie at the level of objects observed with the unaided senses], and the thence resulting structures of particular bodies, and fabrick of the world: besides they say, each of the three mechanical principles, specified in the objection, though but one in name, is equivalent to many in effect; . . . so that it need be no wonder that I should make the mechanical principles so much more fertile, that is, application the production and explication of a far greater number of phenomena, than the chemical. (*Works* 4, 281)

Here Boyle holds that his mechanical hypotheses are richer, more "fertile" in their comparative ability to explain a far greater number of phenomena. Again, what Boyle understands is obviously a system that may use deduction to explain observable phenomena.

So far I have tried to show that Hobbes's and Boyle's epistemology and methodology of the "new learning" were not systematically distinct to any degree that would support Shapin and Schaffer's "radically different" thesis.

Experiments

Aside from the alleged radical differences at the level of methodology, another means that Shapin and Schaffer use to differentiate Hobbes and Boyle is their idea that Hobbes rejected experimental trials as being capable, under any circumstances, of grounding scientific knowledge:

> He [Hobbes] asserted the inherently defeasible character of experimental systems and therefore of the knowledge experimental practices produced. Hobbes noted that all experiments carry with them a set of theoretical assumptions embedded in the actual construction and functioning of the apparatus and that, both in principle and in practice, those assumptions could always be challenged. (1985, 111–12)

In a footnote to this passage, Shapin and Schaffer remark, "The resonance with the 'Duhem–Quine' thesis is intentional. We shall see that Hobbes's particular objections to Boyle's experimental systems provide a concrete exemplar of this 'modern' thesis concerning the impossibility of crucial experiments" (1985, 112). In the context of 1660s physics, underdetermination doesn't seem to have a natural role to play, and this attempt to raise the specter that the deductive underdetermination of belief by evidence is thought to cast on philosophy of science is further evidence of the lengths to which the authors go to torture history into supporting their Strong Programme thesis.

Consistent with their claim that Hobbes was ever the a priori deductivist, Shapin and Schaffer conclude that the inductive, or "inherently defeasible," nature of experimental conclusions was, for Hobbes, a critical and an insurmountable failing of experimental methods. According to Hobbes, they write, "Philosophy [including natural philosophy] was a causal enterprise and, as such, secured a total and irrevocable assent" (1985, 19). This is wrong. The supposed dichotomy between the rationalism of Hobbes and the experimentalism of Boyle is, to a large extent, an artifact of Shapin and Schaffer's highly selective filtration of the historical evidence. Hobbes's texts contradict the contention that he saw no role for experimental method in the generation of scientific knowledge. Let me support this criticism.

It is difficult to credit the claim that Hobbes rejected experiments wholesale when he wrote extensive passages concerning how experiments ought to be performed and what experiments should provide to the natural philosopher. For example, Hobbes said that in examining hypotheses, "You must furnish yourself with as many experiments . . . as you can" (*EW* 7, 88).[11] Hobbes even details what "trials" allow natural philosophers to conclude concerning the "probable" nature of hypotheses:

> A proposition is said to be supposed, when, being not evident, it is nevertheless admitted for a time, to the end, that, joining it to other propositions, we may conclude something; and to proceed from conclusion to conclusion, for a trial whither the same will lead us into any absurd or impossible conclusion, which if it do, then we know such supposition to have been false. But if running through many conclusions, we come to none that are absurd, then we think the proposition probable. (*EW* 4, 29)

In his *Decameron physiologicum* (1678), Hobbes goes so far as to engage two interlocutors over the matter of experimental technique. In the following exchange, Hobbes's concern for the epistemic power of experimental findings is clear. Yet, contra Shapin and Schaffer, there is no wholesale rejection of experimentalism; rather, Hobbes holds with "mean and common" experiments, sufficient to allow a role for empirical inquiry in his epistemology of science.

> A. You see how despicable experiments [are] I trouble you with. But I hope you will pardon me.

> B. As for mean and common experiments, I think them a great deal better witnesses of nature, than those that are forced by fire, and known but to a very few. (*EW* 116–17)

Hobbes's expressed preference for the "mean and common" experiment figures in Shapin and Schaffer's long discussion of Hobbes's dislike of the air-pump trials. Again, however, we find that Shapin and Schaffer are at once selective in their use of texts and Procrustean in getting the evidence to fit their aims.

Although it is true that Hobbes scoffed at the air pump's integrity and made fun of how the seals, in fact, failed to "keep out straw and feathers, but not to keep out air," it was not just Hobbes who fretted about deficiencies of design. Both Hobbes and Boyle were deeply concerned over the implications of the less-than-ideal experimental conditions presented by the air-pump trials. Boyle worried just as long and hard, evidenced by the evolving design of the air pump. He even worried specifically—just as Hobbes did—that the air pump did not evacuate the cylinder completely, writing that "we cannot perfectly pump out the air" (Hall 1965, 332). As the number of demonstration trials increased, Boyle wrote:

> Now although we deny not, but that in our table some particulars do not exactly answer to what our formerly mentioned hypothesis might perchance invite the reader to expect; yet the variations are not so considerable, but that they may probably enough be ascribed to some such want of exactness as in such nice experiments is scarce avoidable. (Hall 1965, 341)

Shapin and Schaffer's telling of the story ignores Boyle's expressed and published worries. This gives the reader the impression that the pumps were a serious embarrassment to Boyle and one that he either did not confront or hid behind his Royal Society trappings.

By ignoring Boyle's concerns and serving up a patently unbalanced account of the experimental trials done with the air pump, Shapin and Schaffer give the mistaken impression that Hobbes deserves credit for bravely raising—where Boyle (and others) ignored—questions concerning the methodological integrity of experimental trials. With Boyle's concerns written out of the story, it is easy to conclude—wrongly— that Hobbes was a lone critical voice, raised against a society of experimentalists who were consciously blind to the shortcomings of their method and who shrouded these shortcomings with new and elitist rhetoric and writing practices.

One last problem. Do Shapin and Schaffer adequately and fairly represent Hobbes's own views on experiments? They tell us that Hobbes is dissatisfied with experimentalism because air pumps leaked.[12] But they do not tell us what we have already seen, namely, that Hobbes's dissatisfaction is not based on a belief that the epistemics of experimentalism is inherently flawed. Rather, we have seen that despite the imperfections, Hobbes still saw a methodological role for experimental trials in gaining scientific knowledge. Shapin and Schaffer missed this, and one wonders why. After all, even the work that anchors Shapin and Schaffer's case study, *Leviathan* (*EW* 3), shows that Hobbes allowed an epistemological role for experimental trials in natural philosophy. Early arguments in Hobbes's *Leviathan* take the categorical stance that true science is deductive and its conclusions are conclusive (part 1, chap. 5; *EW* 3, 35–37; cf. part 1, chaps. 7 and 9; *EW* 3, 51 ff, 71 ff). But later Hobbes writes that scientific reasoning results in knowledge that is probable in nature (part 4, chap. 46; *EW* 3, 664). Shapin and Schaffer never mention these latter passages, however.[13]

New Thinking

What Shapin and Schaffer should have argued is that Hobbes and Boyle were agents during a period of cognitive transition. The old view of knowledge was yielding (at least for some) to the emerging cognitive attitude that defined new epistemological aims for natural philosophy: the old standard required deductive proof, whereas the newer one would be based on experimental probabilities and reasonable belief.[14]

Both Hobbes and Boyle vacillated between the older standard and the newer one. It is by no means clear from the historical record that they were polarized on method in the extreme manner ("radically different") that the discussion and arguments in

Shapin and Schaffer's study portray. With good reason, Hobbes wrote: "Natural Philosophy is therefore but young" (*EW*, Epistle Dedicatory, viii–ix).

Conclusion: Blinkers

My conclusion is that Shapin and Schaffer misrepresent history in their case study of the "Hobbes–Boyle dispute." Their study misrepresents the actual work of Hobbes and Boyle and thereby sustains certain myths such as that this was a major debate of the period; Hobbes's and Boyle's concerns were typical for the period; and Hobbes and Boyle diverged significantly, always, and in every point of methodology.

It might be countered that my "myths" fall directly into Shapin and Schaffer's hands, for one of their aims is to rewrite the history of philosophy of science so as to properly center Hobbes's thought. After all, they not only hold that Hobbes has been undeservedly written out of the history of philosophy of science but also conclude their case study with the puffery: "Hobbes was right."

My response is that I have pointed to central texts from both men showing that neither's views on methodology were fixed and unmoved in the face of the flourishing experimental practices. Furthermore, textual study shows that Hobbes's and Boyle's views of methodology were not always opposed. Killer quotations should not settle the matter, but the quotations I provide to argue against the success of this case study are not obscure and are not isolated; they sit in the body of easily available, widely regarded, published works.

In short: no new philosophy (or, sociology) of science comes out of bad history.

Notes

This paper was read at the first international meeting of the History of Philosophy of Science (HOPOS) group at Roanoke, Virginia, 1996. I acknowledge the critical comments received there and from J. Garrett, J. Maffie, and W. Schmaus. D. Jesseph deserves special thanks for a sharp eye on historical detail.

1. Critics know, too, that Barry Barnes promised to provide the "heir to something better than" philosophy of science but, in the end, fell disgracefully on his own critical sword, underdetermination (see Barnes 1982, 113).

2. Roth 1987 and Roth and Barrett 1990 are notable exceptions to the attitude of inattention.

3. Perhaps even more important, the sandwich theory masks a lack of historical preparation by philosophers of science. If philosophers were true to the naturalism that most of us espouse in our philosophy of science, we ourselves would be delving into the historical underpinnings of these case studies. Attention to the relevant history of science may force us to reject certain case studies of scientific subjects—and, consequently, to close any last avenues of support for the sociology that creates these case studies.

4. The underdetermination thesis is tricky to use in any of the social constructivist or postmodern efforts to be the science of science (see Gale and Pinnick 1997; Pinnick 1994).

5. Sargent (forthcoming) thinks Shapin escapes these criticisms by arguing in more recent works that evidence, rather than theory choice, is determined by social interests. But there is little new here by Shapin; this idea appeared early in Barnes (1982), who borrowed from Hesse

(1974). The underdetermination thesis thus remains a two-edged sword, even with the shift to evidentiary formulation.

6. Detailed analysis of the equicredibility issue must be set aside for another paper, as it runs to the more philosophical (rather than historical) problems endemic to the Shapin and Schaffer case study. I want to focus in this chapter as narrowly as possible on the historical, and not the philosophical, issues. I will simply take the existence of the equicredibility thesis as my point of departure.

7. The book as a whole speaks for itself on the matter that Shapin and Schaffer represent to their readers as a key debate of the period. In fact, this debate goes almost without mention in most historical treatments of this period, and Hobbes and Boyle are regarded as relatively minor figures. See Feingold 1996 for a more typical perspective.

8. By contrast, the Hobbes–Wallis exchanges are prominent. See Bird 1996 for a recent contribution, with all the sharp controversy that the Shapin–Schaffer study needs but lacks.

9. References for Hobbes are to part, chapter, and section of *De corpore* (English translation), and for Molesworth's edition of Hobbes's *English Works*. 11 vols. (London, 1839–45) (hereafter cited as *EW*) by volume and page number.

10. The quotations from Boyle, except where noted, are from *The Works of the Honourable Robert Boyle*, ed. Thomas Birch, 6 vols. (London, 1772) (hereafter cited as *Works*). Also in Hall 1965, 206.

11. In a remarkable about-face with Hobbes's own words, Shapin and Schaffer write: "He [Hobbes] regarded the experimental programme as otiose. It was pointless to perform systematic series of experiments, since if one could, in fact, discern causes from natural effects, then a single experiment should suffice" (1985, 111).

12. We read also that Hobbes was dissatisfied because of the putative politics implicated in the air-pump trials: "For Hobbes civil war flowed from any programme [experimentalism] which failed to ensure absolute compulsion" (Shapin and Schaffer 1995, 152).

13. Gauthier 1996, with commentary by McGuire, supports the view that Hobbes countenanced methodological dualism.

14. Shapiro 1969 and Sargent 1995 make the interesting suggestion that methods in natural philosophy are tied to emerging legal standards and the overlap of barristers and natural philosophers at this time.

References

Barnes, B. 1982. *T. S. Kuhn and Social Science*. New York: Columbia University Press.

Bird, A. 1996. Squaring the circle: Hobbes on philosophy and geometry. *Journal of the History of Ideas* 57: 217–31.

Bloor, D. 1991. *Knowledge & Social Imagery*. London, Routledge & Kegan Paul.

Boyle, R. 1772. *The Works of the Honourable Robert Boyle*. 6 vols., ed. Thomas Birch. (Reprint, Hildesheim: Georg Olms, 1965 facs. ed.)

Feingold, M. 1996. When facts matter. *Isis* 87: 131–39.

Gale, G., and C. Pinnick. 1997. Stalking theoretical physicists: an ethnography flounders. *Social Studies of Science* 27: 113–23.

Gauthier, D. 1996. Hobbes: constructive definition and demonstrative science. Paper presented to Third Quadrennial International Fellows Conference, Castiglioncello, Italy. J. E. McGuire, commentator.

Hall, M. B. 1965. *Robert Boyle on Natural Philosophy*. Bloomington: Indiana University Press.

Harding, S. 1992. Rethinking standpoint epistemology: what is "strong objectivity"? *Centennial Review* 36: 437–70.

Hesse, M. 1974. *The Structure of Scientific Inference*. London: Macmillan.

——. 1980. *Revolutions and Reconstructions in the Philosophy of Science*. Bloomington: Indiana University Press.

Hobbes, T. 1839–45. Sir William Molesworth's edition of Hobbes's *English Works*. 11 vols. London: John Bohn.

Laudan, L. 1984. The pseudo-science of science. In *Scientific Rationality: The Sociological Turn*, ed. J. R. Brown, 41–73. Dordrecht: Reidel.

——. 1990. Demystifying underdetermination. In *Minnesota Studies in the Philosophy of Science*. General ed. Ronald N. Giere. Vol. 14: *Scientific Theories*, ed. C. W. Savage, 267–97. Minneapolis: University of Minnesota Press.

Pinnick, C. L. 1994. Feminist epistemology: implications for philosophy of science. *Philosophy of Science* 61: 646–57.

Roth, P. 1987. *Meaning and Method in the Social Sciences: A Case for Methodological Pluralism*. Ithaca, N.Y.: Cornell University Press.

Roth, P., and R. Barrett. 1990. Deconstructing quarks. *Social Studies of Science* 20: 579–632.

Sargent, R.-M. 1995. *The Diffident Naturalist: Robert Boyle and the Philosophy of Experiment*. Chicago: University of Chicago Press.

——. Forthcoming. The social construction of scientific evidence.

Shapin, S., and S. Schaffer. 1985. *Leviathan and the Air Pump: Hobbes, Boyle and the Experimental Life*. Princeton, N.J.: Princeton University Press.

Shapiro, B. J. 1969. Law and science in seventeenth-century England. *Stanford Law Review* 21: 727–66.

Slezak, P. 1991. Bloor's bluff: behaviourism and the strong programmeme. *International Studies in the Philosophy of Science* 5: 241–56.

15

Reflections on Bruno Latour's Version of the Seventeenth Century

Margaret C. Jacob

Bruno Latour, French anthropologist and would-be philosopher of modernity, has gone back to the history of seventeenth-century England to vindicate his attack on the way "moderns" have constructed "nature." He is particularly vexed by the way modern Westerners have pretended that nature is transcendent and hence that human beings are merely its interpreters. Taking up the narrative offered by Steven Shapin and Simon Schaffer in *Leviathan and the Air Pump* (1985), Latour enlists for his postmodernist enterprise the mid-seventeenth-century confrontation between Boyle and Hobbes. He winds up wearing the chic "postmodernist" label because he embraces deconstruction not simply of texts but also of human subjects: "in France . . . *we take the deconstruction of the subject for granted.*" What people thought, believed, and actually did in the past is irrelevant to the outcome of their actions, then or now.[1] But when you dismiss human agency, as he does, it is possible to misconstrue, to see relations and connections where none may exist. The search for truth, or less falsity, about the past also becomes more difficult.

From the posture of epistemological relativism, the method of his "new anthropology," Latour discovers that fatefully at the moment of modern science's birth in the 1660s, Robert Boyle entered "the closed and protected space of the laboratory." Within it, he did not encounter transcendent nature; rather, he translated "discussions dealing with the Body Politic, God and His miracles, matter and its power" into an account of the vacuum and the air pump. Assent to Boyle's experiments, his artificial construction of "nature," amounted to assent to his elitist politics.[2] Thus the self-deluding perfidy of the "moderns," the duplicitous use of science as the ideological foundation for the modern state, begins at the Restoration with the reestablishment of the church and monarchy in seventeenth-century England. Among those contemporary with the new scientific ideology, only the philosopher Thomas Hobbes, supreme political theorist that he was, saw through it. According to Latour, Hobbes was a relativist, but an honest one. Seeing his contemporaries, among them Boyle and the Royal Society, for what they were, Hobbes observed openly that what counts is never truth, but power; ideas and practices arise because in the state, in Hobbes's "monist macro-structure . . . knowledge has a place only in support of the social order."[3] In *We Have Never Been Modern* (1993), Latour tells us that the modern enterprise of science started to go wrong right at its inception, and historians—with the exception of Shapin and Schaffer—have missed the nature of this profound turning point.

Along with Shapin and Schaffer, Latour believes that Hobbes basically got it right. Sharing the same political goals of order and stability as Boyle did, Hobbes had the decency—when he offered his version of science—not to enlist new natural technologies. No expensive leaking pumps or select gatherings for the more egalitarian Hobbes. Instead, Hobbes stuck with a mathematical or geometrical model of doing science and chose instead to concentrate his energies on society, on guaranteeing that human beings alone, as Latour reads his philosophy, "are the ones who construct society and freely determine their own destiny."[4] Boyle's more daring bid for power permitted him and the Royal Society to invent modern experimental science and then to offer it as a prop for the postrevolutionary, restored state.

These were raw power plays on both sides. Latour maintains that Hobbes and Boyle "agree on almost everything. They want a king, a Parliament, a docile and unified Church."[5] What is at stake in their competing visions of how to do science is actually who will get to do it. Latour characterizes Hobbes's Leviathan as a kind of benign go-between, a "Sovereign [who is] a simple intermediary and spokesperson [*sic*]." Under his gaze, everything is possible, and no one need be shut out from the essentially mathematical game that Hobbes's science would support. Mathematical exercises—not collecting, gathering, or experimenting—constituted the stuff of scientific inquiry—all such pacific activity to be praised by an unthreatened court and its bureaucracy.

In defiance of the historical record, Latour denies the repressive force, the stature above the law, that Hobbes awarded to his political creation.[6] Latour imagines that Hobbes's theory of sovereignty would have permitted a sharing of power between king and Parliament. In effect, it could have offered a foundation for the Revolution settlement of 1689 that Boyle lived long enough to endorse. Latour further maintains that Hobbes's desire to reconfigure the church into a branch of the state amounted to the same thing as Boyle's desire for an Anglican church established by law. Similarly, Latour sees no difference between Hobbes's and Boyle's mechanical philosophy. He ignores the determinism at the heart of Hobbes's materialism and seems unaware of the emphasis on immaterial and spiritual power embedded in Boyle's corpuscularianism. This distortion of past texts arises from the desire to obliterate the profound differences between Boyle and Hobbes on matters of state power and church privileges. With this deconstruction of difference accomplished, it is possible for Latour to render philosophical positions and scientific practices symmetrical, posturing undertaken in the search for ideological advantage. These relativist interpretations of the English past are offered to condemn modernity, its institutions, and its practices.

The involvement of history in the construction of any present-day theoretical position inevitably alerts historians to look at what is being concocted. As a late-twentieth-century American heir of the early modern democratic revolutions, I look with particular attention at any political reading of science that is based on its seventeenth-century origins. The English Revolution of midcentury contributed decisively to the formation of both representative institutions and modern science. The post-1660 compromises that resolved the revolutionary crisis institutionalized Western science's most fundamental characteristics: experimental, socially configured by published exchanges and peer evaluation, dependent on technological innovations, and, in the Baconian tradition, valued for its utilitarian applications particularly in the service of the state.[7] Latour, Shapin and Schaffer, myself, and many other historians since the 1930s agree that events in England from 1640 to 1689 had a major impact on shaping scientific culture in the West. Where we disagree concerns what actually happened.

Boyle's science, which was bequeathed to and vastly elaborated on, by Newton and the Newtonians, reflected aspects of the revolutionary context in which it was first articulated. The debt was expressed in antimaterialist philosophical commitments, as well as in the alignment of the new science with Anglican religious positions. The Protestant understanding of nature, work, and society—as well as the antiabsolutist political stance intended to support the established church—accompanied and reinforced the Boylean experimental legacy. All were integrated into the Newtonian synthesis, into both a clerically constructed physicotheology and the curricula for practical mechanical lectures and demonstrations. To this point, specialists on seventeenth- and eighteenth-century science from Robert Merton in the 1930s to Larry Stewart in the 1980s tend to agree. Only Latour and his allies interpret those centuries and what they may mean for the present in ways radically at variance with the historical evidence.

We need to be sure that all parties to this dispute understand the stakes raised by the Latourian intervention: the gap between the forms of science being advocated, as well as the role that each would serve in relation to church and state. Not least, if we are committed to understanding the origins of democratic institutions and values in the West, we have a special obligation to the historical moment so deeply implicated in the evolution of Anglo-American representative institutions. Largely ignored in Latour's discussion, the long struggle for social and political democracy begins in seventeenth-century England and continues to this day; it can only be assisted by historical understanding.[8] Oblivious of that struggle, Latour relates most dramatically his response to seventeenth-century events as he understands them, by rejecting both the "science" and "nature" he asserts to have been their only interesting legacy.

With Latour's rejection, much is lost. First and foremost is the realization that both the English Revolution and, less obviously, the science prescribed by first the Boylean and then the Newtonian syntheses helped in the evolution of civil society. Placed in the service of Anglicanism, the progress of experimental science hung on the defeat of Stuart absolutism and Catholicism. Put more dynamically, experimental practices invented a new social space that was more than simply, as Latour would have it, a place for the passive "observation of a phenomenon produced artificially in the closed and protected space of a laboratory."[9] The creation of civil society—the zone of relatively free exchange that lies both between and outside the state and the domestic sphere—owes a debt to science. Experimental science requires voluntary associations and practices intended for verification by an independent audience, however gentlemanly or oligarchic its original composition. As both readers and witnesses, the original seventeenth-century audience for science required a new quasi-public space. The interaction of the experimenters with their spectators and consumers could be neither coerced nor predicted; it presumed a *relatively* free market for new rhetorical and experimental technologies.

The machines and devises essential to experimental practice induced not passivity but imitation and application. Access to experimental practices bypassed the inability to comprehend mathematical language. The turn toward applied mechanics visible even during Boyle's lifetime (d. 1691) was vastly expanded by post-*Principia* (1687) explicators and experimenters. The character of eighteenth-century English science thus differed markedly from the more abstract and mathematical forms of inquiry commonplace, for example, in prerevolutionary and absolutist France.[10] The move toward the experimental had consequences well into the early nineteenth century. In the eighteenth century, a culture of applied science arose in Britain and fostered industrial development. If Laplace is symbolically the key figure in French science at 1800, then across the Channel the honor belongs to James Watt, a Newtonian mechanist, experimenter in both mechanics and chemistry, and inventor of the steam engine.[11]

Leviathan Tamed and the Church Secured

Now I wish to shine onto Latour's eccentric readings of seventeenth-century history a differently informed light, one reflecting the Anglo-American democratic and republican tradition that actually grew out of the seventeenth-century English Revolution against the king and established church. In the process I will illustrate the methodological and normative inadequacies of relativism.[12] With history, theory, and prescriptive ideologies on the agenda, simplification will be inevitable. For its sake I will turn to one of the now-famous moments of the seventeenth century when the scientific and ideological legacy of Boyle was first publicly articulated.

On November 7, 1692, the first Boyle lecture, given in the form of a church sermon by the Newtonian and Anglican clergyman Richard Bentley, articulated the relevance of experimental science to the political and religious framework created by the English Revolution. In effect, after Boyle's death, his closest friends and followers brought the full weight of experimental science from Boyle to Newton into open service on behalf of the Church of England. An acquaintance of Isaac Newton, Richard Bentley mounted the pulpit of St. Martin-in-the-Fields in London and proceeded in good clerical fashion to confute atheism. But his physicotheological arguments were distinctive. They rested the foundation of all religion directly on corpuscularianism and Newton's *Principia*. Both proved God's existence from the origin and frame of the world.[13]

Like all marriages of convenience between distant relations, the union of Newtonian science with Anglican interests was a brokered alliance, one made between liberal churchmen and natural philosophers through the agency of a postrevolutionary state. The passion of both sides lay in finding an accommodation with the state, the mortal god—to paraphrase their bête noire Hobbes—who permitted the new liaison of churchmen and scientists. The ardor of churchmen and scientists alike—with Isaac Newton among the most fervent—was directed at the 1689 Revolution Settlement. It had clearly subordinated church to state but had also permitted the church to retain its independence. Dubbed Erastian—a term derived from church–state conflicts in sixteenth-century Germany—the settlement was embodied in a series of laws passed by Parliament and approved by William and Mary. It guaranteed an established Protestant church, but its independent survival depended on a clear subordination of church to state and the end of most clerically controlled church courts. The preservation of an established Protestant church had also required legal toleration of the chapels, schools, and conventicles of non-Anglicans. In 1695, Parliament went further and allowed the act to lapse that had required the prepublication censorship of books and pamphlets. As a consequence of the Revolution of 1688–89, the Church of England survived the threat of Stuart absolutism and the Catholicism of James II. Although weakened, it retained a degree of freedom that would have been denied to it by theorists of political absolutism like Hobbes.

Banished to the sidelines of the post-1689 polity were any essentially democratizing threats. Derived from the midcentury revolution, these could be secular and republican, and hence allied with the lesser gentry and middling sort, or religious and enthusiast, thus the province of the divinely illuminated and possessed found most commonly among the least literate sectaries and their clergy. The new natural religion to which the marriage of Boyleian, Newtonian science, and Protestantism gave birth was procreated by churchmen like Bentley. It was intended to increase reverence for the Creator, the fount of all order and stability. Here, somewhat irreverently, I shall examine the cultural construction called natural religion, just as Hobbes did, as simply an aspect of "humane Politiques."[14]

Although presented as the true, albeit science-inspired religiosity, natural religion also explained to the narrowly defined political nation what the freedoms permitted and obligations required were in the new postrevolutionary state. Part of the requirements included a commitment to an independent church established by law but now stripped of much of its former power. The Anglican Church's diminished authority seemed a small price when the alternative—narrowly averted and still feared—had been the return of Catholicism imposed by a Stuart king intent on absolutism. As Newton confided in his private writings just before the Revolution of 1688, if James II got his way "abrogating ye penal laws and tests we shall absolve ye king from his promise of maintaining ye Church of England according to law, and so have neither law nor promise left for ye support of our religion."[15] Did not French Protestants at that very moment rot in the Bastille, thanks to the theocratic French clergy, who in 1685 had pricked their prince into revoking the Edict of Nantes? Did not the Low Countries fear for their survival as something other than pawns of Louis XIV and his vast army?[16] At the time when Bentley spoke from his pulpit, Bishop Bossuet in France explained that arguments drawn from nature would diminish the grandeur of the Sun King.[17] In eighteenth-century Europe, Newtonian physicotheology became a distinctively Protestant form of religiosity, irrelevant to the Catholic apologists of monarchical absolutism.

For Anglicans of the late seventeenth century, like Boyle and Newton, who had supported the defeat of the English version of absolutism, experimental and Newtonian science fused with Christianity to create an allegory that justified and explained the church's acceptance of its subordination to the needs and interests of the state and to the preferences and freedoms of the county and urban gentry who elected Parliament. They had strong predilections for churchmen like Bentley preaching in the shires on the virtues of order and obedience. Science-inspired churchmen readily agreed. They preached parables that justified and explained hierarchy, order, and authority, both lay and ecclesiastical. Boylean and Newtonian science proclaimed a new and grand natural narrative based on the very structure of the universe, derived from mechanical experimentation and mathematical calculation. The narrative offered a scientific allegory explicating God's active role in nature. Guided by Newton's notes on how to read his *Principia* and speaking from the pulpit both inspired and endowed by Boyle's will, Bentley told the tale intended for the polity and located order, harmony, and design in the very fabric of the universe. Once understood, Newtonian science made order and stability in any realm seem both natural and inevitable. In the course of the eighteenth century, Newtonianism offered a middle way between theocracy and absolutism, on one hand, and the entirely secular, deistic, even atheistic, Enlightenment, on the other.

In laying the metaphysical foundations of the Whig constitution, the ideological union of science with religion from Boyle to Newton also offered an entirely new office to clergymen for whom theocracy was no longer an option. Now they explicated God's purpose in the world through his work. The account offered by the biblical word could be quietly retired. For too long it had legitimated the claims of presbyters and priests to possess "the Supreme Power Ecclesiastical in the Commonwealth"[18]—as Hobbes, the supreme Erastian and the greatest detractor of either Catholic or Calvinist theocracy, put it. From the moment Bentley mounted his London pulpit until the day Darwin published the *Origin of the Species* (1859), English Protestant clergymen of Erastian persuasion—and their continental imitators—could enjoy their local authority, accept their subordination to the needs of the state, and believe that science, properly understood, made a civil and ecclesiastical edifice that was an entirely human construction appear to be imitative of nature and hence providentially instituted. The marriage of "science and religion" grew out of an oligarchic revolution in England that more passionate radicals and republicans throughout the eighteenth century believed had not gone far enough.

Such were the accommodations that English-speaking Protestants, believers in the priesthood of all, made to avoid the final historicism: the admission that having destroyed the divine right of kings and philosophized away the transubstantiated Host, they had also created an entirely new form of civil and ecclesiastical government, one subject to the possibility of constant reformation. As time would tell, not even the order and stability built into the fabric of the Newtonian heavens would save this man-made political system from periodic waves of reform. Because it was an entirely human construction without divine right, representative government would be forever opened to challenge by lesser men—Dissenters, entrepreneurs, industrializers, and eventually even women and workers. The court would constantly have to defend itself against the country. The king in Parliament would be forever subject to pressures from below—of that liberal churchmen who remembered Cromwell or, worse, the Levelers had no doubt.

Bentley and the first generation of Newtonians could not foresee, but they still could fear, what by the 1830s occurred: the Whig constitution would succumb to democratizing tendencies. The accommodation made by the church in 1689 had permitted Dissent to flourish. Between 1776 and 1832, the religious Dissenters and the secular republican and radicals effected other, far more democratic revolutions. But the

American and democratic revolutions of the late eighteenth century, coupled with the parliamentary revolution of 1832, only began a process of democratization that workers, women, blacks, the impoverished, and the utopian have sought ever since to broaden and deepen.

The mid-seventeenth-century parliamentary Revolution laid bare the radical tendencies, dangers, and instabilities for elites that could arise from religious enthusiasm and democratic republicanism. Not surprisingly, both were readily identified as atheism by churchmen charged with protecting the state. In his famous Boyle lecture, Bentley railed against atheism by lashing out against all philosophies that threatened stability by denying God's providential order. He feared the power of lesser men— who, unashamed by the regicide they had once committed and, if given the chance at power again, would reduce the king to a duke of Venice or institute the millennium of the godly. Thus to consecrate the alliance of science with Anglican religiosity, Bentley inveighed against determinism: "The ordinary cant of illiterate and puny atheists that would the world [to be] produced by Fortune, by a fortuitous or casual concourse of atoms."

Sounding at one moment like a relativist who would escape either science or history by detaching words from things, Bentley denounced fortune's being seen as "a real entity . . . [a] physical thing," when it was "but a mere relative signification." Fortune is merely another name for ignorance about causes, and atheists like Hobbes make Fortune "synonymous . . . with Nature and Necessity." To dispense once and for all with these relative signs of which he wanted no part, Bentley proceeded to "give . . . a brief account of some of the most principal and systematical Phaenomena that occur in the World."[19] The message behind the allegory of systematic and harmonious nature concerned the state. The Newtonian system of the world symbolized the order and stability, preordained by God and desired for society and government. Necessity, or fortune, could be replaced by a narrowly circumscribed freedom as guaranteed by the Revolution settlement. Its complete exercise and enjoyment were regarded by churchmen like Bentley as being fit only for Anglican gentlemen.

To thwart revolutionary radicalism, the Newtonians gave religious meaning to Newton's matter theory. Incautiously as the egalitarian ideologies of modern political life would show, Bentley, a child of the first modern revolution, took up Epicurean atomism, as demonstrated by "the excellent and divine theorist Mr. Isaac Newton." Standing in a pulpit officiating at the marriage of the established church to Newtonian science, he endorsed the equality of the atoms, the belief that "all kinds of bodies, though of the most different forms and textures, doth always contain an equal quantity of solid mass or corporeal substance."[20] The stuff of nature had been granted equality by the new mechanical philosophy, the conceptual core of the scientific revolution from Galileo through Boyle and Newton.

With one key element in place in an essentially materialist metaphysics, Bentley thought that he could avoid the democratic implications of mechanical science by recourse to the gambit of Newtonian space. Experimentally configured by Boyle's vacuum, then defined as real by Newton, this space made ontologically real the void found in "the whole concave of the Firmament."[21] At some length, Bentley explained to his long-suffering congregation the specific weights of gold, water, and air, the weight of air at 4,000 miles from the earth, the size of the orbit of Saturn, the weight of the sun, and the diameter of the earth's orbit around the sun, ending up with a calculation of all the empty spaces in the universe—"6,860 million million million." The point was to show that the spaces vastly outnumbered all the matter in the firmation. With each atom of matter, however similar, "many million times . . . distant from any other," Bentley confidently concluded that neither chance nor fortune could ever have permitted all those atoms to unite and thereby to have brought the universe into existence. "Nothing could stick this all together except by mutual gravitation."

In this physicotheology, the force of universal gravitation acquired an actual, immaterial, and ontological reality. The effect of this ontology, the metaphysics that elevated spirit over matter and separated the atoms, was to ensure agency, the force of will, divine and human. More obsessed with the divine than impressed by the human, the first Newtonians had not sought primarily to permit human freedom; (this was later the work of the Newtonian Immanuel Kant). Rather, English churchmen and natural philosophers intended to guarantee providential design and hierarchy and thus to prevent the collective action that might widen human freedom, an attribute that ironically, this ontology so rigorously supported.[22]

What Bentley did not tell his stunned, or sleeping, congregation (except perhaps by his passing reference to the cant of the illiterate) was that if randomness—or better still an orderliness born out of randomness—could be found in the work of the disparate atoms, the gambit would have failed. The intended legitimation of hierarchy and authority, because it now was based on philosophical principles legitimated by experimental science—not on theocracy and divine right—would be ideologically undermined. If nature could be seen to possess knowable movements, if left to their own devices the atoms could be orderly and lawful, hence explainable by science without recourse to clerical interpreters, then what would prevent the access to knowledge and power from widening? Such a conceptual widening of nature's potential inevitably laid the philosophical foundation for the claims of lesser folk. The late-seventeenth-century natural philosophers who officiated at the marriage of "science and religion" knew perfectly well just what the first English translators and explicators of Epicurus, John Evelyn and Edmund Waller, had observed about atomism: "But change and Atomes make *this All* / In Order Democratical, / Where Bodies freely run their course, / Without design, or Fate, or Force."[23]

Despite the best efforts of the Newtonians to prevent it, the new science also fed into materialist doctrines. Simply by locating force in matter, materialists or deists allowed for purposeful action, order, and law amid the appearance of randomness and chaos. Spirit and agency given only vaguely religious associations or divine origins could be returned to nature, as Rousseau's Savoyard vicar so eloquently preached. Hobbes's crude determinism with its deep ambiguity on the issue of human and political freedom, gave way to a metaphysic that allowed for both chance and order in nature, reform, and even revolution in a society of freewilled agents. By the end of the eighteenth century, democratic republicans like Priestley could be materialists and Unitarians, believe in the compatibility of science with their own rational and private religiosity, and have as little use for the hegemony of the established Church of England as they did for slavery. Similarly, whether in the hands of Buffon or Darwin, materialist science could gradually discard religion, its pious if aged partner. The political purpose of their marriage had been all but forgotten since the heady days when Newton's *Principia* had been used to justify the revolution against popery and tyranny.

Originally in the late seventeenth century, the natural philosophical allegory of design and harmony had given natural—and transcendent—meaning to human politics that had actually been deeply subversive of centuries of belief and tradition. In the 1690s and beyond, at the moment when discernibly modern forms of governance emerged in western Europe, Nature as described by experimental science legitimated representative government by contract. Out of the need to stabilize the first modern revolution by retaining oligarchy and requiring the king to rule with Parliament, churchmen and scientists conspired to tell a tale that repudiated arbitrary rule just as forcefully as it opposed the capacity of random atoms to impose order or regulate themselves. Rather than seeking to obfuscate the natural world, as Latour would have it, English natural philosophers sought to expose its laws in order to justify stability, order, and piety in the polity. In their deeply Protestant understanding of the world, only God could be transcendent.

The Comtean Revision

To understand how Latour could so badly misread the intentions of seventeenth-century English agents requires a digression into French intellectual history. By the nineteenth century, European intellectuals were mightily impressed by the ideological service that could be extracted from experimental science. If the edifice of science-inspired religiosity could be replaced by religiously or spiritually satisfying science, a new and potent ideology would result. No longer directed against the divine right of kings but necessary for imposing order in the face of demands for popular sovereignty, science would provide a new architecture and new architects of order and stability. In that insight lay the origins of Comtean positivism.

The political story of the uses to which the marriage of science and religion had been put prior to the French Revolution was known in the nineteenth century by none other than Auguste Comte. Only he told the story in slightly different language from that employed here and from his distinctly undemocratic, but postrevolutionary, perspective. Living in the aftermath of the French Revolution, Comte looked to create a new and transformed version of theocracy. He believed that in the "post feudal–theological" realm in which he lived, the Protestant compromise inherited from the English Revolution should be deconstructed. What he called the Protestant dogma of "liberty of conscience" had, in his view, paved the way for the dangerous doctrine of popular sovereignty.[24] Neither a romantic advocate of the medieval nor a believer in democracy, Comte thought that neither the "popular nor the monarchical doctrines" would suffice. Neither would ensure stability and order.

Comte had been inspired by the ingenious uses to which English Protestants had put science, just as he admired the power accorded to spiritual doctrines and the clergy in the old feudal–theological system. In imitation of the best of both the Protestant and the theocratic, but in an effort to destroy natural religion as overseen by a clergy subservient to the state, Comte offered the "religion of science" and a new, apolitical scientific clerisy. Being an opponent of democracy but recognizing its force, envying "spiritual power, distinct and independent of . . . temporal power,"[25] especially when coupled with a universal doctrine, Comte proposed "positive principles [as] the only basis for true freedom as well as free union."[26]

Gradually divorced from the Comtean fantasy of "moral and mental reorganization," from what Thomas Huxley labeled Comte's "Catholicism *minus* Christianity," positivism reworked—in this century by logicians in Vienna—came to dominate much of Western thinking about science.[27] Positivism evolved into an empiricist philosophy of science capable of embracing various political agendas. Until the 1960s, it could be used to secure the purity of science and its ordering and legitimating role for all knowledge-seeking when undertaken in the secular Western state. But as a result of their peculiarly hegemonic aspirations, positivists were also prone to attributing purity to both science and scientists. It was a claim to which even antireligious Western theocrats had always been especially prone.

Systematically applying the relativist antidote to positivism has proved treacherous because the political history of the relationship between science and religion in its English context before Comte has not always been understood by contemporary French theorists like Latour. As a result, the political stakes raised by relativism have remained obscure. Untutored in history, an antimodernist detractor of positivism like Latour has imagined that Boyle and his heirs were only seeking the power of science and hence that they "agree[d with Hobbes] on everything except how to carry out experiments."[28] Bruno Latour cannot even begin to imagine what was at stake for seventeenth-century men who wanted to avoid absolutism and defeat sectarians and radicals while preserving Christian and Anglican hegemony aided and abetted by their experimental science.

The Political Implications of Contemporary Relativism

It is not just theorists like Latour who can mislead us. Shapin and Schaffer returned to the original seventeenth-century marriage of convenience between science and religion and saw that in its time it had been one of many available alternatives. They tried to give Hobbes his due, to imagine that he may have had it "right." But their strategy can mislead those who know little history. Shapin and Schaffer effectively leveled the historical playing field and imagined that the Erastian and experimental science supported by Anglican churchmen and natural philosophers from Boyle to Newton, as well as the geometrically bound science of the other great Erastian of that revolutionary age, Thomas Hobbes, had been merely two different but equally viable forms of scientific and political life. The goal of the relativist analysis during the 1980s was to expose the polemical circumstances under which experimental science arose and to relate those circumstances to cold war concerns about the closed nature of contemporary science.

Arguing that in our time the scientific "form of knowledge that is the most open in principle has become the most closed in practice," Shapin and Schaffer went back to the mid-seventeenth century, to Boyle of the air pump and Hobbes of *Leviathan*, and recreated the moment when either the Erastianism of the first or the absolutism of the second might have triumphed. In the historical recreation of that moment, they used the relativism at the root of their method to great narrative effect. Yet their method, more epistemologically relativist than simply historicist and intent on recreating the historical moment, also misled them.

Reading late-twentieth-century "doubts about our science" backward, they imagined that in the seventeenth century, contemporaries could see little political or intellectual difference between the rival "games" offered by the experimentally based science of Boyle and the geometrically based science of Hobbes. Either could have been made compatible with the state that evolved after the midcentury revolution; hence Boyle's success can be understood under the general rubric that "he who has the most, and the most powerful, allies wins." Either the science of Boyle or the science of Hobbes could have ensured oligarchy; each in its way possessed tendencies neither more nor less democratic, neither more nor less tyrannical. Although based on very different approaches to knowledge seeking, the Boylean and Hobbesian prescriptions for how and by whom science should be done were presented in the Shapin–Schaffer analysis as equally possible and useful at that time and in that society.[29]

Ignoring Hobbes's absolutism in *Leviathan* and its expressed desire to reduce all independent agencies, whether clerical or judicial, into dependent servants of the state, Shapin and Schaffer left themselves open to the ahistorical, Latourian extrapolation. Not foreseeing it, or perhaps not caring about it if foreseen, they closed the narrative of their microhistory at the moment of choice, ending where they began— with relativism. Wisely given the stakes, Shapin and Schaffer stayed up on the fence they constructed and offered no historical explanation for why in the end Boyle triumphed.[30]

Bruno Latour is no political fence sitter. According to him, the seventeenth-century moment was simply "the settlement of a dispute over the authority of humans and nonhumans," the triumph of "society" over "nature." Latour assures us that in the moment when either the geometry of Hobbes or the experiment of Boyle might have won, all that was at stake was simply "a *mere* distinction between two regimes of representation," two opposing languages, two ways of signifying nature.[31] Both were of equal merit; either would have done; neither implicates us. Relativism—originally intended as the antidote to Big Science and the positivism that supported it—permits Latour to imagine that he, too, can escape history that has nothing to do with him.

Writing from France late in the twentieth century about a microhistorical narrative relating events that occurred centuries before across the Channel, Latour concludes that the experimental science he has inherited has nothing to do with the democratic political institutions currently found throughout western Europe. The seventeenth-century past is without relevance to the twentieth-century present: "We live in a Society we did not make, individually or collectively, and in a Nature which is not of our fabrication."[32] The democratic form of life that slowly evolved out of the original revolution and the experimental science that emerged concomitantly bear no relation to each other, and neither bears any relation to late-twentieth-century Western political institutions. True to the postmodernist vision, Latour cares not at all about democracy, its origins or its promises, its deficiencies, and its challenges.

Sometimes relativist interpretations can badly miss the historical point. To imagine that we are not implicated in the direction that the mid-seventeenth-century English Revolution took permits only a fanciful escape from modern Western history, a renunciation of the democratic political systems that Euro-Americans gradually constructed out of the deconstruction of theocracy and absolutism. By imagining that Boyle's Erastian, constitutional, and oligarchic vision was the equivalent of Hobbes's Erastian, absolutist prescription, relativism promotes a misinterpretation of Hobbes and the nature of his absolutism.[33] Out of it comes the notion that if Hobbes's geometric and nonexperimental way of representing nature had been embraced, it too would have fitted into the constitutional state that developed in late-seventeenth-century England.

Latour's interpretation works only by sidestepping the nature of Hobbes's absolutism. Hobbes wished all learning and knowledge systems to serve the needs of the state. He condemned explicitly "the Vain Philosophy of Aristotle" because its doctrines "serve to lessen the dependance of Subjects on the Sovereign Power of their Country."[34] Even if a philosophy is not vain but true, its teachers should be silenced by the civil authority if their teachings threaten the public quiet.[35] Lawyers also need to recognize that their skills and knowledge exist to serve the sovereign. In England, Hobbes claimed, the king makes the laws and is the sole legislator and supreme judge; the alternative led to the chaos of the civil wars. Hobbes asserted that the relationship of the state to the law is such that (just as Newton feared) only subjects must be made to obey it: "I think the King causing [the laws] to be observed is the same thing as observing them himself."[36]

As his contemporaries well knew, Hobbes also argued that the sovereign had the right to tax, dispossess, or banish any or all who would challenge his authority.[37] Nothing short of brute force would constrain the "the perpetual and restlesse desire for power after power, that ceaseth only in death." Hobbes imagined that even "if a king compel a man . . . by the terrour of his death, or other great corporal punishment, it is not idolatry" to obey him.[38] Hobbes was so horrified by ancient and modern "democraticall writers" whose ideas had fomented rebellion in both England and the Low Countries that he could not "imagine, how anything can be more prejudiciall to a Monarchy, than the allowing of such books to be publikely read."[39] The Hobbesian version of Erastianism would be to deny the clergy any independent power distinct from that of the state.[40] Similarly among the many uses to which geometry can be put, its encouragement of peace in the service of the state particularly recommends it: "The doctrine of lines, and figures, is [not perpetually disputed]; because men care not, in that subject what be truth, as a thing that crosses no mans ambition, profit, or lust."[41] Geometrical science was safe science.

In Hobbes's version of science and its role in "human politiques" rested the irony that the only theocracy allowed would consist of the king and his servants. Had Hobbes's absolutism triumphed, a far more virulent form of positivism than even Comte could have imagined would also have easily been invented. The sociologists as well as the scientists would all be duty-bound civil servants. Then and now faced with such absolute power, antidotes would have been hard to come by. Hobbes could play the relativist game with the best of them—remember in *Leviathan* his biting deconstruction of "ecclesiastiques," comparing them to "fairies" and "elves." But had Hobbes's science and politics been victorious, not even relativism would have been allowed a chance to exorcise it. Men like Latour might scoff in the privacy of their homes but never in civil society. It too would have been stillborn.

For all its oligarchic pretensions, the experimental version of science advocated by Boyle and his associates in the Royal Society ensured a very different polity, along with a more expansive science from what Hobbes wanted. Experimentalism was intended to channel the aggressions and ambitions of the great as well as the lowly. Boyle, Newton, and the next generation of interpreters—Bentley, Clarke, and their followers—were just as afraid of absolute sovereigns and their henchmen as they were afraid of republicans. The vision of science and religion that the Newtonians inherited and expanded gave birth to the physicotheology so beloved by eighteenth-century liberal Protestants in Europe and the American colonies. This middle way worked for a time to prevent the return of absolutism. For its believers, it justified resistance to any pressure for reform.

However imperfectly from our perspective, Erastian churchmen and experimental scientists grappled effectively with their historical moment because they were neither absolutists nor cynical relativists. They believed that words were connected to reality, to their own unstable present, and to a desired future, and in general they recognized that words could never simply be mere arbitrary signifiers. Witnessing the bankruptcy of relativism, the democratic heirs of the Western revolutionary tradition who wish to expand on its promise need also to reconstitute the political and religious history of science. It teaches us that the metaphysical and experimental tradition that first deployed the random atoms, like them, had never been simply of arbitrary significance.

Notes

1. Quoted from an interview with Latour by Werner Callebaut, ed., *Taking the Naturalistic Turn or How Real Philosophy of Science Is Done* (Chicago: University of Chicago Press, 1993), 442, italics in the original. In the same interview, Latour describes the Enlightenment legacy as "asymmetrical rationality."

2. See Bruno Latour, *We Have Never Been Modern*, trans. Catherine Porter (New York: Harvester, 1993), 14–21.

3. Ibid., 24.

4. Ibid., 30.

5. Ibid., 14 and 108: "If you want Hobbes and his descendants, you have to take Boyle and his as well. If you want the Leviathan, you have to have the air pump too."

6. Latour, *We Have Never Been Modern*, 143.

7. The literature on this subject is vast, and here I cite items particularly germane to this critique: James R. Jacob, *Robert Boyle and the English Revolution* (New York: Burt Franklin, 1977); James R. Jacob and Margaret C. Jacob, "The Anglican Origins of Modern Science: The Metaphysical Foundations of the Whig Constitution," *Isis* 71 (1980): 251–67; Julian Martin, *Francis Bacon, the State and the Reform of Natural Philosophy* (Cambridge: Cambridge University Press, 1992); and Larry Stewart, *The Rise of Public Science. Rhetoric, Technology, and Natural Philosophy in Newtonian Britain, 1660–1450* (Cambridge: Cambridge University Press, 1992). One could, of course, argue as Latour does in relation to the French Revolution (*We Have Never Been Modern*, 40), that there was nothing revolutionary about the events in England from 1640 to 1660 and again in 1688–89, but then, one would not be a historian.

8. Throughout this chapter, the voice used differs from "the uprooted, acculturated, Americanized, scientized, technologized Westerner" invented by Latour in *We Have Never Been Modern*, 115.

9. Ibid., 18, citing a 1990 article by Shapin.

10. On French science and its link to aristocratic culture, see Terry Shinn, "Science, Tocqueville, and the State: The Organization of Knowledge in Modern France," in *The Politics of Western Science*, ed. Margaret C. Jacob (Atlantic Highlands, N.J.: Humanities Press, 1994). On the industrial applications of English science, see Margaret C. Jacob, *Scientific Culture and the Making of the Industrial West* (New York: Oxford University Press, 1997), chaps. 5 and 7.

11. Technically, Watt invented the separate condenser; see Jacob, *Scientific Culture*, chap. 6.

12. For further discussion, see Joyce Appleby, Lynn Hunt, and Margaret Jacob, *Telling the Truth about History* (New York: Norton, 1994). For an excellent philosophical and historical discussion of the Shapin and Schaffer book, see Philip Kitcher, *The Advancement of Science: Science without Legend, Objectivity without Illusions* (New York: Oxford University Press, 1993).

13. Richard Bentley, *A Confutation of Atheism from the Origin and Frame of the World* (London: H. Mortlock, 1693).

14. The phrase belongs to Thomas Hobbes, who delineates religions as made by "two sorts of men. One sort have been they, that have nourished, and ordered them, according to their own invention. The other, have done it, by Gods commandment, and direction: but both sorts have done it, with a purpose to make those men that relyed on them, the more apt to Obedience, Lawes, Peace, Charity, and civill Society. So that the Religion of the former sort, is a part of humane Politiques; and teacheth part of the duty which earthly Kings require of their Subjects." Found in Thomas Hobbes, *Leviathan* (1651), ed. C. B. Macpherson (Harmondsworth: Penguin Books, 1968), 143. I used this edition throughout the chapter.

15. King's College, Cambridge, Keynes MS 118, in Newton's hand from the reign of James II, probably 1684.

16. On the French context, see Emmanuel Le Roy Ladurie, "After word. Glorious Revolution, shameful revocation," in Bernard Cottret, *The Huguenots in England. Immigration and Settlement*, trans. P. Stevenson and A. Stevenson (Cambridge: Cambridge University Press, 1991), 290–99.

17. As discussed by Jeffrey Merrick, "The Body Politics of French Absolutism," a paper given at the Clark Library, Los Angeles, April 3, 1993; citing Jacques-Bénigne Bossuet, *Politique tirée des propres paroles de l'Ecriture sainte*, ed. Jacques Le Brun (Geneva: Droz, 1964), 185; written in the 1670s and originally published in 1709.

18. Hobbes, *Leviathan*, part 4, 640.

19. *A Confutation*, 4–8. The definition of fortune closely resembles that given by Hobbes, *Leviathan*, 696.

20. *A Confutation*, 8–9.

21. Ibid., 13.

22. I am reminded of the uses to which Newtonian natural philosophy could be put by Peter Miller, *Defining the Common Good. Empire, Religion and Philosophy in Eighteenth-Century Britain* (Cambridge: Cambridge University Press, 1995).

23. From the 1656 poem by Edmund Waller (which came with a disclaimer by John Evelyn printed in the margin) prefixed to Evelyn's translation, *Essay on the First Book of T. Lucretius Carus, De Rerum Natura* (London, 1656), quoted in Robert H. Kargon, *Atomism in England from Hariot to Newton* (Oxford: Oxford University Press, 1966), 92, italics in original. For a further explication of this theme of the democratizing potential found in the new science, see Margaret C. Jacob, "The Materialist World of Pornography," in *The Invention of Pornography. Obscenity and the Origins of Modernity, 1500 to 1800*, ed. Lynn Hunt (New York, Zone Books, 1993), 154–202.

24. Gertrud Lenzer, ed., *Auguste Comte and Positivism. The Essential Writings* (New York:

Harper & Row, 1975), 14: "The dogma of the sovereignty of the people corresponds to the dogma just considered, of which it is only the political application. It was created as a means of combating the principle of divine right, itself the general political basis of the ancient system, shortly after the dogma of liberty of conscience had been formed to destroy the theological ideas on which this principle was founded.... Both, being devised for purposes of destruction are equally unfitted for construction."

25. Ibid., 22, from Comte's early writings.

26. Ibid., 336, from "The Social Aspect of Positivism, as Shown by Its Connection with the General Revolutionary Movement of Western Europe." For a very good summary of the impact of the Revolution on Comte, see Mary Pickering, *Auguste Comte. An Intellectual Biography*, vol. 1 (Cambridge: Cambridge University Press, 1993), 1–14 and 528, for Comte's repudiation of traditional theocracy.

27. Robert N. Proctor, *Value-Free Science? Purity and Power in Modern Knowledge* (Cambridge, Mass.: Harvard University Press, 1991), 160–70, italics in original.

28. Latour, *We Have Never Been Modern*, 25.

29. Steven Shapin and Simon Schaffer, *Leviathan and the Air Pump: Hobbes, Boyle and the Experimental Life* (Princeton, N.J.: Princeton University Press, 1985), 332–41. On Hobbes's view of the role of knowledge seeking in the state: "For Hobbes there was to be no special space in which one did natural philosophy" (333). For how winning occurs, 342, and on the nature of science late in this century, 343.

30. For some indication of why Boyle might have won, see the important argument by James R. Jacob, "The Political Economy of Science in Seventeenth-Century England," *Social Research* 59 (Fall, 1992): 505–32; reprinted in Margaret C. Jacob, ed., *The Politics of Western Science*. Here I am restating the point about the possibility of ahistorical extrapolations made in my review of *Leviathan and the Air Pump* in *Isis* 77 (1986): 719–20.

31. For Bruno Latour's essay, see Ernan McMullin, ed., *The Social Dimensions of Science* (South Bend, Ind.: University of Notre Dame Press, 1992), 289, italics added. For Latour's proclamation of the end of the Enlightenment, see 288.

32. Ibid., 281.

33. In *Leviathan and the Air Pump*, "absolutism" never appears in the index and, as far as I can see, is mentioned only once in the text.

34. Hobbes, *Leviathan*, 691–92.

35. Ibid., 703.

36. For Hobbes on law and lawyers, see his *A Dialogue between a Phylosopher and a Student of the Common-Laws of England* (London, 1681), 11, 28, and 42 for the quotation. Found in *Tracts of Thomas Hobbes* (London, 1681).

37. Discussed in greater detail in James R. Jacob, "The Political Economy of Science," 513; citing Hobbes, *De cive* (New York: Appleton-Century-Crofts, 1949), 120; and *Leviathan*, ed. R. Tuck (Cambridge: Cambridge University Press, 1991), 106.

38. Hobbes, *Leviathan*, ed. Macpherson, 670–71.

39. Ibid., 368–69.

40. *Leviathan*, 715 (final paragraph of the treatise): "For it is not the Romane Clergy onely, that pretends the Kingdome of God to be of this World, and thereby to have a Power therein, distinct from that of the Civill State. And this is all I had a designe to say, concerning the Doctrine of the Politiques. Which when I have reviewed, I shall willingly expose it to the censure of my Countrey."

41. Ibid., 166.

Part V

Civilian Casualties of Postmodern Perspectives on Science

One function of the university is to provide a tournament ground for the jousting of outlandish points of view. Although many academic "wars" do little damage to civilians, postmodernism is now gaining influence in schools of education and is feeding the media's appetite for increasingly cynical attacks on any institution that honors its own past. Particularly vulnerable to the blandishments of postmodernism are the professional staff at university presses, government and private foundations, and other organizations that mediate between science and the public. Such personnel typically come from the humanities and have little formal training in science. In earlier times, they might have strived to overcome that deficiency. Now they are likely to have encountered the STS or cultural studies critiques in their college career and may uncritically bring such perspectives to bear in their dealings with science. This section describes some of STS's deleterious impacts on mathematics and science curricula, science journalism, and science policy, as well as the general public's understanding of the scientific process.

Noretta Koertge looks at some troubling examples of multicultural mathematics curricula and proposals for making science more "female friendly." She argues that we must resist postmodernist proposals to substitute courses "about" science for courses *in* science and we must reject the corresponding redefinitions of the concept of science literacy. Future citizens need to have basic scientific informa-

tion at their disposal. By having some firsthand experience with the process of structured, critical inquiry exemplified in science, they may also gain a greater appreciation of the importance of rational discourse in an open society.

Norman Levitt analyzes science journalist John Horgan's extraordinary claim that we have reached the end of revolutionary new conceptual breakthroughs in science. From now on, science will consist of either the working out of routine applications of current paradigms or the proliferation of untestable speculations about increasingly incomprehensible theoretical entities. Horgan calls the latter "postmodern science" and claims that physics has already reached this stage. Levitt argues that, despite his adoption of postmodernist terminology, Horgan has a rather naive empiricist conception of the evaluation of scientific theories. Theoretical physics is not doomed to the fate of literary criticism. Horgan also overlooks the prospects for fundamental new insights in fields such as cognitive science and developmental biology.

Meera Nanda reveals yet another instance in which the West has dumped inferior or dangerous goods into Third World markets. This time, the export is shoddy intellectual merchandise—postmodernist and postcolonialist theory. Nanda tells how the progressive people's science movements, in which she has been active, have lost ground to those advocating ethnoscience and ecofeminism. Ironically, these ideas, which sprang from the academic left, have now been taken up by

Hindu nationalists and other cultural conservatives who are quite happy to be given additional arguments for opposing Western influences that might disturb their status quo.

Nanda's report reminds us that the ideas we debate in the "science wars" can directly affect people's lives. Well-meaning moralism is no substitute for intellectual integrity.

16

Postmodernisms and the Problem of Scientific Literacy

Noretta Koertge

Educational policymakers and members of organizations such as the American Association for the Advancement of Science have long been troubled by indications of scientific illiteracy in our society, such as the poor performance of U.S. students on standardized tests and the amount of money spent on extremely dubious alternative health procedures, ranging from the relatively benign old-fashioned homeopathic remedies to new psychological therapies for "recovered memory syndrome" that can be extremely destructive. There is also concern about the salience of antirational and pseudoscientific mythologies in our culture (e.g., angel books, astrology columns, and TV programs on "unexplained mysteries," as well as the persistence and success of the "creation science" movement). The history of science education since *Sputnik* is a story of attempts to ameliorate this situation.

The postmodern approaches to science studies that figure as major players in the "science wars" also have prescriptions for transforming science education and redefining scientific literacy. Although they differ on details and degree, all are intent on radically changing public perceptions of science through an agenda of radical educational reform. In this essay, I show how these various postmodernist accounts of science are now being uncritically absorbed into science pedagogy and argue that many of their proposals are counterproductive.

Science: Golem or Unthinkable Father?

I begin by surveying some of the general recommendations for a new kind of science literacy that are coming out of the sociological and cultural studies approach to science studies. There is no doubt that much of the work is intended to be directly applied to science education. Harry Collins and Trevor Pinch (1993) subtitle their controversial book, which compares science to a golem, "What Everyone Should Know about Science." Part of Cambridge University Press's Canto series, their book is intended for a general audience, including preuniversity students. Research in women's studies is also directed toward revolutionizing science pedagogy. Teachers College Press lists a number of such books, including Sue Rosser's *Female-Friendly Science* (1990) and *Teaching the Majority* (1995).

It is tempting to epitomize the cultural studies perspective on science education with Andrew Ross's dedication of *Strange Weather* (1994): "To all of the science teachers I never had. [This book] could only have been written without them." Perhaps more typical, however, are textbooks of the writing-across-the-curriculum variety in which the student is taught to evaluate articles about controversial issues such as nuclear waste disposal or global warming on the basis of the narrative strategies and rhetorical devices employed, instead of through an analysis of the arguments and the kind of scientific evidence provided.

All the authors I have referred to (even Ross) proclaim that students as citizens need to know more about science, although as we shall see, their picture of science is a very controversial one, indeed. Even more controversial is their denial of the proposition that laypeople need to have a better understanding of the basic content of modern science. Collins and Pinch, in a section entitled "Science and the Citizen," are quite emphatic on this point: The idea that knowing more science would help the public make more sensible decisions "ranks among the great fallacies of our age" (1993, 144). Through their case studies, they claim to "have shown that scientists at the research front cannot settle their disagreements through better experimentation, more knowledge, more advanced theories, or clearer thinking. It is ridiculous to expect the general public to do better" (1993, 144–45).

These commentators also widely agree that students should be explicitly disabused of any ideas they might have picked up about the epistemic desirability of reproducible experiments, controlling variables, statistical analysis, and all the other methodological staples of modern science. Again, Collins and Pinch are quite clear. After describing the bungling efforts of schoolchildren to determine the boiling point of water, they claim that the "negotiation" of their results in the classroom does not differ significantly from the behavior of great scientists working at the frontier: "Eddington, Michelson, Morley . . . are Zonkers, Brians, and Smudgers with clean white coats and 'PhD' after their names" (151).

In tandem with this trivialization of scientific methodology are proposals that would negate central features of the scientific approach. For example, Rosser's recipes for making science more "female friendly" prescribe the inclusion and validation of personal experiences as part of the laboratory exercise and a deemphasis on objectivity (1993, 196). Noting that "well-controlled experiments in a laboratory environment may provide results that have little application . . . outside the classroom" (213), she recommends that students investigate problems of a more "holistic, global scope," using "interactive methods" instead of trying to set up isolated systems or controlling variables. Rosser is convinced that female students would find this new kind of methodology more "friendly." Be that as it may, such recommendations cannot be viewed as friendly amendments to standard conceptions of fruitful scientific methodology.

Ross also calls for a thorough overhaul of scientific practice: "[We need] to talk about different ways of doing science, ways that downgrade methodology, experiment, and manufacturing in favor of local environments, cultural values, and social justice. This is the way that leads from relativism to diversity" (1996, 4). He sharply distinguishes his approach from the descriptive relativism inherent in the sociology of scientific knowledge (SSK) program. Although he approves of the treatment of scientific knowledge as sociologically equivalent to all other belief systems, he feels that SSK does nothing to dictate the necessity to change scientific institutions (11).

From this perspective, although Collins and Pinch indeed want to redefine our understanding of how science operates, they are remiss in not demanding a radical enough change in the practice of science. For example, after concluding from their case studies that the scientific community is constantly "transmuting the clumsy antics of the collective Golem Science into a neat and tidy scientific myth," a description that certainly sounds provocative to the Enlightenment ear, the last sentence of their book praises such mythmaking activities: "There is nothing wrong with this; the only sin is not knowing that it is always thus" (1993, 151).

Appealing to an analogy, Collins and Pinch emphasize that they are not against plumbers—society needs plumbers; they are only against the immaculate conception theory of plumbing (145). The scientific golem is a bumbling giant who knows neither its own strength nor the extent of its clumsiness and ignorance: "It is truth that drives it on. But this does not mean it understands the truth—far from it" (2). They ask the reader to learn about what actually happens in science and "to learn to love the bumbling giant for what it is" (2) It is presumably on the basis of such an attitude of tolerant superiority and epistemological condescension that SSKers such as Collins and Pinch deny so vehemently that they are antiscience.

However, much of the cultural studies commentary on science is explicitly hostile to not only the technological applications of science but also the values of pure scientific inquiry. Mary Daly, a radical feminist philosopher at Boston College, provides a parody of traditional philological scholarship in her *Wickedary*. She describes as necrophilia the essential message of science under patriarchy and states that "phallotechnology" has "rapism as its hidden agenda and destruction of life as its final goal" (1987, 217). One never knows how seriously to take Daly—there is certainly dark humor lurking in her promise to take the "dick" out of "dictionary." But there are others who quite seriously contend that traditional science overemphasizes the stereotypical masculine virtues, such as a respect for objectivity and a preference for methodical, analytic modes of problem solving, while at the same time magnifying all the stereotypical masculine vices. Thus Sandra Harding claims that in Bacon's influential writings, "both nature and inquiry appear conceptualized in ways modeled on rape and torture—on men's most violent and misogynous relationships to women—and this modeling is advanced as a reason to value science" (1986, 116).

In a chapter entitled "Unthinkable Fathering" (a reference to Brian Easlea's book *Fathering the Unthinkable*), Jane Caputi, a professor of American studies, claims to show a "compelling connection between incest and nuclearism" ranging from "the nuclear-father figure" and "the predominantly masculine character of the abusive cohort" to "the cult of secrecy, aided by psychological responses of denial, numbering, and splitting (in both survivor and perpetrator)" (1993, 118). Kathy Overfield sums up this perspective succinctly: "Male science furthers the capitalist, imperialist tradition in which it was begotten: it exploits, rapes, destroys" (1981, 247).

Given such a horrifying picture of science, it is not surprising that the ensuing recommendations for science education call for extreme measures. Until science is changed, the argument goes, anyone who is not into rape and necrophilia will have to enter science as an undercover agent. Hilary Rose describes the telos of science and technology as one of nuclear annihilation (1987, 265). "Women who manage to get jobs in science," she observes, "have to handle a peculiar contradiction between the demands on them as caring laborers and as abstract mental laborers" (278). "Thus a woman scientist is cut in two. . . . Small wonder that women, let alone feminists, working in natural science and engineering are rarities. It is difficult enough to suppress half of oneself to pursue knowledge of the natural world as a woman" (279).

Let us now look at some specific pedagogic proposals that have come out of the postmodernist critiques of science. I find some of the proposals to be sensible but hardly novel and others, though controversial, certainly worthy of discussion. Some appear to me to be silly but largely harmless; others strike me as totally misguided and likely to have effects that even the critics would find undesirable. The main point I wish to establish is that postmodernist conceptions of science are rapidly diffusing out from the graduate humanities seminar into the K–12 science classroom. If as others in this volume argue, those conceptions are seriously flawed, we have reason to be concerned about their impact on education.

Postmodern Mathematics

It is easy to see the line of transmission in our first example, which might well be called the "case for menstrual mathematics." The trajectory begins with the Belgian philosopher Luce Irigaray's speculations about the history of fluid mechanics. Irigaray's prose—in either French or English translation—is difficult to understand; her influence on educators has come primarily through an essay by Katherine Hayles, an English professor and contributor to the *Social Text* issue on the "science wars." Here is Hayles's summary:

> The privileging of solid over fluid mechanics, and indeed the inability of science to deal with turbulent flow at all, she [Irigaray] attributes to the association of fluidity with femininity. Whereas men have sex organs that protrude and become rigid, women have openings that leak menstrual blood and vaginal fluids. Although men, too, flow on occasion—when semen is emitted, for example—this aspect of their sexuality is not emphasized. It is the rigidity of the male organ that counts, not its complicity in fluid flow.
>
> These idealizations are reinscribed in mathematics, which conceives of fluids as laminated planes and other modified solid forms. In the same way that women are erased within masculinist theories and language, existing only as not-men, so fluids have been erased from science, existing only as not-solids.
>
> From this perspective it is no wonder that science has not been able to arrive at a successful model for turbulence. The problem of turbulent flow cannot be solved because the conceptions of fluids (and of women) have been formulated so as necessarily to leave unarticulated remainders. (1992, 17)

The speculative nature of Irigaray's musings should be obvious to anyone, although they might seem plausible to those inclined toward psychoanalytic historiography. The more one knows about the history of science, however, the more preposterous these assertions seem. For starters, fluids have hardly been "erased" from science. Thales, the first pre-Socratic discussed by Aristotle, thought that the whole world was formed from one element, water. Aristotle made water one of his four elements and designated "the moist" and "the yielding" as primary qualities. In early modern science, Descartes's cosmology was based on the vortex motions of a continuous fluid, and to refute his account, Newton had to develop an alternative treatment of vortices. Eighteenth-century explanations of electrical phenomena posited either one or two electrical fluids. Imponderable fluids such as "caloric" or the "ether" were prominent in the ontology of nineteenth-century physics, and many of the most important theoretical advances in chemistry dealt with the properties of solutions. Furthermore, Irigaray's description of the mathematical models used in hydrodynamics is completely distorted. (See chapter 5 in this volume for a remarkably restrained dissection of Hayles's attempt to improve on Irigaray's account.)

Unfortunately, these and other equally erroneous interpretations of the history of science and mathematics are now being cited in the professional pedagogical literature. Consider, for example, the speculations about why girls find mathematics and physics so counterintuitive, found in a recent book published in collaboration with the National Council of Teachers of Mathematics:

> We might ask again . . . why adolescence marks the increased separation of women from mathematics. In the context provided by Irigaray we can see an opposition between the linear time of mathematics problems of related rates, distance formulas, and linear acceleration versus the dominant experiential cyclical time of the menstrual body. Is it obvious to the female mind–body that intervals have endpoints, that parabolas neatly divide the plane, and, indeed, that the linear mathematics of schooling describes the world of experience in intuitively obvious ways? (Damarin 1995, 252)

In a related vein are recent Internet discussions of an e-mail request for introductory physics texts that treat sound and light before presenting mechanics, the premise being that female students would find the study of wave phenomena more congenial than the motions of those darned old rigid bodies! (But might not the menstrual body be able to relate to the ebb and flow of the tides or the undulations of harmonic oscillators?)

Similar explanations have been offered for the difficulty some Navajo children have with elementary school mathematics:

> The Western world developed the notion of fractions and decimals out of a need to divide or segment a whole. The Navajo world view consistently appears not to segment the whole of an entity. (Bradley et al. 1990, 8)

> Non-Euclidean geometry, motion theories, and/or fundamentals of calculus may be naturally compatible with Navajo spatial knowledge. Math classes should begin with these notions and continue deemphasizing the segmentation of notions into smaller parts. (7)

Some of the suggested pedagogical remedies are quite astounding, and not all come from humanities departments or schools of education. For example, after discussing the Navajo case with approval, a college teacher of mathematics exclaimed, "In other words, for some students, it might be appropriate to teach calculus as elementary mathematics, and fractions in college!" (Shulman 1994, 9). Leaving aside the conceptual difficulty of introducing the definition of the slope of a line without recourse to quotients, one might at least expect that someone who was truly sensitive to the needs of Navajo schoolchildren would worry about the consequences of not explaining to them how to compute sales tax or how to figure out how many quarter-pounders there are in a pound until they go to college!

These proposals to rearrange the order of presenting mathematical ideas and to emphasize those allegedly most accessible to students of a particular gender or ethnic group are sometimes accompanied by the more radical claim that the very truths of mathematics are dependent on culture. For example, the following are four explicit objectives of a class offered in a mathematics department that can be used to satisfy a general education requirement:

> After taking this course the student will be able to:
>
> 1. Describe the political nature of mathematics and mathematics education.
> 2. Describe gender and race differences in mathematics and their sociological consequences.
> 3. Examine the factors influencing gender and race differences in mathematics.
> 4. Critically evaluate eurocentrism and androcentrism in mathematics. (Kellermeier 1995a, 40)

The course bibliography includes some useful historical material on great women and non-European mathematicians, but the syllabus is heavily skewed toward so-called ethnomathematics and readings on "Understanding the Politics of Mathematical Knowledge" and "Fighting Eurocentrism in Mathematics" (42).

John Kellermeier describes his course as "a first step toward changing students' perceptions of mathematics, and making them aware of the sexism, racism, and elitism in mathematics" (1995a, 39). He provides many heartfelt comments from students who were relieved to learn that it was not their fault that they had previously not done well in mathematics, and one student reports that she is actually contemplating registering for another math course! (38).

Other things being equal, it is certainly good to relieve math anxiety and increase students' self-confidence. But I also wonder what difficulties this student will experience when she once again encounters proofs employing the "rigorous application of a form of deductive axiomatic logic" that she has now been taught is the result of a "Eurocentric bias in mathematics" (36). It is one thing to introduce students to the Mozambican method of squaring up the foundation of a house (make sure the diagonals are equal) (35) and quite another to tell them that ethnomathematics is in conflict with academic mathematics (36).[1]

It sometimes appears that these reformers are less interested in removing barriers that impede students' ability to learn mathematics or science than they are in using required courses in these areas as a political platform. In articles entitled "Queer Statistics" and "Women's Studies in a Statistics Classroom," John Kellermeier argues for the importance of using word problems that "reflect cultural diversity." Here are some examples:

> A recent study suggests that 80% of women with anorexic and bulimic eating disorders were sexually abused as children. Suppose 10 women with eating disorders are interviewed. . . . What is the probability that at least 7 of these women would have a history of sexual abuse? (1994, 29)

In sixty-five percent of all rapes, the victim knows her assailant. If we interview twenty women who were raped, what is the probability that no more than four of them were raped by strangers? (1992, 24)

Research shows that 36% of lesbigay people have been threatened with physical violence because of their sexual orientation. If 150 lesbigay people are surveyed, what is the probability that at least 40 of them have been threatened with physical violence? (1995b, 102)

Kellermeier correctly points out that the content of word problems is a constituent of what he calls the "hidden" curriculum, the "overt" part being the statistical concepts we intend to teach. He is trying to utilize the "hidden" part for progressive ends. But as every teacher of any formal subject such as introductory logic, the probability calculus, decision theory, or finite math knows, one of the most difficult things for students to learn is to ignore the specific content of a word problem and to concentrate instead on its formal structure. Many students find this difficult to do. For example, logician Andrea Nye begins her feminist interpretation of the history of logic with a story about the difficulty she had in representing sentences with Ps and Qs. When confronted with the example "Jones ate fish with ice cream and died," Nye, who had come to philosophy from literature, says she found her mind wandering off into speculation about why Jones ate such a bizarre dish and why death was the consequence (1990, 2). And Shulman tells the story of a female student who was even distracted by an example in which a jacket cost only $25 (12).

Would such a student be helped or hindered by a steady diet of statistics exercises that deal with rape, female battering, and the correlation between bulimia and sexual abuse? If the data in Kellermeier's word problems are factually correct, then a significant proportion of the women students he is trying to accommodate have had first-hand experience with the traumas that figure so prominently in the homework sets and might well find them disturbing. Kellermeier does not mention any such difficulties, but he does report that most students were very aware of the "hidden" curriculum in the "queerstatistics" class and that some reacted strongly to it. The following is one of the more moderate student responses he received: "The content of the problems is a touchy topic with some people and in a way I do not see why every question had to be about 'it' but I did not have a problem with them" (1995b, 106).

Kellermeier felt that the experiment using a thematic approach to word problems was successful because it heightened students' awareness of homophobia and might help gay students in the coming-out process. But again, one wonders about the effect on students who have to confront puzzles about statistical significance while being distracted with feelings of ambivalence about sexual orientation. Be that as it may, it is telling that Kellermeier makes no attempt to find out whether his students learned more or less about statistics with this new approach. This is surely a case in which pedagogy is being driven by ideology. However progressive Kellermeir's political agenda may be, it does not belong in a statistics class. Either the political proposals will not be critically discussed, in which case the students will rightly object to a form of indoctrination, or they will be openly discussed, but without the benefit of informative readings and at the expense of the statistics syllabus. One need only imagine parallel situations featuring statistics describing the mothers of crack babies or the *y* of gay men to realize the undesirability of politicizing the "hidden" *m.*[2]

,o accident that there is intense interest in new strategies for mathematics educaᵤᵤn. Introductory classes in mathematics and related formal disciplines such as logic and statistics serve as prerequisites for many career paths, such as business, nursing, criminology, and physical therapy, as well as majors in both the natural and social sciences. There is a strong correlation between the number of math courses taken in college and a graduate's starting salary. So math competence is of great importance in American society, yet it is an area in which women and minorities seem to have a disproportionate amount of difficulty.

All parties to the debates about science studies are concerned about these under-represented groups, but they support two quite different philosophies about how to remedy the situation. One is to enrich the exposure to the fundamental concepts of mathematics for those women and minority students (and others) who need them. Here the message is simple: It is not your fault that your mathematical education thus far is deficient, but it is you who will suffer lost opportunities unless something is done about it. We're here to help you learn the basics as efficiently as possible.

The other approach is to try to adapt both mathematics itself and mathematics pedagogy to the students' commonsense belief systems, to tell them that the reason they find ordinary math difficult is that they come from a different culture and to teach them to view the widespread utilization of mathematical and analytical reasoning as a form of cultural imperialism and social injustice.

Female-Friendly Science

Feminist suggestions for science pedagogy are another mixed bag. Let us look briefly at one of the seemingly more moderate and systematic approaches, the "Model for Transforming the Natural Sciences" (1993, 193), proposed by Sue V. Rosser, a former senior program officer at the National Science Foundation. Although Rosser presents her proposal as a modest one, it is immediately evident that it is science itself that is to be transformed, not just science education.

The initial recommendations are sensible—one should include the names of women scientists who have made important discoveries and provide wherever possible hands-on experiences for the students instead of making them rely on textbook descriptions. Rosser then moves on to the familiar feminist injunction to "use less competitive models and more interdisciplinary methods to teach science" (205). (It is anybody's guess what that might entail, but it sounds innocent enough.)

The last phases of Rosser's planned transformation, however, are explicitly based on culture studies' critiques of science. For example, when Rosser recommends not having laboratory exercises in introductory courses in which students must kill animals, what is surprising is not the recommendation itself—there has already been a dramatic decrease in the use of animal subjects—but the reason Rosser gives: "Merchant and Griffin . . . document the extent to which modern mechanistic science becomes a tool men use to dominate both women and animals. Thus many women may particularly empathize with animals" (202).

Rosser goes on to criticize all of modern biology for being reductionistic because of its emphasis on cell and molecular biology. From the beginning, she says, students should be encouraged to develop hypotheses that are "relational, interdependent, and multi-causal rather than hierarchical, reductionistic, and dualistic" (213), and curricula should include "fewer experiments likely to have applications of direct benefit to the military" (194). The goal is to arrive at a new kind of science, one that is "redefined and reconstructed to include us all" (214).

Some of Rosser's proposed reforms show little awareness of the actual practice of science education. Neither I nor my colleagues were able to come up with any examples of student experiments that seemed to be strongly associated with the military. The example Rosser gives, the calculation of trajectories that could be used to describe the motions of bombs and rockets, can equally well be illustrated with space age examples involving *Starship* Captain Kathryn Janeway. And for some time now, the killing of laboratory animals—as opposed to the dissection of them—has been very uncommon in student labs. Other of Rosser's proposals are totally impractical. Since multifactor models are notoriously difficult to evaluate, it hardly seems wise to encourage beginning students to concentrate on them. And the issue of reductionism in modern biology is an extremely complex and contested one. In any case, one can hardly afford not to teach cell biology.

In addition, Rosser's citation of Griffin's prose poem as an authoritative source on the history of science reveals the extent to which her proposals are based on the simplistic negative images of science so common in feminist critiques.[3] Rosser describes her approach as an application of women's studies, whose central message is that the overall impact of science on the lives of women (and minorities) has been oppressive.[4] Claims that science and technology have eliminated drudgery and many of the dangers of childbirth and provided improved contraceptive devices are said to be outweighed by science's contributions to warfare, the decline of midwifery, and the devaluation of more intuitive, holistic approaches to problem solving.

The methods and values of science are claimed to be intrinsically sexist. Instead of viewing science as an important component of the Enlightenment (a term with positive connotations), many feminists see the rise of modern science as responsible for "the death of nature" (as the historian of science Carolyn Merchant argues in a book with that title). For a variety of reasons (some would posit "womb envy"; others cite the "object relations" successor to Freudian theory), men are seen as more inclined to treat living things as "objects" and to value them only to the extent that they can dissect and control them. Science—with its emphasis on analysis, abstraction, quantification, and prediction—is then viewed, to repeat Mary Daly's terminology, as a paradigm case of patriarchal, phallocratic necrophilia.

It then seems inevitable that the content of science must be sexist. Dozens of books describe cases purporting to show that sexist bias seeps into scientific theorizing. Many of the examples deal with theories of reproduction. Perhaps the most famous one is the saga of the active sperm and passive egg, about which it is claimed that gender stereotypes about human males and females influenced not just the rhetoric of popular accounts of the heroic little sperm valiantly swimming upstream hoping to penetrate the big, fat noncooperative egg but also contributed to the tendency to emphasize research on DNA to the neglect of embryology. (For a thorough debunking of this feminist legend, see chapter 4 in this volume.)

Sophistication versus Cynicism in the Classroom

What unifies the various commentators on science education that I have called postmodern are the convictions that science has too much authority in our society and that there is too much respect for scientific reasoning within the general population. To combat these alleged tendencies toward science worship or science boosterism, these commentators believe that science education must be transformed so as to present students with a less heroic and less idealistic picture of scientific inquiry. Once students learn that scientific knowledge is constructed just like every other segment of our cultural beliefs and once they realize that the results of scientific experiments are the product of social negotiations, just as are the appraised values of the damage caused by a tornado, then, the social constructivists believe, students will grow up to be less admiring of scientific findings and better able to cope as citizens in a complex technological world.

Since allegedly there can be no compelling evidential basis for scientific claims, the discerning citizen needs to ask not whether the claim is well supported but, rather, whose political interests are served by such a claim. Since science supposedly has a long history of oppressing women and minorities, young people in these groups are thought to have a special need for inoculation against a naive belief in the products and procedures of traditional science.

What are we to make of such a program? Even if we reject extreme versions of social constructivism, are there at least some beneficial practical reforms coming out of these critiques? First, we all can agree with the postmodernists—and C. P. Snow—that citizens need to know more about science, but only if what they are told is reasonably accurate and only if it is not presented as a substitute for learning more of the content of science. Of course, students need to learn about scientific controversies like the interesting cases Collins and Pinch discuss in their book. But the study of scientific debates is hardly a novel pedagogical innovation. Old textbooks, such as Holton and Roller and the Harvard Case Histories, gave rich accounts of historical controversies that clearly illustrated the strengths and weaknesses of both positions and the difficulties in settling the matter. The most conventional of the current science books offer alternative accounts of why the dinosaurs became extinct, how the moon was formed, how life began, and how best to model the bonds in benzene. Let us concede to Collins and Pinch, however, that students also need to learn more about professional rivalries in science and the politics of science journalism (more on this later), but the picture must not be reduced to a cynical cartoon of real science. Students must also learn about the role of evidence and rational debate in scientific conflict resolution and also successful theory construction and refinement and how they were achieved.

Second, let us agree with the cultural critics of science—and Sir Karl Popper—that students need to learn more about the fallibility of scientific theories. We also freely grant that the hypotheses that spring most readily to the scientist's mind often reflect the commonsense beliefs, metaphysical systems, and ideological predilections of the times. But what we must add to the cultural studies picture and give prominence to is the old Popperian point about the refutability of false scientific theories and the possibility of learning from our mistakes. However obnoxious various outdated scientific theories of reproduction, intelligence, and human differences may be, it still was possible to discredit them through the routine application of the ordinary self-corrective devices of scientific inquiry. These methodological procedures do not always operate as efficiently as we would like, and they also can be subverted by political pressure. But there is no historical warrant for building into the curriculum a cynical, worst-case-scenario view of science.

Third, let us welcome all those who support the proposition that science education at all levels should become more accessible to and welcoming of women and minorities. Professional scientific organizations have long been concerned about the gender gap and racial gaps in scientific literacy. On average, members of these groups are more prone to math anxiety and score lower on standardized math tests, are less likely to take nonrequired science courses in high school, are less inclined to major in science or engineering, and are more liable to drop out if they do choose a science-based career. Sociological studies of attitudes toward science also report interesting gender differences in the patterns of attraction to pseudoscience and belief in paranormal phenomena (Harrold and Eve 1995). In accord with their explicit political commitments, postmodernist commentators on science emphasize the problem of the alienation of such groups from science. But too often their solution is to give each political identity its own special ethnoscience, tailor-made in ontology, methodology, and content to match the ideology and interests of that community. They believe that by denying the transcultural validity of scientific results, they will somehow undermine the power of transnational corporations and military alliances. Somehow feminist science will empower women, and ethnomathematics will improve the lot of the Navajo people.

But here their misunderstanding of the nature of science can have tragic consequences for the very people they want to liberate. It is not easy to generate truth claims about the world, and the scientific process does not always succeed in doing so, but when it does, what wonderful stones they provide for David's sling! To give just one example: I submit that DNA testing—a product of the reductionistic, nonholistic science that cultural critics of science deplore—promotes more justice to rape victims than do the special pleadings of feminist psychologists about the veracity of female plaintiffs. At the same time, DNA testing can promote justice for those African American men who have been falsely accused of rape.[5] The science underlying DNA testing is universal science, not ethnoscience or genderized science, and that is why it can be used to arbitrate disputes between groups in such a manner that the politically less powerful can prevail. What a tragedy to educate members of disadvantaged groups in ersatz science and mathematics, all in the name of multiculturalism. And how dangerous it is to suggest to the powerful in our society that science is part of their rightful patrimony because after all, as Sandra Harding put it, "science is politics by other means" (1991, 10). A denigration or subversion of the ideals of science can never be politically progressive in the long run.

It is sometimes tempting to respond to the postmodernists' litany of the fatal flaws in science with a naive defense of science as sacred. Even though I am firmly convinced that the ideals and institutions of science provide the best approach we have for understanding the world, I do not deny for a moment that they are both intrinsically imperfect and subject to improvement. The whole point of science education in a liberal arts tradition is to help students benefit from what has been learned before while simultaneously acquiring the ability to question and improve on the currently received views.

There are quick ways to foster a critical stance in young people, well illustrated by the success of various forms of cynicism. It is easy to make students feel sophisticated— at least temporarily—by giving them a hedonistic explanation of altruism, a vulgar Marxist analysis of the rise of atomism, or a slick psychoanalytic interpretation of scientific metaphors. What is more difficult is to help them attain a genuine sophistication about the content, methods, and institutions of science.

Traditional concerns about scientific literacy have focused on content and methods. The writings of postmodernists forcibly remind us to pay more attention to the importance of protecting scientific institutions. The epistemic quality of scientific results depends not only on the methods and ideals of scientific inquiry but also on the institutional conditions under which those ideals can best be realized. Scientists themselves, as well as traditional sociologists of science such as Merton, have written about the relationships between scientific societies and society at large and how the integrity of science can be threatened by governmental secrecy, patenting policies, funding patterns, and pressures from the media. Students can profit from studying these and countless other sociological dimensions of scientific practice, but as an addition to and never at the expense of the central components of science education.

There are many reasons for retaining an emphasis on content and methodology: First, as the many factual blunders in the discussions of Dolly the cloned sheep illustrate, one simply can't have even a rudimentary understanding of science policy issues without some basic knowledge of the science in question.[6] This is an argument for general education in a broad range of scientific disciplines.

There is also a strong argument for providing students with a more sophisticated understanding of at least one science, and it may be because of a deficiency of this kind that so many postmodernist analyses completely miss the boat. Although we can present many features of the scientific approach by drawing parallels with commonsense reasoning, the most distinctive epistemological features of science can be illustrated only by working with fairly sophisticated examples. Suppose we want students to ask why scientists sometimes become convinced of the reality of the entities posited by their theories even when they can't observe them or manipulate them. This question cannot be adequately answered by either analogies to everyday life (does my pen pal really exist?) or skeptical philosophical arguments about the underdetermination of theory by evidence.

Instead, students need to be familiar with scientific examples in which mutually relevant, but distinct, theories are confirmed by vast numbers of independent experiments. Heinz Post (1968) describes an atomic physics course in which he presents literally dozens of methods for determining the value of Avogadro's number, ranging from the crude method of measuring the spread of an oil drop over water, to indirect calculations based on the mean free path of a molecule, to direct calculations from

Einstein's theory of Brownian motion, to the simple measurement of the volume of helium produced by a known number of alpha-disintegrations. It is the amazing convergence of these results—what the philosopher William Whewell called the "consilience of inductions"—over time and over disciplinary specialties that undergirds scientific confidence in the reality of theoretical entities.

During the century in which these methods were developed, there were many debates about the existence of atoms, and the detailed pictures presented by atomists changed dramatically. But eventually it became impossible to fault the proposition that whatever the little critters were like, there were 6×10^{23} of them per mole, for this result was invariant with respect to the experimental procedure used, or theoretical system, or disciplinary specialty. If students are to appreciate the epistemic power that scientific systems sometimes attain and the standards against which junk science, pseudoscience, and immature science must be judged, they need to encounter examples of science at its best. And this requires learning a fair bit of some science.

What those who would prioritize the study of the social aspects of science fail to realize is that it is impossible to analyze the functioning of scientific institutions without thoroughly understanding the aims and methods of scientific inquiry. And understanding the process of science in turn requires a detailed knowledge of the structure of some actual scientific theories and of the nature of the specific data-gathering processes that are used to evaluate them and how both the theoretical and experimental components develop over time. It is no use trying to make a virtue out of ignorance by thanking the science teachers we never had or by telling kids that they no longer need to study the content of science but need simply to learn what the sociology of scientific knowledge says about it. After all, what if science studies turned out to be the real golem?

Notes

1. I find Kellermeier's discussion of this example quite amazing. First, he treats Mozambican carpentry practice as if it were exotic. Yet I know from casual experience that this same method is used by farmers of Scots-Irish descent in southern Illinois and the Amish in central Indiana. Kellermeier then claims that the Mozambican concept of rectangle is different from the one presented in math textbooks, implying that the difference between the operational definition and the formal one given in math classes is a form of incommensurability. But on this analysis, Euclid also had an exotic concept of rectangle, for he defines it as any parallelogram containing a right angle (Euclid's *Elements*, book 2, definition 1). The most common definition today is provided by the *Oxford English Dictionary*, which describes a rectangle as "a plane rectilineal four-sided figure having all of its angles right angles, and *therefore* its opposite sides equal and parallel" (my italics). But in none of these cases are there discrepancies between the properties said to hold true of rectangles. At most, there might be formal differences in which propositions are to be taken as definitions and which ones as theorems. However, since the carpenters in neither Mozambique nor the Hoosier heartland seem to be committed to any particular axiomatization of Euclidean geometry, this formal difference is unlikely to constitute a psychological barrier to their children.

2. There is a long history of incorporating tendentious material in story problems. Lewis Carroll's anti-Semitic syllogisms in his book on symbolic logic were probably misguided attempts at humor. Norman Levitt tells me that Daniel Harvey Hill's 1854 algebra book is full of problems portraying Yankees as mean, greedy, hypocritical, and altogether dishonorable by the standards of Southern chivalry.

3. The first meditation in Griffin's 1948 book, entitled "Matter: How Man Regards and Makes Use of Woman and Nature," is an outline of the history of science punctuated by comments about the status of women. Her allusions are to genuine historical documents, but any connections between the juxtaposed items are provided only by proximity and poetic devices. Here is one example:

1619   Kepler publishes his third law, *De Harmonice Mundi*.
1619   The first black slaves are introduced in America.
(She is asked if she signed the devil's book.
She is asked if the devil had a body.
She is asked whom she chose to be an incubus.)
1622   Frances Bacon publishes *Natural* and *Experimental History for the Foundations of Philosophy*.
1622–1623   Johann George II, Prince Bishop, builds a house for the trying of witches at Bamberg, where six hundred burn. (16)

There is no pretense of an attempt to present a reasoned historical interpretation of the items. The book, though moving, contains no research on which to base a theory of science pedagogy.

4. For an extended critique of the commentaries on science coming out of women's studies, see Patai and Koertge 1994. Rosser fully endorses the central tenets of feminist critiques of science: "I believe that scientific research and theories cannot be gender (or class, race, or sexual preference) neutral as long as our society is not neutral on these issues" (1990, 23). She is nevertheless deeply committed to immediately attracting women and other underrepresented students into science. This distances her considerably from more extreme critics, whom Spender describes as arguing "that science *is* men's studies and cannot be modified and that a 'woman-centered-science' would be so radically different that it would no longer be invested with the meaning of 'science' as we understand it" (Rosser 1990, 237, italics in original).

5. When remembering the history of African American men being lynched after being accused by white women, who were sometimes their lovers, the term *rape victim* takes on a new connotation.

6. Yet the Dolly episode also refutes the Collins–Pinch thesis that science policy issues are so complex that laypeople can never evaluate the merits of the competing arguments and hence must rely on analyses of the political factors involved. Emerging from the flurry of media attention were many excellent accounts of the cloning procedure, the difficulties in transferring the technology to other species, how clones compare with identical twins, and why the technique had scientific and technological value. Anyone with a minimal degree of biological literacy—and who was so motivated—could learn enough to enter into the ethical/policy debates without being swayed by scenarios equating clones with an array of synchronous goose-stepping robots. The confidence that one can procure the scientific information needed to make personal decisions should be another important goal of scientific literary.

References

Bradley, Claudette, Charlotte Basham, Melissa Axelrod, and Eliza Jones. Language and mathematics learning. *UME Trends* 2(5): 7–8, 1990.

Caputi, Jane, *Gossips, Gorgons & Crones: The Fates of the Earth*. Santa Fe: Bear & Co., 1993.

Collins, Harry and Trevor Pinch. *The Golem: What Everyone Should Know about Science*. Cambridge: Cambridge University Press, 1993.

Conant, James B., ed. *Harvard Case Histories in Experimental Science*. Vols. 1 and 2. Cambridge, Mass.: Harvard University Press, 1957.

Daly, Mary, *Websters' First New Intergalactic Wickedary of the English Language*. Boston: Beacon Press, 1987.

Damarin, S. Gender and mathematics from a feminist standpoint. In Walter G. Secada, Elizabeth Fennema, and Lisa Byrd Adajian, eds., *New Directions for Equity in Mathematics Education*. Cambridge: Cambridge University Press, 1995, pp. 242–57.

Easlea, Brian, *Fathering the Unthinkable: Masculinity, Scientists, and the Nuclear Arms Race*. London: Pluto Press, 1983.

Euclid. *Elements*. Vol. 1, ed. by Thomas Heath. New York: Dover, 1956.

Griffin, Susan, *Woman and Nature: The Roaring Inside Her*. New York: Harper & Row, 1978.

Harding, Sandra, *The Science Question in Feminism*. Ithaca: Cornell University Press, 1986.

———, *Whose Science? Whose Knowledge? Thinking from Women's Lives*. Ithaca: Cornell University Press, 1991.

Harrold, Francis B. and Raymond A. Eve, eds., *Cult Archaeology & Creationism: Understanding Pseudoscientific Beliefs About the Past*. Iowa City: University of Iowa Press, 1995.

Hayles, Katherine, Gender encoding in fluid mechanics: masculine channels and feminine flows. *Differences* 4: 16–44, 1992.

Holton, Gerald and Duane H. D. Roller, *Foundations of Modern Physical Science*. Addison-Wesley Publishing, 1958.

Jackson, Allyn. Feminist critiques of science. *Notices of the American Mathematical Society* 36(6): 669–72, 1989.

Kellermeier, John. Writing word problems that reflect cultural diversity. *Transformations* 3: 24–30, 1992.

———. Women's studies in a statistics classroom. *Feminist Teacher* 8: 28–31, 1994.

———. Mathematics, gender, and culture. *Transformations* 6(2): 35–53, 1995a.

———. Queer statistics: using lesbigay word problem content in teaching statistics. *National Women's Studies Association Journal* 7(1): 98–108, 1995b.

Nye, Andrea. *Words of Power: A Feminist Reading in the History of Logic*. New York: Routledge, 1990.

Overfield, Kathy. Dirty fingers, grime and slag heaps: purity and the scientific ethic. In *Men's Studies Modified: The Impact of Feminism on the Academic Disciplines*, Dale Spender, ed. New York: Pergamon Press, 1981, pp. 237–48.

Patai, Daphne and Noretta Koertge, *Professing Feminism: Cautionary Tales from the Strange World of Women's Studies*. New York: Basic Books, 1994.

Post, H. R. 1968. Atomism. *Physics Education* 3: 1–3, 1900.

Rose, Hilary. Hand, Brain, and Heart: A Feminist Epistemology for the Natural Sciences. In *Sex and Scientific Inquiry*, Sandra Harding and Jean. F. O'Barr, eds. Chicago: University of Chicago Press, 1987, pp. 265–82.

Ross, Andrew, *Strange Weather: Culture, Science, and Technology in the Age of Limits*. London: Verso, 1994.

———. Introduction. *Social Text* 46/47 14(1,2): 1–13, 1996.

Rosser, Sue V., *Female-Friendly Science: Applying Women's Studies and Theories to Attract Students*. New York: Teachers College Press, 1990.

———. Female Friendly Science: Including Women in Curricular Content and Pedagogy in Science. *Journal of General Education* 42: 191–220, 1993.

———, ed., *Teaching the Majority: Breaking the Gender Barrier in Science, Mathematics, & Engineering*. New York: Teachers College Press, 1995.

Shulman, Bonnie Jean, Implications of feminist critiques of science for the teaching of mathematics and science. *Journal of Women and Minorities in Science and Engineering* 1: 1–15, 1994.

Whewell, William. *Novum organon renovatum*. 3rd ed. London: 1858.

17

The End of Science,
the Central Dogma of Science Studies,
Monsieur Jourdain, and Uncle Vanya

Norman Levitt

John Horgan's recent book, *The End of Science* (1996), is a well-publicized and highly pugnacious addition to the growing literature of analyses of science by nonscientists. Horgan, an editor of *Scientific American*, has extravagant ambitions whose grandeur is evoked by his title. He announces nothing less than the retreat of science into the margins of intellectual life, although he conveniently leaves vague the timescale over which this is supposed to happen. But Horgan is unbending in declaring that the end is in sight and that it is ineluctable. In this regard, he is at one with the essence of radical science studies and less circumspect than his purely academic counterparts in his dismissiveness. But because his thesis is grounded in assumptions that starkly contradict the epistemological bomb throwers who have come to dominate the science studies movement, it is unlikely that he will be welcomed into the fold by everyone. This paradox can be explained, as I will argue, by noting that Horgan's hostility toward science is rooted in a kind of cultural anxiety and that this is also the source of the intellectual distemper that one finds in the superficially distinct manifestos of the science studies movement.

In hopes of avoiding implicit accusations against those whom I don't want to accuse, let me use the term *science studies* narrowly to denote its most vociferous, fashionable, and moralistic faction, whose ostensible radicalism is both political and epistemological.[1] There will be little distortion if I do so, since the ambition and aggressiveness of this sect has all but completely appropriated the label. It is by now widely known that thus defined, science studies has a central dogma—social constructivism[2]—to which all its communicants adhere, albeit occasionally with equivocation and an eye on possible lines of retreat. Set forth bluntly but not, I think, unfairly, the dogma holds that science is a well-entrenched species of ideology, that its theories and constructs are shaped and dictated by "social" forces, which is to say that they encode social, political, and economic ideas to the point that "nature" plays only an auxiliary role—possibly no role at all—in the determination of scientific truth.

It would be going too far to say that everyone who does science studies is wedded to this dictum in its strongest form, yet the notion dominates the field in that it commands automatic deference and constitutes the trump card in debate and polemic. If a science studies scholar wishes to evade the full force of the doctrine, a great deal of tact is required, along with repeated genuflection.[3] Ultimately, the constructivist dogma is the icon of the faith that science studies is on to something new and that it has something vital, indeed earthshaking, to say about the way science works.

The fascinating thing about Horgan's thesis and its elaboration is how deeply its premise contradicts the central constructivist dogma of science studies while simultaneously allowing Horgan to brandish a skepticism—sometimes growing into scorn—toward scientific practice and scientists themselves that echoes the dominant tone of the science studies subculture. Horgan's principal reason for prophesying the senility of science is that it has already made all the great discoveries that human cognitive limitations and the finite range of human technology will allow. Contemplating the great scientific narrative synthesis that encompasses the Big Bang, stellar evolution, the condensation of the solar system, the origins of life and its Darwinian evolution via the molecular mechanisms of DNA, and, finally, the emergence of the human species, Horgan remarks: "My guess is that this narrative that scientists have woven from their knowledge, this modern myth of creation, will be as viable 100 or even 1000 years from now as it is today. Why? Because it is true" (1996a, 16).

This passage will, I think, earn Horgan at least some transient fame, if only because it scorns the contemporary academic diffidence that insists that a proposition cannot be true but merely "true." Such sentiments clearly court the rage of the science studies faithful. It's hard to think of a more succinct and uncompromising contradiction of the constructivist view that the narrative Horgan extols is ephemeral, destined to live and die with the culture that brought it forth. Indeed, this celebration of the enduring accuracy of extant scientific models goes further than many cosmologists and evolutionary biologists, schooled in Popperian propriety, would ever go. Nonetheless, Horgan exploits this encomium to what science has already achieved as part of a strategy that substitutes for the arrant relativism of the constructivists a vision of the limitations of science that is hardly less hostile to scientific self-esteem. For him, science is senescent not because its master narratives are wrong or because such narratives are necessarily phantoms but because science has reached the limit of its capacity to generate further narratives.

Part of Horgan's argument rests on the rather arbitrary notion that a scientific discovery deserves the highest reverence only if it forces an entire society to revise its metaphysical and ontological assumptions, to recast its notions of reality, and to redefine its sense of humanity's place in the cosmos. Clearly, the stunning revision in astronomy that began with Copernicus and culminated with Newton belongs in this category, as does Darwin's great work. It is less clear to me that either relativity theory or quantum mechanics has the same status, even though, by the standard of pure intellectual magnificence, they are unsurpassed. In my view, the cultural implications of these foundation stones of twentieth-century physics are relatively muted simply because comprehension of these ideas falls off so rapidly as one moves away from the world of professional physics and mathematics. Occasionally popularizations—some good, some dreadful—stir the cultural stew a bit, but these have only a brief effect.

It seems to me, however, that the development of molecular biology has indeed had resounding cultural effects, not because of the grandeur of the founding ideas, which to my prejudiced mind simply don't approach the majesty of either relativity or quantum mechanics, but instead because of the exponentially proliferating consequences of biotechnology. Quantum mechanics, too, has generated far-reaching technological developments, but few members of the lay public perceive the connection. By contrast, in the case of biotechnology, almost everyone is at least vaguely aware of the direct link to our understanding of molecular genetic mechanisms.

If we disallow Horgan's postulate that the kind of heroic science he believes is coming to an end is necessarily marked by profound effects on social consciousness, some of the force is removed from his argument. But the core remains to be dealt with: henceforth, few, if any, revolutions in thought will emerge to shake up the scientific cognoscenti, let alone the public, as the great scientific epiphanies of the past have done. It is interesting to watch how Horgan deploys this purported insight to ease himself partway into the camp of science studies and its postmodernist allies, notwithstanding that he begins with a flat denial of their primary article of faith. The gimmick that enables this remarkable bit of footwork is the notion of "ironic science."

Briefly described, ironic science is the kind of speculative (often mathematical) thought that takes over a scientific field when its theoretical questions outrun the possibility of empirical—that is, experimental or observational—confirmation or refutation. Primary examples, in Horgan's view, are superstring theory, attempts to find comprehensible explanatory models for quantum mechanics, cosmological models of the early universe, and certain aspects of evolutionary history. Since this kind of science, on Horgan's view, falls outside the reach of Popperian "falsifiability," it is not really science at all but, rather, the mere creation of fable, mythmaking akin to literary creation or, for that matter, literary criticism. It may intrigue, entertain, or perturb us, but it is not true science, for it tells us nothing reliable about the real world.

In part, Horgan is playing the old logical–positivist game of declaring utterances "meaningless" because no testable predications follow. But he turns this sword against science, including the most mathematical "hard" science, a reversal of the positivist program which in itself certainly deserves to be called ironic. More important to contemporary sensibility, he is mimicking the postmodernist posture, thereby aligning himself in a crucial way with the constructivists, for whom all science is "just a narrative," a discourse to whose elements the label *truth* can be affixed only with an ironic smile. Indeed, Horgan uses "postmodern science" as a synonym for "ironic science." This maneuver also allows him a partial reconciliation with constructivism: If the ironic science practiced by so many eminent thinkers lacks accessible means for testing it empirically and is hence undisciplined by physical reality, a gap is opened for the influence of those factors emphasized by constructivist theory—politics, ideology, and the impulse to echo social myth.

Horgan's flirtation with postmodernism has gone even further since the publication of *The End of Science*. He sprang to the defense of the hapless editors of the postmodernist journal *Social Text* when they fell victim to Alan Sokal's now-famous hoax and included his mock paean to the supposed affinity between postmodern philosophy and cutting-edge physics (1996) in their issue devoted to defending science studies against its critics. In an op-ed piece for the *New York Times*,[4] Horgan asserted that the episode actually revealed Sokal to be the true naïf, a purblind scientist unaware of how far his discipline has wandered into the realm of irony. By contrast, Horgan singled out for praise Ilya Prigogine, naming him as one of the few scientists who have consciously embraced the transformation of their calling into just another branch of postmodernist yarn swapping. To a degree, *Social Text* has returned the compliment: in an essay attempting to extenuate his own gullibility, one of its editors cited Horgan as an authority on quantum mechanics (Robbins 1996).

Using slick and convenient imagery—the rock-solid empirical science of the past dead-ending into the ironic science now being born—Horgan has succeeded, if only for a precarious moment, in straddling the enormous gap between the orthodoxy of scientists and their admirers, on the one hand, and the orthodoxy of science studies, on the other. I think that ultimately he is fated to tumble into that gap, for his posture is inherently unstable. Nonetheless, his audacious performance is worth studying, not for the soundness of its ideas but rather for what it reveals about the parlous cultural position of contemporary science and for what it tells us about the roots of antiscience. I shall defer that analysis for a while, however, to explore why I believe Horgan is mistaken as a prophet.

Why Horgan Is Wrong

To state it again, Horgan's central thesis is that science has, or is about to, run out of problems that are both soluble and profound, in which "soluble" implies that the solution must be testable by practicable experiment and observation, and "profound" means that the solution entails a sharp qualitative revision in the way the experts, if not the population at large, think about the world. I believe that Horgan is wrong in the simplest sense: such problems do exist. Needless to say, they are staggeringly difficult. It could not be otherwise if the demand for profundity is to be met. My suggestion that they are solvable presupposes at least a modicum of hopefulness. It also implies a subtle but important revision of the notion of what it means to validate a solution. For Horgan, this seems to be formulaic; a theory is proposed; it has novel and surprising consequences; excited scientists rush off to their labs or observatories or wherever; and within a reasonably short time, the novel phenomena are detected as predicted.

This is indeed one model of theoretical success, and an extremely important one. The confirmation of various predictions of general relativity is a classic example. But there are important variants. Einstein provides one of these as well in his analysis of the photoelectric effect (the work for which he actually won his Nobel Prize). In cases like these, there is an anomaly, a phenomenon inexplicable in terms of extant theory. Here, the principal work of experimentalists occurs in advance of theory; their task is to show that the anomaly is real and essential, neither a random error nor an artifact of biased experimental procedure. The successful theorist then produces an explanation that at once accounts for the anomaly and expands—or modifies—existing theory so as to integrate the new explanation. The point is simple: confirming evidence may appear before, as well as after, the theory it confirms (or, more likely, there may be confirmation both before and after the theory is formulated). Simple though it is, Horgan's failure to grasp it leads him badly astray.

Consider physics, for instance, where Horgan bluntly derides superstring theory and a number of other approaches to the unification of basic physical laws. He cites those working in this area as leading exemplars of his notion of ironic science, even if (like Edward Witten!!) they are said to be "naive" in not recognizing their status as postmodern ironists. Here, Horgan's own naïveté is exhibited, for he fails to understand the key motivation of the work he criticizes. The two great achievements of twentieth-century physics are relativity and quantum mechanics, each almost miraculously successful in its own predictive sphere. The problem is that they are inconsistent with each other at a deep conceptual level; they both can't be "right" in the literal sense. Heroic attempts have been made to unify them but thus far only partial and tentative success has been won. Superstring theory represents an ambitious and promising, though exceedingly difficult, program to build a unified model for the foundations of physics. If it succeeds in creating a coherent picture that reconciles relativity with quantum mechanics in a fundamental way, this will represent an intellectual triumph almost beyond praise. The important measure of its success will be (as always) logical economy and conceptual simplicity.

Of course, "conceptual simplicity" is another notion with which Horgan has difficulty. Mathematical abstraction makes him squirm. The 10-dimensional space-time of superstring theory, for instance, is, according to Horgan, so bizarre as to be an abomination. But this queasiness reflects the limitations of Horgan's education. For mathematicians and theoretical physicists, the notions that appall Horgan for their supposed lack of intuitive clarity are, in fact, quite intuitive, very accessible, and often rather tame. Superstring theory (or any alternative) will likely succeed if the account it provides turns out to be unified, elegant, economical, and conceptually simple in the judgment of the mathematically literate. That these virtues will not be apparent to Horgan's circumscribed intuitions (and those of most people) may be unfair, but this unfairness is of little weight in deciding whether such an achievement is to be ranked with those of Newton, Maxwell, Einstein, and Heisenberg.

Moreover, although it would be wonderful if immediately testable new predictions were to follow from a fundamental reformulation of physics, this is not the crucial criterion. The most important requirement is to impose conceptual unity on a field whose foundations are, at present, ad hoc to a great extent. This might not be "evidence" in the naive sense of brand-new lab reports, but it counts as evidence in a more general and, in some ways, more important sense. There is an enormous predisposition in favor of theoretical proposals that achieve this kind of unity, and the history of science, as well as post hoc philosophizing, shows this to be essentially sound. Horgan's jibes at superstring theory and the like make sense only if this principle is disregarded.

As I see it, the central, long-term project of theoretical physics is to produce a coherent foundational point of view, one that not only unites (perhaps with subtle modifications) relativity and quantum mechanics but also provides a model that dispels the seeming paradoxes that have bedeviled our understanding of the latter. This is essentially a mathematical problem. An elegant, logically economical mathematical model that meets these so far unmet stipulations will, quite appropriately, strike physicists as having a strong claim to be viewed as a faithful model of physical reality. Novel empirical confirmation won't be needed for the simple reason that an ocean of confirmation—what is usually called *retrodictive confirmation*—already exists in the form of the experiments and technology that verify both relativity and quantum mechanics within their respective predictive domains.[5]

In connection with superstring theory in particular, Horgan is simply wrong about some of the assertions that he uses to justify the imputation of "irony" to this frontier of physics. The experts in the field, including Witten, contend that at least some predictions specific to the theory can be tested by devices that exist or are feasible to build, and some experimentalists agree.[6] (The same might well hold for various conceptually intriguing attempts to reformulate quantum mechanics in a spirit antithetical to the "Copenhagen interpretation," which Horgan also regards as "ironic.")[7] Thus, the bottom drops out of Horgan's central argument.

Turning to biology, one is struck by the peculiar dogmatism of Horgan's belief that the elucidation of DNA's structure not only constituted the last great revolution in the field but the last possible one as well. Together with the Darwinian evolutionary paradigm, it supposedly forms the broad conceptual envelope marking the limits of the field, thereby reducing all current and future work to an important but unexciting mopping-up operation that can never attain the same intellectual grandeur. I think this is wrong and that Horgan, though he flirts with the issues that render it dead wrong, is too enthralled with his own hypothesis to look at them carefully. Several times, he alludes to the central problem of developmental biology—"how a single fertilized egg becomes a salamander or an evangelist" is how he puts it—but treats it as one of those intricate but unexciting technical problems that are all "the end of science" will allow us.

There is a profound error here, although it is perhaps extenuated by the fact that the topic comes up in Horgan's discussions with evolutionary theorists, not embryologists. The fact is that in forming a large-scale model of how heredity operates, the work that has been done to date, brilliant and significant as much of it is, has dealt with only the relatively easy aspects of this central question. Our understanding of how genes correlate to somatic features of a complex organism is still, frankly, primitive. We are still not much past the "gene-for-characteristic-X" stage. The really deep question is not what somatic variations the various alleles of a gene correspond to but, rather, the egg-to-salamander question: the difficult morphogenetic question of how the genotype, taken as a whole, determines somatype in the mature organism. The ideal sought is nothing less than a method of transcribing the finite code of the genotype into an accurate description of the animal or plant coded for. In other words, we want to be able to look at a cow genome, say, and, merely on the basis of what is found at that level plus a general description of the environment, derive an accurate picture of a large, placid ruminant that lives in herds and says "moo." We have no clear idea how to do this. We haven't even invented the conceptual categories that would be required, those morphological ideas that would enable us to describe systematically the physical form of creatures and the geometric dynamics of morphogenesis. Even if those were on hand, we would have to crack the coding scheme that allows the transformation of the discrete information embedded in the chromosomes into a morphological process. How vast this problem is may be grasped when we consider that its infinitely simpler cousin, the protein-folding problem, is still a great conundrum; we don't yet understand how an individual gene determines the geometric structure of an individual protein.

A huge conceptual apparatus will have to be built if the morphogenetic problem is to be solved. Much of this will undoubtedly call on computer modeling using techniques now undreamt of. Even more important, it is almost inevitable that challenging mathematical ideas will be involved. Developmental biologists will have to become as mathematically adept as cosmologists. Progress will depend on many breakthroughs, each of which will have to be at least as intellectually deep as the crystallography that went into finding the structure of DNA, and probably much deeper. It is certain that once found, or even partially achieved, the solution will rank, contra Horgan, with the greatest of historical scientific revolutions and may well have many startling things to say about the human condition.

Another problem that is enormously difficult but by no means hopeless is that of artificial intelligence. Let me formulate this in a way that will sidestep philosophical feuds over what intelligence or consciousness is: Is it possible to build a "machine"—hardware, software, or some combination thereof—that will pass the famous Turing test? To keep up to date with technology, let's imagine this as a device that can create the image of a recognizable human on a TV monitor and animate that image realistically while synthesizing appropriate speech. The test subject, a real person, "converses" with the simulacrum and also, via an audiovisual hookup, with the control, another real person. If the subject (on average over many tests and subjects) can't reliably distinguish which of the two respondents is the real human, the device will have passed the Turing test.

It is obvious that if such a gadget is ever built, the philosophical, ethical, and cultural ramifications will be stupendous, easily ranking with those that attended the Copernican and Darwinian revolutions. On the other hand, if it is somehow shown that it is impossible to contrive a Turing test–passing machine, the implications will be hardly less cataclysmic. Either case represents precisely the kind of scientific leap that Horgan categorically disallows.

Other examples from various fields—cosmology, evolutionary biology, cognitive neuroscience—could easily be cited. Each involves problems of a very high order that will certainly be worked on and might very well be solved. Success in any of these areas would necessitate epochal conceptual developments and would carry intriguing philosophical implications. In other words, they would constitute the kind of heroic—and definitely nonironic—science that Horgan declares at an end. Since Horgan, an experienced science writer, is not particularly ignorant—certainly not by a lay journalist's standards—his emphatic insistence on the impossibility of what is clearly possible invites inspection of his motives, acknowledged or otherwise.

Horgan, Science Studies, and Cultural Anxiety

In the first section, I noted the affinities that link John Horgan's ideas to academic work in science studies, as well as the differences. To my mind, however, the affinities are more interesting, but not on account of their manifest theoretical content. Rather, it seems to me, the important connections are found on the level of personal, social, and cultural attitudes. In pursuing this point, I shall be openly disrespectful of both Horgan and the science studies camp, because in neither case are we dealing with bodies of thought that are compelling or even interesting in themselves. They are important only because they reflect wider and sometimes inarticulate attitudes toward science.

The practitioners of science studies, for their part, hardly bother to hide their disdain for the object of their scrutiny. Says one critic, "Long enshrined as a kind of apex of rational knowledge production, so powerful as to remain largely immune to the vicissitudes of social change, science is up for deconstruction just like all the rest of the Western canonical fare" (Franklin 1996, 142). With allowance for variations in emphasis and mode of special pleading, this sentiment echoes and reechoes throughout the science studies literature. Horgan, however, tries to take an elegiac tone: science, that once-Olympian enterprise, has now fallen into stagnation, and all its grandeur lies in the past. But as many reviewers of *The End of Science* have noted,[8] resentment runs so strongly in Horgan's soul that he repeatedly lapses into inadvertent self-disclosure.

In his anxiety to establish the epochal nature of his own notions, particularly "ironic science," Horgan becomes notably churlish and self-aggrandizing:

> Much of modern cosmology, particularly those aspects inspired by unified theories of particle physics and other esoteric ideas, is preposterous. Or, rather, it is ironic science, science that is not experimentally testable or resolvable even in principle and therefore is not science in the strict sense at all. Its primary function is to keep us awestruck before the mystery of the cosmos. (Horgan 1996a, 94)

Even worse, Horgan's sneering descends into the silliest kind of ad hominem attack against most of the scientists he describes. He ridicules their appearance, their mannerisms of speech, even their office furnishings.[9] If they disagree with him, they are fools or hypocrites, and he has no compunctions about turning mind reader in order to claim that the victims (that's the right word) of his interviews really do agree with him even when they say they don't. This is poor writing and wretched journalism. But it's fascinating because it suggests that the overt disparity between Horgan's ideas and the constructivist central dogma of science studies is, finally, of little importance. Both doctrines emerge, I think, from a deep pool of resentment that has become central to contemporary culture and is manifest at various cultural levels, as much in UFOlogy and homeopathy as in pretentious academic science studies programs—or books by disgruntled science journalists.

Whence does this arise? Clearly, the discontents of the modern world generate a degree of distrust of science, in that science is implicated in all aspects of modernity, good or ill. But for those associated with "high" culture in some fashion, other elements infuse the brew. In speaking of the "end" of science, Horgan echoes the mortuary tone of much current discourse. We have had the "end" of this and the "death" of that from all sorts of writers. Corpses seemingly litter the ground; God, the self, the author, ideology, representational art, nonrepresentational art—the list is endless. This litany reflects a culture that, to many, itself seems dead in the water.

I invite the reader to play a mental game suggested by the philosopher John Searle[10] to account for the sullenness of current intellectual life. Think of various fields of art and thought, and compare the outstanding figures of the turn of the century with those at work today or in the recent past. Is there a novelist we would wish to match with Tolstoy or Henry James, a composer with Brahms or Debussy or Mahler or Strauss, a painter with Degas or Picasso or Cézanne, a sculptor with Rodin, a poet with Yeats or Hardy, a playwright with Shaw or Wilde or Ibsen or Chekhov, a philosopher with

Nietzsche or Russell? The composer Krzysztof Penderecki (once an aggressive experimentalist) has it about right: "We live in a decadent time because in the arts there is absolutely nothing new happening. It's not a period of discovery. It's no longer possible to find something which will shock other people, because everything has already been done" (quoted by Schwarz 1996, 33).

This sense of stagnation, moreover, applies to wider matters than those that merely engage the highly literate. At the end of what has been a century of unprecedented murderousness (facilitated, of course, by scientific ingenuity), the best we can say for ourselves is that we have survived. What may be thought of as the central political project of modern Western culture—the leveling of great disparities in power and wealth, the diminution of hereditary privilege, the scouring of superstition and obscurantism from public discourse—seems to be running out of steam. The egalitarian dream, most fully articulated in various socialist movements, is largely moribund, notwithstanding its volubility in narrow venues of identity politics. The sense that the moral progress of civilization is a live possibility, if not historically inevitable, has faded to the point of invisibility, and attempts to revive it have a somewhat frantic quality. Yet, if we live in a cul de sac culture, we nonetheless find one grand exception to the general air of desuetude: the natural sciences.

I shall merely assert what can easily be argued: From the conceptual point of view, the sciences are in an unprecedentedly robust state of health, strength, and vigor. Theoretical understanding from biology to physics is deeper and sharper than it has ever been. Overall, there is greater unity and greater cross-fertilization among the various scientific disciplines than has ever been seen. The monistic, reductionistic point of view that forms the main philosophical current of science seems increasingly to be vindicated by a string of breakthroughs.[11] The major puzzles that certainly exist are regarded with delight by young practitioners eager to unravel them. This is, I think, self-evident to most scientists and leads to a touch of smugness on their part.

From the point of view of many nonscientific intellectuals, however, what seems like such a happy circumstance to scientists looks far less like a cause for celebration. The asymmetry between the scientific and nonscientific cultures does not console but, rather, irritates. If the general culture seems headed into a stagnant slough where irony infuses everything and all values are transient or self-negating, why shouldn't science be drawn into the overall disaster? The attitudes that generate this kind of querulousness are generally called "postmodern." Under its vaunted irony one usually finds diffuse rage at having been born too late for intellectual and artistic self-confidence. The fact that such self-confidence permeates the sciences, where it is so casual that scientists are hardly conscious of it, leads to bitterness and sullen thoughts. In the case of science studies, the resulting ill will is amplified by an ethical and political commitment to the values of the left.[12] The epistemological radicalism of the genre is matched by a self-ascribed political radicalism, in which traditional socialism shares space with radical feminism and a partiality for the supposed virtues of nonwhite and non-Western peoples. In this community, the "privileging" of "Western" science as a uniquely accurate way of finding out how things work is, of course, seen as an aspect of Western white male hegemony and is therefore to be deplored.[13] These feelings are all the more ferocious because, in a practical sense, the left has little sense of what to do or where to turn in the real world where power and authority are contested. Consequently, it retreats into the comforts of sweeping denunciation. A Wagstaffian "whatever it is, I'm against it" mood flourishes.

By contrast, Horgan's own resentment, though just as strong, doesn't seem to flow from overt political commitments but simply from the sense that the special authority of science (and of scientists) is unjust because it leaves little room for intellectuals like him to claim to be doing work of comparable importance or validity. Unlike the science studies faithful, he doesn't daydream about a new scientific order in which the power to oversee science would be handed over to feminists, radical environmentalists, variously oppressed ethnic groups, and so forth. What he seems to want instead is the personal authority to condescend to science and scientists, to tell them, in effect, that much of what they are doing is self-deluded nonsense.[14]

If we compare Penderecki's remark with Horgan's "end of science" thesis, the commonalities leap out at us; the culture is played out and what once was vital is now moribund. The difference, however, is that Penderecki is talking mostly about music, in which as a composer, he has spent years trying to find a congenial style for himself. Horgan, however, is not a frustrated scientist but a frustrated amateur metaphysician. He has ascribed a role to science and is furious that scientists won't accept it. He insists that many of today's most inventive scientists have, unawares, been practicing ironic science all along. Unlike Monsieur Jourdain, they are not pleased to learn of this unwitting accomplishment. This traps Horgan in a feedback loop. The more he insists on the "ironic" quality of contemporary science, the more contemporary scientists dismiss him. The more he is ignored, the more he persuades himself that science is sinking irretrievably into irony.

Speaking of Monsieur Jourdain, it is certainly possible to discern something of his brusque philistinism in John Horgan. But Chekhov's Uncle Vanya perhaps provides a better prototype. Horgan seems mired in a swamp of resentment and thwarted ambition. He has spent years as a science journalist, serving, as he admits, the vanity of scientists and their desire for favorable PR:

> I therefore did my best, in writing later about the ideas of Hawking and other cosmologists, to make them sound plausible, to instill awe and comprehension instead of skepticism and confusion in readers. That is the job of the science writer after all.
>
> But sometimes the clearest science writing is the most dishonest. (Horgan 1996a, 93)

Now he has convinced himself that he has something of great philosophical importance to say, which reveals, among other things, the fatuity of the supposedly great scientists he has served so faithfully. The comical, pathetic, and self-defeating tone of Horgan's book derives from this. He rails at the scientists who were once his idols, but the shots he fires off go far wide of their intended targets.

There is a strong taint of Vanya-ism in the science studies movement as well, although there it has political trappings. Many people in the field exhibit a desperate need to find reasons for believing that it is they who these days generate the stunning insights and the seismic revolutions in thought. They contrive to see scientists much as Horgan sees them, though without using his special vocabulary: scientists are oblivious, blinkered, laboratory automata who are unaware of what they are really doing. Horgan has them doing ironic science; the constructivists have them following the unperceived imperatives of a social ideology. In either case, scientists are to be brought low and put in their place.

Vanya-ism of this kind is not limited to card-carrying intellectuals. The prestige and insularity of science—and perhaps the Serebryakov-ism of some scientists—evoke it in many ordinary people, who can't help yearning for the authority to define what is to be taken as knowledge rather than myth and for the insight to comprehend or even construct a sweeping vision of the universe. The popularity of various modes of pseudoscience and parascience, almost all of which are characterized by pugnacity toward orthodox science, is evidence that Vanya is a ubiquitous archetype. His incarnation in a science journalist like John Horgan and in the questionable scholarship of what insists on describing itself as science studies should warn scientists that the task of maintaining a culture hospitable to science is difficult and unremitting.

Notes

1. To give a reasonable sense of the range of opinion representing what I have chosen here to call "radical" science studies, let me mention some names: Paul Forman, Simon Schaffer, Karin Knorr-Cetina, Steve Fuller, Sandra Harding, Donna Haraway, Steven Shapin, Bruno Latour, David Bloor, Barry Barnes, Steve Woolgar, Evelyn Fox Keller, Stanley Aronowitz, Sharon Traweek, Trevor Pinch, Sheila Jasanoff, Harry Collins, and Hilary Rose. For my purposes, any point in the area roughly defined by this selection of representative instances (the "convex hull," to take a metaphor from my home discipline) would be counted as a radical position. Here, the term *radical* refers chiefly to epistemology rather than left-wing politics; however, a majority of those listed count themselves as radical in this secondary sense as well (although some refuse that label).

A number of scholars, however, pursue "radical" science studies, in the sense that they link their scholarly efforts to hopes for egalitarian social chance while harboring doubts about the epistemological radicalism and postmodern enthusiasms allied with it. Moreover, the political radicalism of many epistemological radicals is, in some politically radical circles, viewed as more problematical than helpful (see, e.g., Nanda 1997).

2. The term *social constructivism* deserves some scrutiny. In essence, what I describe here is the social constructivism of the so-called Strong Programme associated with the Edinburgh school of sociology of science. Sometimes, however, the term is used to denote a far less radical doctrine, more concerned with the social context in which institutional science arises and endures than with showing that scientific knowledge is merely the transcription of sociopolitical notions and processes. This label is sometimes eschewed by theorists whose doctrines seem to call for it. In this case, one suspects that the rejection of the specific designation is an artifact of academic competitiveness, in which originality is at a premium and being seen as a mere follower of someone else's school is not a good career move. The trick here is to devise a minor variant of social constructivism, one that offers the same emotional and political satisfactions while nominally sailing under a different banner.

3. See the long, two-part review article, "Truth and Objectivity," by Paul Forman, in two successive issues of *Science*, which is chiefly concerned with upbraiding backsliders from the constructivist faith. This makes clear how far a science studies scholar risks loss of face for failure to play, with sufficient reckless enthusiasm, what Collins and Yearley call "epistemological chicken."

4. John Horgan, "Science Cut Free from Truth," *New York Times*, July 16, 1996. See as well the subsequent replies of Sokal (July 22), Berndt Mueller (July 20), and Sidney Coleman (July 23) in the letters-to-the editor section.

Horgan's ambivalent attitude toward science studies is also on view in a recent book review by him ("It's Not Easy Being Green," *New York Times Book Review*, January 12, 1997). He seems to agree with scientists who assert that much of this work is "pretentious gibberish" yet finds virtue in some of it. For instance, he cites with approval Sandra Harding's notion of "strong objectivity," although it's not clear that he knows much about this highly politicized formula beyond the innocuous label Harding has attached to it (see, e.g., Harding 1992).

Some commentators, by contrast, are not shy about comparing Horgan himself with the more egregious postmodern thinkers. John Casti's "Lighter Than Air" declares that Horgan employs "terms like 'law,' 'discovery,' 'fundamental,' and even 'science,' more as they might be employed in a journal of deconstructionist literary criticism or, perhaps, as they would be propounded by certain continental philosophers whose names I shall pass over with the silence of the grave."

5. Superstring theory, to the extent that it has been developed thus far, has been criticized in the physics community for positing phenomena whose direct testing seems impossible in practical terms, now or in the foreseeable future. Horgan is certainly aware of this, and it underlies his critique. However, he fails to understand how much even the critics will be obliged to concede if superstring theory (or any rival) achieves the kind of conceptual unity that has been so elusive for 70 years.

6. This observation is based on remarks by Witten and other physicists (including experimentalists) made in my presence.

7. See Cushing 1995 for a discussion of this point, especially in regard to David Bohm's elaboration of the Louis de Broglie "pilot wave" model.

8. For some sardonic reviews of *The End of Science*, see Horvitz 1997, Goodstein 1996, or Hayes 1996, as well as Casti 1996.

See also the review "La fin de la science?" by Bruno Latour, one of the proponents of "radical science studies" listed in n. 1. Latour is scornful of Horgan, not because Horgan sneers at science, but because he does not sneer properly following Latouresque style: "However, the author [Horgan] is probably right if, by 'the end of science,' we mean the end of a facile argument that jumbles together the unending progress of budgets and of knowledges, the fight against religious obscurantism, the dream of mastery, divine omniscience, the fat budgets of the Cold War, certain edifying histories of Galileo or Newton or Mendel, and a flagrant incomprehension of the social world" (1997, 97). Latour takes care to commend, in place of Horgan's notion, the attitude of his current collaborator, Isabel Stengers. This is in keeping with the fact that of the prominent scientists canvased by Horgan, Latour specifically singles out for beatification another of Stengers's partners, Ilya Prigogine (*Entre le temps et l'éternité*). This does not strike me as the most inevitable choice (see J. Bricmont, "Chaos in Science or Science in Chaos?"). Needless to say, unlike Horgan, Latour nowhere actually addresses any germane scientific issues.

9. "His dimly lit office, lined with dark, heavily varnished bookcases and cabinets, was as solemn as a funeral parlor" (Horgan 1996a, 63, describing Sheldon Glashow's office).

10. Searle made this suggestion in response to a question during a lecture at Rutgers University in 1991.

11. See Weinberg 1995 for a revealing discussion of reductionism and the appropriateness of its association with the scientific worldview. Alternatively, see chapter 3, "Two Cheers for Reductionism," in Weinberg 1992. Let me quote a sentence from the latter piece in antici-

pation of some of the objections likely to erupt in reaction to the red-flag word, *reductionism*. Says Weinberg: "For me, reductionism is not a guideline for research programs, but an attitude towards nature itself." A similar theme is sounded in Robert A. Frosch's brief "Reductionism and the Unity of Science" (2).

12. It would perhaps be more accurate, and more fair, to those with left-wing views (including mine) to note that postmodern academic fashion seems to champion one particular variety (and a hothouse variety at that) of left-wing values.

13. See Gladwell 1996 and Eichman 1996 (wry accounts of a debate between Alan Sokal and some of the *Social Text* editors victimized by his prank) for amusing anecdotal evidence of this peculiar solicitude for "non-Western ways of knowing" when they come into conflict with well-confirmed science.

14. Yves Gingras, a sociologist of science, suggests (personal communication) that Horgan's indignation arises from the recent tendency of some scientists (mostly physicists) to turn out popular work with strong quasi-theological, or even unabashedly theological, pretensions. Although I see this as a possible irritant (I'm not overfond of the genre myself), I don't agree that it's the main source of Horgan's bile.

References

Casti, J. 1996. Lighter than air. *Nature*, August 29, 769–70.

Collins, H. M., and S. Yearley. 1992. Epistemological chicken. In *Science as Practice and Culture*, ed. A. Pickering, 301–26. Chicago: University of Chicago Press.

Cushing, J. T. 1995. *Quantum Mechanics: Historical Contingency and the Copenhagen Hegemony*. Chicago: University of Chicago Press.

Eichman, E. 1996. The end of the affair. *New Criterion*, December, 77–80.

Franklin, S. 1996. Making transparencies: seeing through the science wars. *Social Text* 46–47 (Spring–Summer): 141–55.

Forman, P. 1995a. Truth and objectivity: part 1, irony. *Science*, July 28, 565–67.

———. 1995b. Truth and objectivity: part 2, trust. *Science*, August 4, 707–10.

Frosch, R. A. 1997. Reductionism and the unity of science. *American Scientist*, January–February, 2.

Gladwell, M. 1996. A matter of gravity. *New Yorker*, November 11, 36–38.

Goodstein, D. 1996. The age of irony. Review of *The End of Science*, by J. Horgan. *Science*, June 14, 1594.

Hayes, B. 1996. The end of science writing? Review of *The End of Science*, by J. Horgan. *American Scientist*, September–October, 495–96.

Harding, S. 1992. After the neutrality ideal: science, politics and "strong objectivity." *Social Research* 59: 567–87.

Horgan, J. 1996a. *The End of Science: Facing the Limits of Knowledge in the Twilight of the Scientific Age*. New York: Addison Wesley.

———. 1996b. Science set free from truth. *New York Times*, July 16, 17.

———. 1997. It's not easy being green. Review of *The Idea of Biodiversity*, by D. Takacs. *New York Times Book Review*, January 12, 8.

Horvitz, L. 1997. The enemies of science. *Boston Book Review*, January–February, 28–29.

Latour, B. 1997. La fin de science? *La Recherche*, January, 97.

Nanda, M. 1997. The science wars in India. *Dissent*, Winter, 78–83.

Prigogine, I., and I. Stengers. 1988. *Entre le temps et l'éternité*. Paris: Fayard.

Robbins, B. 1996. Anatomy of a hoax. *Tikkun*, September–October, 58–59.

Schwarz, K. R. 1996. First a firebrand, then a romantic. Now what? *New York Times*, October 20, sec. 2, 33.

Sokal, Alan. 1996. Transgressing the boundaries: toward a transformative hermeneutics of quantum gravity. *Social Text*, Spring–Summer, 217–52.

Weinberg, S. 1992. *Dreams of a Final Theory*. New York: Pantheon.

———. 1995. Reductionism redux. *New York Review of Books*, October 5, 39–42.

18

The Epistemic Charity of the
Social Constructivist Critics
of Science and Why the
Third World Should Refuse the Offer

Meera Nanda

If we forgive too much, we understand nothing.

Ernest Gellner

Ethnoscience: Gift or Charity?

They say it is impolite to look a gift horse in the mouth. It is indeed doubly impolite if the gift was intended as a token of respect and solidarity. Yet it is precisely this rather delicate task of returning a well-intended gift that I have taken upon myself in this essay. I hope to carry out my assignment politely but firmly, and always with utmost regard for the generosity of the spirit behind the gift.

The gift I want to return is the cluster of theories that forbids outsiders from evaluating the truth or falsity of any beliefs of other people in other cultures from the vantage point of what is scientifically known about the world and, conversely, allows the insiders to reject as ethnocentric and imperialistic any truth claim that does not use locally accepted metaphysical categories and rules of justification. These theories hold that because modern "Western" science is but one among the many ways of understanding the world and is as embedded in its own cultural context of production as other knowledges are in theirs, it cannot serve as a transculturally valid source of knowledge. All sciences are ethnosciences, and none is more universally true than any other.

This gift has many names, many givers, and many presumed beneficiaries. It is variously called *ethnoscience,* situated knowledge, anti–Northern Eurocentric, or *postcolonial* science—labels that derive their force from their parental rubric of social constructivist theories of science.[1] Its most generous sponsors are the self-consciously left and often self-described postmodern academics from North American and European universities (and increasingly also from non-Western universities as well), who see any claim of universality of modern science as the West's ploy for "disvaluing local concerns and knowledge and legitimating outside experts," as Sandra Harding put it (1994, 319).

The most aggressive consumers of ethnoscience are the equally "left" postcolonial intellectuals and activists associated with cultural/religious and other "new social movements" that aim to purge their cultures of all alien (mostly Western) elements. These intellectuals and movements openly and stridently reject the calls of earlier modernist/anticolonial "people's science movements" in favor of postmodernist/postcolonial "alternative science movements." Whereas the former sought to assimilate modern science into local settings as a means of cultural change and economic development, the latter see modern "Western" science as a source of all that ails non-Western societies and seek alternative "ways of knowing" grounded in their own civilizations. The Western and Third World critics of the universality of science are united in reversing the terms of respect in Sandra Harding's statement quoted earlier; that is, they want to value local concerns and knowledge and delegitimize outside experts, assuming all the while that the local and the outside are irreconcilable and that the knowledge of the "outside" experts—that is, modern science—is nothing more than an imposition on reluctant local knowers.

What unites—and justifies—all theoretical versions of science as ethnoscience is a well-elaborated theoretical edifice of the social construction of science. Social constructivist theories hold as their first principle that the standards of evaluation of truth, rationality, success, and progressiveness[2] are relative to a culture's assumptions and that the ways of seeing further vary with gender, class, race, and caste in any given culture. Such relativization has some "liberating" effects for both the sponsors and the intended beneficiaries of ethnosciences. To begin with, it frees both of them from ever having to say that others, or they themselves, though acting rationally by local standards, could be holding false beliefs and that these beliefs can or should be corrected in the light of what we—Westerners and non-Westerners—have collectively learned about the world through the methods and institutions of modern science in the last 300 years. The gift of ethnoscience simply does not allow for any transcultural reasons for accepting some facts of the matter as justified beyond reasonable doubt. And in any case, the gift givers and takers have convinced themselves that such boring "facts" about the material world are not "morally salient" to how we live our lives.[3]

Such a conception of science as ethnoscience lies at the heart of the postcolonial project of "provincializing Europe" (Chakrabarty 1992, 20), with the goal of recovering and giving voice to the silenced sciences of the colonized. The postcolonial critics argue that because modern science is merely one ethnoscience among others—its "universals" put in place by the violence and rapaciousness of colonialism and not by its superior claims to validity—non-Western cultures can never gain a true knowledge of the natural world as *they* experience it unless they develop their own "alternative universals" grounded in indigenous categories, cultural idioms, and traditions.[4]

Once the once-colonized peoples understand that social location and cultural meanings make a difference to scientific knowledge, the argument goes, they can at least hope to "decolonize their minds" by repudiating the knowledge brought in by the colonizers: to overcome the colonial Kipling, the colonized must learn to repudiate the rationalist Descartes as well.[5] The gift-givers see Cartesian rationalism and the entire tradition of the Enlightenment as having prepared the grounds for colonialism by disassociating scientific knowledge from the rest of the culture and holding it up as a gold standard, so to speak, in the international exchange of ideas. It is much better to stop worrying about any such universal abstractions and instead try to understand why other people hold the beliefs *they* think are true and what meanings and purposes their own truths have for *them*. Charity begins with giving up objective truth and embracing hermeneutic truth.[6]

Why should anyone want to refuse so generous a gift, least of all someone like myself whose own native Indian culture was berated for so long by the British rulers as irrational, mystical, and superstitious? How can anyone urge ex-colonial people to refuse this poultice of relativism when they are still so obviously smarting from the indignities of colonialism and when they need to affirm their identities to resist the seductions of the fast-encroaching McWorld? And how can anyone even think of speaking up for the old discredited modernist myth of bringing science to the masses in Third World societies when it is Western science that has long denied them their "epistemological rights" to live by their own lights?[7]

My reason for urging a rejection of ethnoscience is this: What from the perspective of Western liberal givers looks like a tolerant, nonjudgmental, therapeutic "permission to be different"[8] appears to some of us "others" as a condescending act of charity. This epistemic charity dehumanizes us by denying us the capacity for a reasoned modification of our beliefs in the light of better evidence made available by the methods of modern science. This kind of charity, moreover, enjoins us to stop struggling against the limits that our cultural heritage imposes on our knowledge and our freedoms and to accept—and in some Third Worldist and feminist accounts, even celebrate—these cultural bonds as the ultimate source of all "authentic" norms of truth, beauty, and goodness. Moreover, such an injunction to prefer authenticity over truth, or at least consider authenticity as a determinant of truth, severely limits people's science movements in non-Western societies that strive to challenge the claims of local standards of truth and morality. For these movements wishing to popularize science, there is poison in the gift of ethnoscience. It is mainly on behalf of these movements that I wish to return the "gift."

But even as I do that, I don't wish to minimize the importance of what the gift givers have accomplished. They have challenged the smugness of the colonizers who defined themselves as enlightened, rational, and free against the backdrop of the innately emotional, superstitious, and unfree "Other." In the past, such a bipolar worldview has given credence to theories of racial and civilizational essences, which were in turn used to justify the unspeakable evils of genocide, slavery, and colonialism. The nonessentialist sentiment that informs social constructivist theories of knowledge, and postmodernism in general, is indeed a gift for all those whose difference has been read as aberration.

Yet, the antiessentialist argument that best demonstrates the historical contingency of the supposedly innate characters of civilizations is entirely inappropriate when applied to the purpose and norms of knowledge. Countering the imperialists' factually erroneous claim that the natives are inherently irrational by insisting that rationality itself is a Western obsession, or that people in non-Western cultures participate in "forms of life" in which rationality has a different point for them altogether, is only a Pyrrhic victory, for it leaves the non-Western peoples so fundamentally and incommensurably different—truly an "other"—that they could conceivably have no use of the facts of the natural world uncovered by the rationality of modern "Western" science. By defining the very nature of rationality and truth as internal to social practices, social constructivist theories do indeed give the natives their "permission" to be different—but, then, so did apartheid.

Understandably, the gift givers are dismayed to have their generosity interpreted as charity, ethnosciences seen as antiscience, and their invitation to be different read as intellectual apartheid. This is not what they meant at all, they assure us Third World ingrates. The thoughtful among them take seriously the charge of condescension and deny that respectfully understanding others' beliefs implies a suspension of critical judgment. Such a project only demands, they claim, that every society should use criteria that are internal to its own "specific historical tradition," for the criticism of its own knowledge and values.[9] But if Western sciences must be used to criticize non-Western practices, they insist, the critics must acknowledge that Western science is not a god's eye view but a situated, ethnocentric, Western view of the world (see Rentlen 1988).

The gift givers thus assure us that they are not against modern science per se, but only against its universal pretensions. They wish science to confess its culture, take on an ethnic middle name ("Western") and become one among many other ethnosciences. Thus provincialized, science is deemed acceptable for certain limited and purely instrumental purposes, with no claims to truth, worldviews, and social values. "Modern 'Western' science, yes, but modern universal science, no!" could well summarize the prevailing ethnoscience/postcolonial position. This position, incidentally, turns Joseph Needham's words on their head: "Modern universal science, yes, Western science, no!" (1969, 54).

The rest of this essay can be read as my reasons for why Joseph Needham's old-fashioned "modern universal science" is better suited for the purpose of an honest, internal self-critique of non-Western cultures than is a science that confesses its culture by acquiring the new ethnic middle name "Western." To that end, I first demonstrate that treating science as "Western" has already contributed to a rather uncritical adoration of the nation and its traditions in many parts of the Third World. The growth of local tyrannies, each justifying itself by culturally authentic standards is not a far-fetched fear at all, and the proponents of ethnosciences cannot ignore the unintended—and frightening—consequences of their theories. Next, I try to explain exactly how, at what point, and through what logic the gift of ethnoscience turns into charity and the worthy opposition to colonialism ends up justifying intellectual apartheid. To that end, I tease apart the arguments of the Strong Programme (SP) of the sociology of science, the ur-text of constructivism, to show how it grants "others" rationality but only by weakening rationality to the point that a critical evaluation of the knowledge claims of any culture falls from view altogether. Finally, I offer a sociological argument as to why such a weak and nonevaluative rationality is inadequate for the rapidly modernizing Third World societies in which the certainties and securities of traditions are fast disappearing.

Ethnoscience in Action

How is the post in *post*colonial related to the post in *post*modern science critiques? My answer is simple: the postmodern elements of the constructivist science critique strengthen the *pre*modern elements of postcolonial societies.

Even though the proponents of ethnoscience speak in more Durkheimian and often in (vulgarized) Marxian and strongly feminist dialects, they share many standard postmodern views of the impossibility and undesirability of objective knowledge. Postmodernism in its various versions is skeptical of the representational view of knowledge in which facts are supposed to correspond to elements of a pregiven reality. Postmodernists see the knower, knowledge, and reality all as active constructs of one another, all bearing the marks of the socio-historical context. Thus, knowledge both participates in the construction of reality and is itself a construction.[10]

It is precisely the adoption of a nonrepresentational, postmodern view of knowledge that distinguishes postcolonial critics from all previous cross-cultural critics of science. In the postcolonial lexicon, roughly speaking, modernism implies colonialism, whereas postmodernism implies freedom from coloniality or postcoloniality. The modernist, objective science is seen as having directly legitimated colonial rule by treating local knowledges of other cultures as distorted representations of reality against the benchmark of scientific knowledge. Conversely, the postmodern view of knowledge is seen as "returning" to the many "Others" the capacity to have true knowledge, for it does not hold up Western science and Western view of reality as universally valid. On the contrary, it allows for understanding how cultural stratagems and even violence "play a decisive role in the establishment of meaning, in the creation of truth regimes, in deciding whose and which universals win" (Chakrabarty 1992, 21).[11] If the balance of forces were different, the argument goes, the non-Western representations of "reality" would be universally accepted as "true," for to paraphrase Chakrabarty again, "truth is but a dialect backed by an army" (21).[12] The postcolonial enterprise seeks not only to make visible the force that accompanied the universalization of "Western" science but also to release the Third World from the thrall of Western modernity and enable it to seek its own alternative universals.[13]

Thus the post of postmodernism and the post of postcolonialism make common cause in their desire to go past the Enlightenment vision of the progressive growth of knowledge leading to a flourishing of human capabilities. (Their objection is obviously not to the flourishing but to the idea that the growth of scientific knowledge has anything positive to do with it). The crucial difference is that when post*modernism* repudiates the Enlightenment, it declares itself free of the need for *any* ancestors at all; it refuses to seek shelter from contingency in any firm foundations of a priori or empirical truths. The post*colonial* critics of the Enlightenment, however, *only* repudiate alien ancestors that they wish to replace with their own ancestors and the voices embedded in the folk traditions: the postcolonial is post-Western but definitely not postnational, at least in science studies.[14] But the two posts are so vehemently united in what they don't like—something they call "Enlightenment Reason"—that for the most part, they overlook the rest of each other's projects. This sincere and yet incomplete alliance between the two is the source of much confusion, with the consistent postmodernist taking great umbrage at being lumped in with the nativists, and the postcolonials, in turn, repudiating the total skepticism of the postmodern.

It is my contention that the epistemic charity of the postmodern and the postcolonial science critics lies in the constitutive role they assign to social relations and cultural narratives in providing the norms of truth. Because they see nothing—not truth, not beauty, not goodness—that is not fully social, they see the free play and autonomy of local webs of meanings as the supreme priority, not to be constrained by any "transcendent" goal. But such a view of knowledge is problematic on at least three counts: (1) It allows social relations and cultural meanings, as they exist today with all their inequities and oppressions, to set limits on what we can know about the world. (2) Simultaneously, it disables any critique of the existing relations and meanings based on knowledge not derived from these same social relations. (3) Last but not the least, it delegitimizes and denigrates intellectuals and movements that bring modern science and scientific temper to bear on local knowledges. As we see in the following scenarios, under the prevailing contexts in most of the Third World, such a logic ends up strengthening those upholding the status quo, be they traditional cultural elites or the modern state. The losers in all these cases are the internal critics—people's science movements, human rights, and democracy movements—that attempt to challenge the existing cultural mores by using the "alien" worldview of science.

    Scenario 1, India: Frederique Apffel Marglin, a well-known anthropologist from a well-known American university, with the full endorsement of India's foremost social scientists, recently declared that the eradication of smallpox from India using the modern cowpox-based vaccine was an affront to the local custom of variolation, which included inoculation with human smallpox matter accompanied by prayers to the goddess of smallpox, Sitala Devi. Despite her own admission that the traditional variolation is at least 10 times more likely to actually cause the disease as compared to the modern vaccine, Marglin persists in deriding the introduction of modern vaccine in India by the British (and the later support of mass-vaccination programs by the government of independent India) as an imposition of "Western logocentric mode of thought," which treats health as a binary opposite of illness, over the "Indic" nonlogocentric, binary-denying view, which treats the goddess Sitala as both the disease and its absence (F. A. Marglin 1990). Marglin defends those who resisted the modern vaccine in the name of the goddess as fighting for a form of life that does not distinguish between natural and supernatural forces. She is joined by a radical economist, Stephen Marglin, who issued a moratorium on using the impersonal, value-free *episteme* of modern science as a resource for improving on the locally embedded, situated knowledge, or *techne*, of other cultures on the grounds that different beliefs of different people create their own truths (S. A. Marglin 1990a).

    In calling for "decolonizing the mind," the Marglins are only affirming and adding to a long line of influential Indian intellectuals, from Mohandas Gandhi to the neo-Gandhians like Ashis Nandy and hybrids of Gandhianism and a vulgarized, pop-postmodernist like Vandana Shiva who has been writing giddy requiems for modern "Western" science for quite some time. These nativist proponents of "alternative" or "patriotic" science have declared modern science to be "intrinsically" violent, colonizing, exploitative, and patriarchal, a "pathology" that must be replaced with indigenous knowledge.[15] These alternative/patriotic science movements are self-consciously pitted against not just modern science but also Western civilization itself, for they see the distinctive features of modern science, especially its emphasis on value freedom and knowability of the world, as Western cultural traits parading as universals.

    The members of these movements wish to construct a new universal science that draws its legitimacy from the Indian (which mostly translates into Hindu) civilization.[16] Not surprisingly, these intellectuals tend to brand all modernizing forces—from international development agencies and the Indian state to social movements challenging the indigenous cultural mores—as agents of an alien and oppressive ideology. They reserve their most pointed barbs for people's science movements, which they accuse of "trampling on ordinary people's epistemological rights . . . by gallantly attempting to replace the village sorcerer or *tantrik* with the barbarism of modern science's [methods]" (Alvares 1992, 230).[17] True "patriots" like Nandy have not even spared those who protested a recent incidence of widow immolation (*sati*), branding them as modernized Westernized elites who denigrate authentic folk practices (Nandy 1988). Not surprisingly, such nativist, antimodernist ideas have found a sympathetic audience among right-wing Hindu fundamentalist parties (Nanda 1997b).

Because the people's science movements (PSMs) are one of the chief targets of the "patriots," a brief description of these movements may be in order here (for a more detailed description and defense, see Nanda 1997a). Among all new social movements in India, the popular science groups are the largest, with some 15,000 to 20,000 activists throughout the country. Scientists, engineers, intellectuals, political activists, environmentalists, and often industrial workers and farmers who get involved with these movements bring with them diverse motivations and a varied, often contradictory, mix of ideologies, ranging from Marxism, nationalism, and appropriate technology to sustainable development. Apart from consciousness-raising for ecologically sensitive and nationally self-reliant development policies, segments of the science movements were also (at least in the 1980s) involved in popular science education and the propagation of what was popularly referred to as "scientific temper." Informal education in basic scientific concepts, explicitly aimed at demystifying folk beliefs and traditional myths, used to be the central goal of large, nationally known groups like Kerela Sastra Sahitya Parishad, Kishore Bharati (now defunct), Eklavya, and many of their informal affiliates in institutions of higher learning and research. (In the interest of full disclosure, I must reveal that I myself was an active member of one such group based jointly at the Indian Institute of Technology and the All India Institute of Medical Sciences, both in New Delhi.) The objective of these modernist elements in PSMs was to develop a critical scientific temper that would subject inherited traditional knowledge claims to the test of empirical evidence. To that end, these groups promoted mass literacy in modern scientific ideas—from the germ theory of disease and the genetic basis of unity of life to the Newtonian laws of physics—and tried to relate these ideas to the everyday life of people through all available cultural means. (For a detailed description of the activities of Kerela Sastra Sahitya Parishad, India's best-known science movement, see Zachariah 1994.)

The modernist, Enlightenment elements of the Indian people's science movements set them apart, for some time at least, from the American and European "science for the people" movements of the 1960s. Undoubtedly, the Indian people's science movements were greatly influenced by the New Left and Science for the People movements in America, and share their antipathy to claims of reason and progress. This antipathy, coupled with the movements' deeply nationalistic views of Indian intellectuals, informs the Indian science movement's opposition to the Green Revolution and other technological innovations. But unlike the American New Left, the Indian left was still not entirely cut off from its Enlightenment roots. Indian science movements of the

1960s through the 1980s still had strong critical rationalist elements derived from the Indian Renaissance of the nineteenth and early twentieth centuries when important Indian reformers such as Jawaharlal Nehru and eminent Indian scientists, including J. C. Bose, P. C. Roy, and Meghand Saha, tried to bring scientific ideas to bear on social attitudes and traditions.[18]

Over the last couple of decades, as Indian intellectuals have taken a constructivist turn in their thinking about science and a postmodernist turn in social theory more generally, the Enlightenment heritage of the people's science movements has been discredited and silenced. The modernist, critical rationalist element of the science movement was indeed the first to be attacked as the turn toward culturalist and constructivist ideas became apparent in the early 1980s. The attack was initiated by Ashis Nandy, Vandana Shiva, and other prominent Indian intellectuals allied with Nandy who counterposed the scientific temper against a "humanistic temper," as if the former were an enemy of the latter.[19] Those of us who tried to defend the scientific temper as humanistic were labeled as "traitors" to our culture and as "comprador elites" interested only in securing privileged positions. The decline and the near disappearance of the Enlightenment tendencies—of critically evaluating and challenging our cultural traditions on scientific grounds—in people's science movements can easily be traced to the "scientific temper" debate, which was an opening salvo in the Indian "science wars." The arguments against "Western" science as an imposition of an alien cosmology have been taken up and internalized by various segments of the postcolonial Indian left, including feminists and environmentalists.[20]

What is striking about the Indian science war—which I call the First Science War, the current Sokal affair being the Second Science War—is the close affinity between the theoretical vocabulary used by the Indian and Western critics of science and the Enlightenment. The original scientific temper debate was couched in the theoretical terms derived from the anti-Enlightenment elements of Mahatma Gandhi's writings at home, reinforced by (much misunderstood) readings of Thomas Kuhn and the emerging discipline of sociology of science abroad. The emphasis in the self-identified postcolonial literature has shifted to the works of Foucault, appropriated through Edward Said's well-known *Orientalism*.[21] The Indian brand of ecofeminism represented by Shiva justifies itself as a "standpoint epistemology" of Third World women, an idea directly borrowed from Sandra Harding's work.[22] In general, the Indian critics of science tend to accept uncritically the social and cultural constructivist views of science that prevail in the American academe as the only valid and "progressive" theories of knowledge. Indeed, the close working relations between the postcolonial and the postmodern (using these terms somewhat loosely) science critics is evident in how they use each other's scholarly output. Works by Ashis Nandy and Vandana Shiva, which distort the Indian reality to fit the social constructivist theory, are read in science studies departments as "authentic" Third World voices.[23] What has completely disappeared is the fact that there were, and still are, many progressive Indian intellectuals, reformers, scientists, and laypeople who have used "Western" science for a critical evaluation and, in some cases, revitalization of our own traditions. The complex and intimate relationship between the once colonized people and science has been reduced to a massive false consciousness enforced by the colonial and neocolonial powers.

Scenario 2, Pakistan: If the people's science movements are the favorite target of derision for the alternative science movements in India, it is the working scientist, (including most prominently the late Abdus Salaam, the Pakistani Nobel physicist), who bear the wrath of the state-sponsored movement for "Islamic science" in Pakistan and other Islamic countries, notably Egypt and Saudi Arabia. In Pakistan, the chief proponents of specifically Islamic epistemology, or *ilm*, that integrates knowledge and Islamic values, is the group of scholars around Ziauddin Sardar, a Pakistani émigré living in Britain, and Munawar Ahmad Anees, a U.S.-based biologist and Islamist. Like their counterparts in India, though without their nonreligious language (for the most part), Sardar and his associates seek not just want to "Islamicize" science but rather to create an entirely new universal science in which the facts of nature would be different, derived solely from the conceptual and ethical categories of Islam. They find attempts by Abdus Salaam and other scientists to bring modern science to bear on specific values and problems of Muslims as misguided, if not actually a crime against Islam.

These critics insist that modern science cannot be reconciled with *Ilm* for the simple reason that it embraces radical doubt and dares to question the Qu'ran. Explicitly citing the work of feminist science critics and those of the new post-Kuhnian sociologists of science, the proponents of Islamic epistemology call for the repudiation of skepticism in favor of faith as a source of knowledge, and they demand a continuous reintegration of facts and Islamic values at each stage of inquiry (see the collection of papers in Sardar 1988 and, more recently, Anees 1993). (In turn, postcolonial feminist critics of science, notably Sandra Harding, cite Sardar and associates among the "progressive" postcolonial critics of science.) Given the active state sponsorship of religious ideologies, the Islamic science movement is having serious and deleterious effects on the practice of science in Pakistan, as seen by the constant threats against allowing the theory of evolution or Einstein's relativity to be taught in schools, and the state sponsorship of "scientific miracles" (see Hoodbhoy 1991). The call for *Ilm* does not end with a call for Islamic epistemology alone. Recently, demands for specifically Islamic (and also Hindu, Confucian, and African) conceptions of human rights have also been put forth (for a critical view, see Afshari 1994).

Scenario 3, China: Students and workers gathered at Beijing's Tiananmen Square on May 4, 1989, to demand democracy in the name of science. The protesters were not only explicitly evoking the legacy of the original May Fourth Movement of 1919— the science and democracy movement, often described as "Chinese Enlightenment"— they were also inspired by their own teachers, a disproportionately large number of whom included theoretical physicists and natural scientists, including Fang Lizhi, a renowned physicist now in exile in the United States. Through their long struggle against the Chinese Communist Party's condemnation of Einstein's theory of relativity as "Western bourgeoisie physics," Fang Lizhi and other dissident physicists came to value liberal and pluralistic social values. These scientists who inspired and participated in the Tiananmen Square protest saw an "epistemological connection between science and democracy and dissent" (Fang and Link 1996, 44; see also Miller 1996).[24] It is this connection between antiauthoritarianism and doing good science that explains why the protesters at Tiananmen Square demanded democracy and science together. Tragically for the dissidents, the Chinese government saw it differently and sent in the tanks. The Deng regime, though anxious to cultivate modern science and technology for economic development, treated any attempt to relate scientific ethos to antiauthoritarian politics as a sign of the "spiritual pollution" of China's socialist values.

In all these cases, "civilizational logic" is embraced by those defending the status quo—new nationalistic social movements in India and state-sponsored intellectuals in China and Pakistan. All these defenders of civilizational knowledge make similar moves. First, they reject the neutrality ideal of modern science. They claim that perspectiveless, bias-free knowledge is impossible; that the value of freedom is itself a cultural construction of the West in the interest of cultural imperialism; and that, of all values, other civilizations prefer not freedom but a "subordination of reason to other instrumentalities"—namely, the generation and regeneration of the cosmos and society in the Indic civilizations, faith in the Islamic civilization, and socialism in Communist China. They then claim the "epistemological right" of other civilizations to reject the Western value of value freedom to do science in tune with their own values. This "right," interestingly enough, is justified not only in the name of "self-determination" but also in the interest of discovering "many more universal laws of nature that delinked sciences could discover if they were permitted to develop out of civilizational settings different from . . . European projects" (Harding 1994, 325). These new "non-Western universalities," so runs a constant refrain in these scenarios, may be the key to human survival for generating more humane and ecological knowledge.

It now becomes easier to see where the postmodern connects with the postcolonial in science studies: social constructivist theories of science make available to their postcolonial comrades the central plank of their case for separate but equal sciences. This central plank states that social relations and cultural meanings constitute the logic, content, and goal of all knowledges alike, from witchcraft to molecular biology. Without this constitutive role of social relations, the idea of alternative sciences that are rational according to their own socially embedded rules and values simply ceases to make sense. We now turn to the parental program of all such theories that make knowledge constitutively social and cultural, the "Strong Programme."

Constructivist Condescenions

The constructivist giveth and the constructivist taketh away. The Strong Programme (SP) endorses the rational unity of humankind but denies as "myth" the idea that there could be a universality of knowledge; that is, it denies that there could be facts about the natural world that knowers in different times and places could recognize as more rational to hold on the grounds that they are based on better evidence, are better approximations of the truth, or have withstood serious attempts to refute them. The constructivists, then, confer on the "Others" the ability to think, but then take away the ability to choose, on occasion, the knowledge of aliens over the knowledge of ancestors.

The recognition of the rational unity of humankind is an advance over the earlier theories of anthropology and sociology (especially those by Levi-Bruhl and Peter Winch) that saw the "primitive" people as prelogical and simply not interested in a rational pursuit of explanation, prediction, and control of the material world (for a critical review of some of the earlier theories, see Horton 1993). In contrast, the SP and allied constructivist programs grant not only that all of us are mapmakers of our physical world but also that all of us are equally competent and skilled mapmakers: All mapmakers in all cultures relate to the physical world using the same perceptual apparatus; all seek evidence to back their assertions; all reason logically; and all come up with maps equally useful in coping with the world they live in, depending on how different communities measure success in coping. There are no nonsocial and non-conventional grounds for preferring the way modern science maps the world in its theories. The only reason that anyone in a non-Western society could conceivably prefer the latter is due to the intellectual hegemony of the West backed by its military and economic might.[25]

The underlying claim is not simply that all people everywhere are biologically endowed with the ability to reason: nothing less would be expected and accepted from any even marginally respectable theory of knowledge. The underlying claim is, rather, that all cultures are equally rational in arriving at and holding the beliefs they actually hold and that no sense can be attached to distinguishing between "mere beliefs" and "facts" backed by good, falsifiable reasons. As Barry Barnes, one of the original architects of the SP-based sociology of science, put it, "wherever men [and women] deploy their cultural resources in the authentic task of explanation and investigation [of the material world], what they produce deserves the name of knowledge" (1977, 24).

Accordingly, the famous symmetry tenet of SP admonishes sociologists, anthropologists, and all other "scientific" students of knowledge to treat the alien beliefs they come across as: "empirical events to be described . . . and explained without making any practical distinction between knowledge and belief, or between our 'genuine' knowledge and their mistaken 'knowledge,' or between sound and unsound inferences . . . or between objective and biased judgments. For sociological purposes, all *evaluative* distinctions and dualism are ignored" (Barnes 1991, 325; italics in original).[26]

SP admits that sometimes an anthropologist will encounter societies whose members find credible a belief that the anthropologist finds incredible and objectively false (e.g., the belief that smallpox is caused and cured by the goddess Sitala). The alien observer is then required to lay aside her own incredulity and explain why the members of these societies should believe in Sitala using the same sociological factors that she would deploy to explain why she herself and her tribe find a belief certified by modern science (e.g., smallpox is a contagious disease caused by a virus). What a thoroughly symmetrical sociologist or anthropologist of knowledge *cannot* assume is that the latter belief is a better approximation of reality or that it is has better evidence to back it up or even that it has withstood serious efforts to falsify it. All these traditional reasons for a comparative evaluation of belief and the criteria on which they are based are said to be purely local, contextual, and internal to the cultures under question.

Again, SP's admonition to treat the beliefs of "ancestors, aliens and deviants as cultural variants and not as difference in basic natural rationality" (Barnes 1977, 24) is an advance over the frankly racist theories of an earlier era that read difference in beliefs as a sign of primitive, prelogical, or mystical thinking. By treating knowledge everywhere as motivated by the same interests—explanation, control, and prediction—the SP affirms the rational unity of humankind. And there is no denying that as an initial methodological principle, it makes sense to set aside one's own evaluation of the truth or falsity of beliefs others hold, to understand why they find it reasonable to hold these beliefs. But note that such generosity is an internal matter of outsiders looking in, trying to improve their own understanding of the insiders and, in the case of the Western anthropologists, also to make amends for the sins of their imperialist forebears. Such methodological neutrality does nothing for the actual lives of the supposed benefactors—the insiders—whom it binds to their own terms of reference and their own norms of truth.

A closer look at the SP's conception of rationality helps explain this paradoxical gift that imprisons us in our own webs of beliefs even while it retrieves our basic humanity from racist denigration. The key to this paradox is another paradox hidden in the very name "Strong Programme": its illusion of strength derives precisely from the weak version of rationality with which it operates. To put it bluntly, SP and the allied social constructivist programs break the connection between rationality and any sense of progress in improving the reliability and validity of knowledge. The place evacuated by all traditional surrogates for validity—Karl Popper's falsifiability, Robert Merton's institutional norms, and Thomas Kuhn's paradigmatic consensus—is left open to contingencies of power, social interests in maintaining power, and other pragmatic uses of knowledge.[27]

Rationality, as most broadly understood, is characterized by four dispositions: suspicion toward received authority, a commitment to continually refining one's own understanding, a receptivity toward new evidence and alternative explanatory schemes, and a dedication to logical consistency (see Lewis and Wigen 1997). The whole point of calling a belief rational is not that it is guaranteed to be true but only that it is arrived at through a process of inquiry that allows the inquirers to be so oriented to evidence that they can change their beliefs in ways that make it more likely that they are true. This sense of cognitive progress—a sense that through science we have learned how to learn from our experience so that we can give up erroneous views—is how the scientific worldview has traditionally been distinguished from prescientific worldviews. Note that this notion of cognitive progress requires at a minimum that there be an external world that is in some sort of a determinate state about which we can be right or wrong, and it also requires some criteria to determine whether we are moving closer to or further from understanding this state of the world. This is the strong sense of rationality (see Jarvie 1984).

The SP's main goal is to show through anthropological and historical case studies that the social institutions and practices of modern science are no better oriented to evidence that would enhance the likelihood of its findings being any closer to the facts of the matter than are the knowledge-seeking practices of any other culture. Thus we find constructivists spending a lot of time dispelling as myth the idea that there is any difference between traditional and scientific modes of legitimation: normal scientists operating within a paradigm are no different from knowers in traditional societies, in that both accept the existing fund of knowledge handed down by their intellectual ancestors and both act consensually, with one paradigm reigning at any given time.[28]

One result of this supposed similarity in traditional and modern scientific knowledge systems is that the former is no better at critically monitoring theories, as Popperian and Mertonian rationalists assumed. Constructivists claim that the institutions of modern science and scientific ideas are as integrated with the rest of everyday common sense, cultural frameworks, and social relations as are the knowledge-seeking practices of any other premodern society. This total interpermeability of knowledge and society implies that all systems alike struggle to maintain the coherence of their beliefs, a coherence that serves to maintain the existing balance of political and social power. In a version of holism (or "finitism" in SP-speak), presumably derived from Duhem, Wittgenstein, and Kuhn, the strong programmers insist that "existing beliefs may always be defended . . . current experience can be described as consistent with any extant body of ancestral knowledge, even as offering inductive confirmation of it. Equally, it can be made out as refuting any specific body to knowledge or item thereof" (Barnes, Bloor, and Henry 1996, 78).

It couldn't be otherwise, the strong programmers tell us, because cognition is social: something will be recognized as evidence for anything new, different, and challenging only under certain preexisting contextual conditions. Rationality is multiple and diverse, and all forms of logic are in the final instance, sociological or ethnological.[29] Thus the time-honored link between rationality and objective truth is broken: "What makes a belief true is not its correspondence with an element of reality, but its adoption and authentication by the relevant community," as one of the central planks of the SP would have it (Fine 1996, 234). Thus the SP finds all distinctions between truth and what is *taken* as true by any community to be sociologically irrelevant. Knowledge is accepted belief, and "from a sociological perspective, there is no value in a fundamental distinction between science and ideology" (Barnes 1982, 107). (One can't but wonder what Fang Lizhi and other Chinese dissidents, who, at great personal cost, challenged the party ideology of natural dialectics in the name of Einsteinian physics, would make of this sociological "insight.")

Hurrah! then, for the ancestors who shall always be with us, standing between the evidence of our senses and what sense we can make of it. Hurrah! also for traditions standing guard over our unruly, unexpected experiences of the material world, channeling them along paths sanctioned by our hallowed "forms of life." But thumbs down for the real world: it, too, shall always be there, but as a silent source of our primitive causes, a world that is "silent, "indifferent," and "tolerant" to the extent that "we may say what we will of it, and it will not disagree" (Barnes 1991, 331) or, in one word, that is inconsequential to our beliefs.[30] Thumbs down, too, for the project of cultural change through growth of knowledge and rationality, for "to favor one paradigm rather than another is . . . to express a preference for one form of life rather than another [which] cannot be rationalized in any non-circular way" (Barnes 1982, 65).

The co-making of facts and values, or the total permeability of the constitutive values of science (logic, evidential reasoning) and the contextual values (social, ethical, aesthetic, and metaphysical) of the rest of the web of belief, is exactly what the postcolonial intellectuals invoke to support their respective civilizational epistemologies. Thus we have here the intellectual labor of self-consciously "progressive" constructivist critics of modern science in the West aiding and abetting the ideas that support the far from progressive status quo in Third World societies. Although the

PSMs in India and the dissident scientists in Pakistan and China we encountered in the previous section argued that certain traditions were standing in the way of a better understanding of the world on which more rational ethics and social policies could be erected, the constructivists tell us that all traditions alike play an equally constitutive role in all knowledge. In other words, by accepting modern science, we in the Third World are not moving any closer to a truer understanding of the world but are only exchanging Western cultural traditions for our own and sinking deeper into a colonial mind-set—exactly the accusation those upholding the status quo make against all modernists.

So the reason that we—the dogged advocates of scientific temper in non-Western societies—must reject the privilege of having our traditional knowledge considered at par with science is clear: the project of different and equal sciences for different people completely negates our project of science for all people. We prefer our much-maligned universalistic project because we are not interested in a supposed cognitive equality of different cultures but, rather, in substantive equality for all people in terms of healthier, fuller, and freer lives. We prefer the cold, objective facts of science to the comfortable, situated knowledge of our ancestors for the simple reason that we refuse to subordinate what is good to what is ours.

But our gift givers may well interject at this point and insist that they, too, speak for the good, for a substantive equality, freedom, and well-being for all. As has become abundantly clear in the debate sparked by Paul Gross and Norman Levitt's *Higher Superstition* and Alan Sokal's "transgression," constructivists feel that their critics have misread their legitimate efforts to identify the traces of culturally shaped human subjectivity in the enterprise of science as antiscience. It is likely, then, that the constructivist proponents of ethnosciences will ascribe my preference for science over ethnosciences to my scientistic mind-set, a carryover from my days in the lab.[31]

Indeed, the constructivists could easily point to Donna Haraway's opposition to nativism, relativism, and all myths of organic belongingness. Didn't she openly urge a blurring of all dualities rooted in essential identities, including the duality of "us" and "them," in favor of an oppositional consciousness? Didn't she assert that she would "rather be a cyborg than a goddess?" (Haraway 1991, 181). Or take Sandra Harding. Hasn't she made it clear that she is not ruling out modern science for Third World people but is "only" urging that modern science give up the pretension of being the only game in town and take its place as one among equals in the "collage" of knowledges from all different cultures? (Harding 1996, 22). Furthermore, the gift givers may ask, don't Haraway and Harding (and Vandana Shiva as well) argue for "situated knowledge" and "standpoint epistemologies" as offering not just contextually true, relativistic knowledge but as actually a stronger, more objective kind of knowledge? They contend that partial perspectives and standpoints don't deny objectivity per se but only a certain kind of objectivity that claims to transcend the lived experiences of ordinary people and the cultural heritage of working scientists. A stronger objectivity emerges, both Haraway and Harding assure us, when different knowledge systems enter into a critical engagement with one another: new and unexpected connections emerge when reality is seen from different perspectives. Thus, those who favor a cognitive egalitarianism defend it in the name of better science and not in the name of traditional values.

It would be a tragedy indeed if we were to reject such a promise of more objective and more egalitarian science out of a misunderstanding. But there is no misunderstanding: the proponents of situated knowledges and standpoint epistemologies may want a better, stronger, truer, less biased science, but they cannot deliver it. They cannot deliver it because they have made a certain politics—what Haraway calls "oppositional consciousness"—rather than truth the goal of science. It is not the truth content of ideas but their location—geopolitical and class, gender and racial—that legitimates beliefs (for indeed, truth claims cannot be separated from location).[32] This means that the outcome of the promised critical engagement of different partial perspectives is decided by political concerns rather than facts, which may well go against the preferred political outcome (Nanda, 1997b). In addition, a critical look at the world from different partial perspectives may be a good initial step, but in the end, a choice must be made regarding which perspective is closer to the facts. That this choice can even be made on nonpolitical and culturally neutral grounds is denied by the constructivists.

It is precisely because of the devaluation of truth in favor of politics that situated knowledges won't work for a critique of cultural values in non-Western societies. The facts with which we counter some of our oppressive cultural values are, in the constructivist account, inseparable from Western cultural values. The non-Western critics cannot use these facts without being accused of bringing in alien cultural values and thus being "comprador Westernized elites," an accusation that all enforcers of the status quo make against internal critics. This accusation blunts and disables the critique and silences the critics. In sum, when science is divorced from truth and all criteria of truth are made internal to the social context, we end up with ethnosciences that fail to give us a stronger, better science and society and instead end up legitimizing all kinds of ethnonationalism and nativisms.

The Scientific Temper and Political Progress

Aristotle observed that "all human beings seek out not the way of their ancestors, but the good" (quoted in Nussbaum 1993, 242). Two millennia and more later, our relativists tell us that the way to the good lies only through the ways of the ancestors. Barnes and Bloor, for instance, teach us that "faced with a choice between the beliefs of his own tribe and those of the other, each individual would typically prefer those of his own culture" (Barnes and Bloor 1982, 27). This choice, as their symmetry principle insists, has nothing to do with the objective truth of the belief in question, for the tribe also provides the norms and standards for justifying the belief as true. They go further and "show" that it is nothing more than hubris on the part of modern scientists to think that they can escape the ways of their own tribe. Scientific "truths" are true only in the web of belief spun by the tribe of Western scientists.

But, then, what about the iconoclasts, the reformers, and the rebels who, in all societies at all times, insist on putting the good above the ways of the tribe? What about the people's science movements in India? What about the May Fourth Movement in China? What about the modernizing states? And what about working scientists who believe that they make progress only by challenging the basic assumptions of their paradigms? All these are instances of a cosmopolitan ethos that seeks the good justified by the best and the strongest standards of rationality and not just by the norms of one's own tribe. Indeed, these cosmopolitan movements see their own tribes connected to the rest of the humanity through a series of concentric circles of family, identity groups, and nations. The cosmopolitan ethos of science and popular science movements seeks to widen the circumference of their own tribes by making all other tribes and their stock of knowledge a part of their own community of dialogue and concern (see Nussbaum 1996).

Such cosmopolitan movements are conspicuous by their absence in the constructivist reflections on science in non-Western societies. In all the feminist, postcolonial, and other studies in the constructivist vein that I have encountered, the only acknowledged relationship between "Western" science and the non-Western people is that of imperialism, with all its ugliness and exploitation. Science is seen as an imposition on reluctant and/or brainwashed non-Western natives, whose otherwise coherent and basically humane world is threatened by the West for ulterior motives of profit and dominance. Not surprisingly, the only social movements whose existence is acknowledged and celebrated are the nativist, cultural nationalist, "civilizational science" movements that resist modern science in the name of cultural authenticity.

Unacknowledged though they are, constructivists seem well aware of the threat which the very existence of cosmopolitan, popular science movements poses to their entire theoretical edifice of science as ethnoscience. For why else would there be so many theoretical maneuvers to insist that the cosmopolitan aim of science and science movements—to advance the good by the best, most rigorously obtained truth—is impossible, undesirable, or, if possible, not very important anyway? It is these arguments I now examine very briefly.

Replace objectivity with solidarity, and rationality with sentimentality, exhorts Richard Rorty, for objective and transculturally valid knowledge is impossible and, in any case, not morally relevant to the advancement of liberties, human rights, and other goods that cosmopolitan social movements seek (Rorty 1993). Rorty's position, incidentally, is not that far apart from the SP view on this matter, encapsulated neatly in Barry Barnes's words, "cultural change is not to be understood as the consequence of increase in rationality" (1976, 124). The SP, as we saw, argues that facts cannot change values, for given the right kind of social interests (solidarity?), it is possible to save any values by making adjustments in other parts of the web of beliefs. Thus, Rorty claims that objective knowledge about human beings and their social world does nothing to correct moral intuitions about human beings and that such knowledge has contributed nothing to the emergence of a human rights culture in the West or anywhere else. Adding to these views are anthropologists like Stephen Marglin, who claims that the facts of modern human biology are irrelevant and culturally inappropriate to changing the accepted beliefs regarding female genital mutilation in some African cultures: Beliefs simply create their own truth, and any attempt to counter cultural beliefs with alien truths is an attack on the cultural autonomy of the believers. ("Ah, but they won't be Aztecs any more," is S. A. Marglin's stock reply to the idea that scientific beliefs can change values [1990, 13].)

The relationship between factual knowledge and moral knowledge is too complex an issue to be dealt with adequately here. But a dismissal of the progress in our knowledge of the nature of human beings and human societies as irrelevant to widening the realm of human liberties is a travesty of the entire history of the Enlightenment in the West, which saw "industry, knowledge and humanity as linked together by an indissoluble chain" (David Hume, quoted in Gay 1969, 26). In the Newtonian worldview of the Enlightenment philosophies, an improved understanding of the order of nature (cosmos) was considered inseparable from a more rational social order (polis). To be sure, Rorty and other critics of modernity fully appreciate the close correspondence and mutual constitution of the cosmopolis of the Enlightenment (Toulmin 1990). They want to deny only that such a grounding of culture in nature is relevant to today's world. For the postmodern world of ours, they argue, knowing what we do of the atrocities committed in the name of bringing the social order in tune with some purported laws of nature, we need to sever the cosmos from the polis; that is, we need no grounding of our ethics in some objective facts of nature. As Rorty would have it, we need hope more than we need knowledge, solidarity more than objectivity.

Fitting human beings and their lived relations into the Procrustean bed of what the latest scientific theory says about human abilities and needs is no doubt dangerous, and Rorty and others are right to oppose it. The danger is compounded many times over if the social order so coerced has traditionally conducted its affairs according to very different rules, derived from a very different cosmopolis. But it is intellectually dishonest to use this extreme model of coercive instrumental rationality to stand in for all the different ways that scientific rationality and knowledge actually operate in different societies at different times. Such a blanket rejection of the moral relevance of science, furthermore, leaves out the real cases of the slow and patient work of education and persuasion carried out by democratic popular science movements. The experience of these movements shows that a reinterpretation of mythologies and traditional knowledges can and does change views on important social issues or, at a minimum, puts in place the psychic processes needed for questioning the authority structure upholding the traditional values.

There are indeed sound theoretical reasons for the scientific temper—that is, the demand for publicly testable, good reasons for cultural norms and cultural authorities—and its corrosion of the ancestral cosmopolis and opening of the way for new demands for civic equality and liberty. Insofar as cultures and traditions have a cognitive component—that is, to the extent they justify themselves on the basis of some theory of the nature of things—they will remain open to competition from other explanatory theories in the same domain. Coercive and exploitative though Western colonialism has been, and uneven and painful though the experience of modernization has been for most ex-colonies, colonialism and modernity have introduced alternative explanatory frameworks in which old facts can be reinterpreted. In such a situation of increased competition of ideas, it is simply not easy for the traditional legitimation—"we do it because our ancestors did it"—to carry much weight (for a similar defense of Western ideas, see Appiah 1992). Modern science will remain "morally relevant" to non-Western societies, and also to Western societies, as long as it threatens the traditional legitimation of ideas, as long as it helps people stand back from, critically reflect on, and lose their faith in the ways of their ancestors. This is why modern science is, correctly, perceived as a threat by all defenders of the traditional worldview. And this is why all the "radical" critics of science who purport to show that in principle and in practice, modern science is "no different" from traditional legitimation based on authority, faith, and material rewards are wittingly or unwittingly serving the interests of traditional authorities everywhere.

So, to answer those who think that it is an impossible and potentially dangerous quest to try to change ancestral values by bringing in new facts or by offering a new understanding of old facts: If scientific facts are understood as a corpus of absolute truths that demand obedience from all aspects of social life, then indeed, they must be treated with as much skepticism as one would any other dogma. If, by contrast, scientific facts are treated as facts held as a result of more open and critical inquiry, they are indeed the necessary elements of a self-critical, authority-defying free society. Scientism no, but scientific rationality yes.

But this answer cannot and does not satisfy those who see skepticism of science and its impulse to distance itself from personal biases, desires, and fears as an undesirable Western and patriarchal import that leads to a split between the subject and the object of knowledge. From this perspective, which is shared by nearly all the prominent advocates of situated knowledges and postcolonial science, popular science movements in Third World societies are not just blinded by science; they are unethical as well for imposing an "alien" rationality on non-Western people. Thus, for instance, Shiva, citing the "new philosophies of science" (which she ascribes to Kuhn, Harding, Keller, and Merchant), declares that the very idea of objectivity as a goal of knowledge is Western, capitalist, patriarchal, and therefore "inherently" foreign to Indian cultural values (Shiva 1988, esp. chap. 2). Non-Western cultures, according to Claude Alvares, a self-proclaimed "Luddite" and a prominent Indian intellectual, don't value the ideal of value freedom or objectivity but, rather, prefer to "subordinate reason to other instrumentalities" (1988, 71).

This is not all: postcolonial critics routinely deny that the kind of individual liberties and freedoms lauded in the West are much valued by people in other societies, a sentiment well expressed by Stephen Marglin, who asserts that "outside the West, adaptation [to the environment and material circumstances] may commend itself rather than control" (1990b, 11). In all these cases, as Ashis Nandy makes it clear, whenever values of modern science "become incompatible with cultural traditions, the latter should have priority over the former . . . [because] they are close to the ways of real life people . . . and more restrained by participatory politics and the democratic process" (1994, 12).

There are two problems with this cultural absolutism. First is this matter of an unacknowledged but a potentially damaging family quarrel between those who would take social constructivism in a more postcolonial direction and those who would pull it in a more inessentialist, contingent, postmodernist direction. The tenets of SP work against any cultural essences. To their credit, the fathers of the SP clearly and unambiguously deny that there are any differences, good or bad, in the natural rationality of different peoples and different cultures; they "only" deny that one can give good reasons for picking one over any other. Yet repeatedly, the postcolonial critics turn the constructivist critiques of modern science into an argument for the superiority of nonmodern, non-Western sciences. Despite this obviously incomplete and opportunistic use of constructivist ideas, I have not found even a single instance in the literature in which the constructivist theorists critically engage with their postcolonial colleagues the issue of separate, civilizational sciences. Indeed, the trend is in exactly the opposite direction. I suspect that the constructivist critics of science in the First and Third Worlds are appropriating each other's theories and "findings" equally opportunistically: the former need the latter to provide "case studies" for cultural variation in knowledge, and the latter need the former for legitimacy for their separatist ideas.

Beyond this internal contradiction, there remains the issue of where the postcolonials draw the boundaries of their cultures. The postcolonial understanding of culture is suffocatingly narrow and parochial. It simply leaves no room for the possibility of what Anthony Appiah calls "cosmopolitan patriots," those who are situated existentially in the Third World but equally grounded in the Popperian World 3 of objective knowledge that exists transnationally and transculturally. Indeed, the most important reformers, from Frederick Douglass in the United States to Vivekanand in India, have been those with a double consciousness of their own local traditions and those of other societies. Moreover, it is not clear whether any culture in the world was ever as different from all others in basic interests and values as the postcolonials portray their benighted local cultures.

In the contemporary world, however, the kind of parochialism that postcolonials wish to defend is simply unviable. In today's world, when all cultures are in a constant and intimate contact with one another, when even the remotest cultures are a part of a much bigger plurality, all cultures are entitled to call on the values and traditions of others to cast a critical look on their own practices. Values of one part of a plurality can enter an internal critique of another part, or as Martha Nussbaum and Amartya Sen put it, "internal criticism can have a long reach" (1989, 321).

Finally, some people concede that internal cultural criticism using "external" criteria of science is possible, and perhaps even desirable, but they insist that it nevertheless has very limited value. Such a thesis is advanced by, among others, Bryan Turner, a sociologist who appears to have no sympathy with the cultural absolutism of the nativist critics of modernity but who is equally critical of rationalists who stress cognitive changes as a source of cultural change. Turner asserts, correctly, that "people do not adopt or reject belief systems on the rationalistic grounds that they are not intellectually coherent. Beliefs are adopted or rejected because they are relevant or not to everyday needs and concerns" (1994, 10).

I agree with Turner's views on the limits on the weight that scientific ideas alone can bear in the process of cultural change. Culture has cognitive content, but learning a culture is not like solving a math problem or doing an experiment. Better logic and evidence can only go so far in challenging the basic cultural frames we are born into. Besides, the need that all societies have for pretense may set limits on the utility of strictly intellectual orientations toward cultural change. Strategic ambiguities seem to be needed to hold a society together. It is also true that unless material realities change, a mere change in beliefs themselves will not go very far.

Precarious and dependent (on other enabling conditions) though the actual domain of science is in the life of a society, it is nevertheless far more powerful than any other cultural innovation. Once modern science enters a social system for even a reason as selfish and exploitative as colonialism, it cannot be safely quarantined. Whatever space and prestige it is allowed, science has the potential to cause reverberations throughout the rest of the frame of presuppositions that assign human life a place in the natural order. True, these frames of meanings don't simply crumple and disappear, for they are not, to paraphrase Clifford Geertz, "experimental science in search for laws, but rather webs of significance, in search for meanings" (1973, 5). Cultural frames are able to absorb scientific facts and theories by making adjustments in other beliefs or by changing the significance of beliefs (e.g., religion, after all, continues to coexist with science in Western societies by treating the scriptures as symbolic rather than literal truths).

And yet, gradually, and especially when science is not ideologically under attack as it is currently in most parts of the Third World, the spread of scientific learning and style of thinking can alter the plausibility structure of the cultural frames. Goddesses and witches simply cease being plausible. Those parts of nature and humanity that prescientific cultural frames took out of the play of human ability to know— those "demon-haunted worlds, regions of utter darkness"[33]—have become knowable and have lost their terror and awe. Women don't need to be burned as witches when the terror of disease and unexplained weather can be better explained by microbiology and meteorology. Likewise, values that were at least understandable given a certain explanation of human origin and nature (e.g., the origin of the untouchables from the feet of the god and the Brahmins from the head) can no longer claim to be ethical in the face of the findings of molecular biology showing the unity of all life. Granted that scientific facts and scientific reasoning are not sufficient for this humanistic rearrangements of meanings—witness the continuing appeal of magical beliefs and irrationalities in scientifically advanced Western societies. But scientific thinking is necessary for any such shift to take place. It is this necessity that the people's science movements in India and the May Fourth intellectuals in China, for instance, were trying to embrace and popularize.

What do the science popularizers in non-Western societies see in science that makes it their ally in social change? What is distinctive about science that these movements value over the ways of knowing sanctioned by their natal traditions? Simply that all beliefs to which we consent should be backed not by authority but by appropriate and independent evidence (independent not of all presuppositions, which is impossible, but independent of the belief in question). Scientific rationality in social life is basically a demand that social authority be backed not by power but by reason.

This is an idealized picture of science, the constructivists tell us. An insult to the "Other," the multiculturalists charge. It is making a new idol out of science, the local priests and their postcolonial allies cry out, accusing the people's science movements of hypocrisy and treason.

Science movements are guilty as charged. But so what? What other way of knowing has made an icon out of iconoclasm? What other way of knowing has set reason and not authority as an ideal to strive for, even if the idea is everywhere imperfectly achieved? The constructivist friends of the Third World can keep their gift of ethnoscience. The people's science movements are after the icon of iconoclasm, something that constructivists believe to be pure illusion. And that spirit of science no one has to "give" to us, for it is, as J. D. Bernal put it, "a treasure . . . for all to spend and increase" (Bernal 1965, 924).

Notes

1. Following the standard practice in science studies, I consider as social constructivist all theories of science that seek to "investigate and explain the very content and nature of scientific knowledge . . . knowledge as such, as distinct from the circumstances surrounding its production" (Bloor 1991, 3).

2. The traditional virtues of scientific inquiry—truth, rationality, success, and progressiveness of beliefs—are bundled together as TRASP and declared irrelevant to explaining how scientific beliefs are justified (see Collins 1981).

3. "The emergence of the human rights culture seems to owe nothing to increased moral knowledge and everything to hearing sad and sentimental stories" (Rorty 1993, 118).

4. Sandra Harding (1994) provides a useful review of various postcolonial critiques of universalism of modern science. According to Harding, science's universality is "as an empirical consequence of European expansion" and must not be located in the supposed validity of scientific claims (319). Harding goes on to claim that if allowed to break from the West, different cultures will discover many more alternative universal laws of nature that start out from their own indigenous cultural idioms (325). A similar argument for recovery of the non-Western "others" by repudiating all social categories framed in and by the history of Europe runs through the so-called subaltern studies tradition of postcolonial critique. For a representative work, see Gyan Prakash, "Writing Post-Orientalist Histories of the Third World: Perspectives from Indian Historiography," *Comparative Study of Society and History*, April 1990, 383–408.

5. For a scathing critique of the tendency to link positivism with cultural imperialism and hermeneutics and postmodernism with postcolonialism, see Gellner 1992. The idea of Descartes preparing the ground for Kipling is from Gellner, 30.

6. For statements on the shift from "getting reality right" to an explication of cultural meanings of science, see Martin 1996 and Franklin 1995.

7. A recent and rather sad instance of such condemnation of science for the people appears in Richard Lewontin's critical views of Carl Sagan's lifelong passion for popularizing science. Lewontin sees Sagan's enterprise as misguided and hopeless, for it supposedly pits the powerless masses against the "rationalizing materialism of the modern Leviathan," which is a special preserve of university-educated elites. Lewontin's critical remarks acquired a special poignancy for they appeared shortly after Sagan's death (see Lewontin 1997).

8. "The aim of relativist teachings is to give permission to diversity and difference . . . on the grounds of the coequality or noncompatibility of divergent forms" (Shweder 1989, 99).

9. As Linda Nicholson puts it, "The postmodernist need not abandon the distinction between legitimate and illegitimate claims to power as she or he need not abandon . . . criteria of truth. The difference . . . is rather that the former and not the latter denies the possibility of such criteria external to any specific historical tradition. From within any given tradition, we always have means to allowing certain discursive moves and not others, of admitting the legitimacy or not of specific claims" (1992, 87).

10. For a clear explication of the connections between the strong program in sociology of science and postmodern philosophy, see Murphy 1990.

11. Among many texts linking scientific rationality with colonialism, see Ronald Inden, "Orientalist Constructions of India," *Modern Asian Studies* 20 (1986); and Shiva 1988.

12. Dipesh Chakrabarty deploys the army imagery ostensibly for language: "If a language is, as has been said, but a dialect backed up by an army, the same could be said of the narratives of modernity that almost universally today point to a certain 'Europe' as the primary hbitus of the modern" (1992, 21). Applying this metaphor to truth, as I have done, is not an unfair reading of Chakrabarty, for in the rest of his paper, he does critique "reason" precisely

on the ground that it was installed as obvious and universal because it was backed by the imperialist powers.

13. Thus, almost universally the postcolonial scholars insist that they are not recommending a nativist embrace of the premodern. Rather, they justify their attacks on modern science in the name of an alternative modernity that will be grounded in the cultural categories of the non-Western societies. Among the most prominent proponents of this view is Sandra Harding (1994, esp. 325). For another recent statement from a very different vantage point, see Arif Dirlik, "Modernism and Anti-Modernism in Mao Zedong's Marxism," in *Critical Perspectives on Mao Zedong's Thought*, ed. Arif Dirlik et al. (Atlantic Highlands, N.J.: Humanities Press, 1994), 59–83.

14. The postcolonial critiques of science are different from postcolonial literary studies and history in that the latter are more consistently postmodern; that is, they at least try to carry their antiessentialism further and deconstruct all essences of nationhood, just as they deconstruct the essences of the East and the West. As Aijaz Ahmad's critique of Edward Said shows, they don't succeed in going past the nation very well, either (see Ahmad 1992). In science studies, however, even the attempt to transcend the nation is missing.

15. For a representative sample, see Nandy 1988 and Sardar 1988. The best-known work of this tradition in the academic left circles in the United States remains Shiva's *Staying Alive* (1988).

16. For a recent statement of such a civilizational imperative, see Nandy 1994. For a mildly critical assessment of the civilizational analysis of science, see Guha 1988. In treating scientific knowledge as inherently and uniquely Western, these radical intellectuals are not very distant from the position taken recently by Samuel Huntington, whom most radicals would consider as an archconservative (see Huntington 1996).

17. Likewise, Vandana Shiva derides liberal feminists as having "colonized minds" (1988, 47).

18. For a recent and sympathetic account of the Indian Renaissance, see Zaheer Baber, *The Science of Empire: Scientific Knowledge, Civilization and Colonial Rule in India* (Albany, N.Y.: State University of New York Press, 1996). See also Irfan Habib and Dhruv Raina, "Copernicus, Columbus, Colonialism and the Role of Science in Nineteenth Century India," *Social Scientist* (India), 164 (1991): 51–66.

19. I refer here to the famous "scientific temper" debate that raged among Indian intellectuals for nearly a year in the early 1980s. The debate was sparked by the publication of "A Statement on Scientific Temper" drafted by a group of critical rationalists including scientists, humanists, and political activists. The dissidents lined up behind Ashis Nandy's "Counter-Statement on Humanistic Temper" which saw scientific temper as an affront to the folk traditions of the Indian people. The debate appears in the periodical *Mainstream* from July 1981 to June 1982.

20. Here I can only offer my own observations as a long time participant-observer of the Indian left. After the scientific temper debate, cultural nationalism—justified by ever more sophisticated theories—became the dominant ideology of the Indian left. The Indian left has sacrificed the need for a frontal attack on the oppressive status quo in favor of idle and over-blown—but much safer—attacks on an omnipotent and evil West.

21. Foucault's and Said's influence on Indian intellectuals is critically examined in Ahmad 1992.

22. For a critical reading of Shiva, see Meera Nanda, "History Is What Hurts: A Materialist Feminist Perspective on the Green Revolution and Its Ecofeminist Critics," in *Materialist Feminism: A Reader*, ed. Rosemary Hennesy and Chrys Inghram (New York: Routledge, 1997).

23. I was shocked and saddened to see the edited volume by Ashis Nandy, *Science, Hegemony and Violence: A Requiem for Modernity* (Delhi: Oxford University Press, 1988), being read

as a standard Third World text in science studies circles in the U.S. universities. This particular volume is the compilation of essays presented at one of the many antiscience seminars in the aftermath of the "scientific temper" debate. What these essays hide, and what their Western readers will never know, is that there was an active opposition to these ideas and that these essays distort and deny the history of the nascent Enlightenment in India.

24. For an illuminating history of the May Fourth movement and a moving testimony to the tremendous appeal of "Western" science and democracy for non-Western people, see Schwarcz 1986.

25. For a recent call to give up all comparative evaluation, see Watson-Verran and Turnbull 1995. The authors argue that because all knowledge systems are local, "Western contemporary technosciences, rather than being taken as definitional of knowledge, rationality or objectivity should be treated as varieties of knowledge systems . . . [and compared with other local systems] on an equal footing" (116). Likewise, Sandra Harding contends that different ethnosciences must be seen as historicized to different cultures' projects, and their truths seen as "caused by social relations as well as by nature's regularities and the operations of reason." In the very next breath, however, Harding assimilates "nature's regularities" into "the social/political history . . . [that is internal to] the image of nature's regularities and the underlying causal tendencies." In other words, the criteria for choosing among different "knowledges" are "historicized" completely to the sociopolitical context (Harding 1994, 314).

26. The Symmetry tenet of David Bloor's formulation of SP states that sociology of scientific knowledge "would be symmetrical in its style of explanation. The same type of causes would explain, say true and false beliefs" (1991, 7).

27. "So what is science? Nothing but a *space*, one that acquires its authority precisely from and through episodic negotiations of its flexible and contextually contingent borders and territories. Science is a kind of spatial marker for cognitive authority, empty until its insides get filled and its borders drawn amidst context-bound negotiations over who and what is 'scientific'" (Gieryn 1995, 405, italics in original).

28. For an illuminating discussion on intertheoretic competition in science and other knowledge systems, see Horton 1993 and also Laudan 1990.

29. The ethnoknowledge idea is most clearly stated in Barnes and Bloor 1982, 21–44.

30. For a persuasively argued case why SP treats nature as inconsequential to our beliefs, see Pinnick 1992.

31. In personal conversations with those swayed by ethnosciences, I have been accused of being Westernized and alienated from the culture of my birth. These sins are attributed to my class background (which most interlocutors incorrectly assume to be upper class) and to my training as a scientist (which most interlocutors, incorrectly again, assume to be a liability in appreciating the advances in social studies of science).

32. What else could Haraway have meant when she wrote, "I am arguing for politics and epistemologies of locations, positioning, and situating, where partiality and not universality is the condition of being heard to make rational knowledge claim"? (1991, 195). Or what could Evelyn Fox Keller have meant when she wrote, "Far from being "value-free," *good science* is science that effectively facilitates the material realization of particular goals, that does in fact enable us to change the world in particular ways. . . . In this sense, *good science* typically works to bring the material world in close conformity with the stories and expectations that a particular we bring with us as scientists embedded in particular cultural, economic and political frames"? (Keller 1991, 5, italics in the original).

33. The phrase is from the Isa Upanishad, a 600 B.C. Hindu scripture, quoted here from Sagan 1995, 114.

References

Afshari, R. An Essay on Islamic Cultural Relativists in the Discourse of Human Rights. *Human Rights Quarterly* 16: 235–76, 1994.

Ahmad, A., *In Theory: Classes, Nations, Literatures*. London: Verso, 1992.

Alvares, C. Science, Colonialism and Violence: A Luddite View. In *Science Hegemony and Violence: A Requiem for Modernity*, A. Nandy, ed. Oxford: Oxford University Press, 1988.

———. Science. In *The Development Dictionary: A Guide to Knowledge as Power*, Wolfgang Sachs, ed. London: Zed Books, 1992.

Anees, M. A. Islam and Scientific Fundamentalism. *New Perspectives Quarterly* Summer, 1993.

Appiah, K. A., *In My Father's House: Africa in the Philosophy of Culture*. New York: Oxford University Press, 1992.

Barnes, B. Natural Rationality: A Neglected Concept in the Social Sciences. *Phil. Soc. Sci.* 6: 115–26, 1976.

———, *Interests and the Growth of Knowledge*. London: Routledge Kegan and Paul, 1977.

———, *T. S. Kuhn and Social Science*. New York: Columbia University Press, 1982.

———. How Not to Do Sociology of Knowledge. *Annals of Scholarship* 8: 321–36, 1991.

Barnes, B. and D. Bloor. Relativism, Rationalism and the Sociology of Knowledge. In *Rationality and Relativism*, M. Hollis and S. Lukes, eds. Cambridge, Mass.: MIT Press, 1982.

Barnes, B., D. Bloor, and J. Henry. *Scientific Knowledge: A Sociological Analysis*. Chicago: University of Chicago Press, 1996.

Bernal, J. D., *Science in History*. London: C. A. Watts, 1965.

Bloor, D., *Knowledge and Social Imagery*. 2d ed. Chicago: University of Chicago Press, 1991.

Chakrabarty, D. Postcoloniality and the Artifice of History: Who Speaks for "Indian" Pasts? *Representations* 37, 1992.

Collins, H. M. What is TRASP?: The Radical Programme as a Methodological Imperative. *Philosophy and Social Science*, 1981.

Fine, A. Science Made Up: Constructivist Sociology of Scientific Knowledge. In *The Disunity of Science: Boundaries, Contexts, and Power*, P. Galison and D. Stump, eds. Stanford: Stanford University Press, 1996.

Franklin, S. Science as Culture, Cultures of Science. *Annual Review of Anthropology* 24, 1995.

Gay, P., *The Enlightenment: An Interpretation*. New York: Norton, 1969.

Geertz, C., *The Interpretation of Cultures*. New York: Basic Books, 1973.

Gellner, E., Concepts and Society. In *Rationality*, Bryan Wilson, ed. New York: Harper and Row, 1970.

———. *Postmodernism, Reason and Religion*. New York: Routledge, 1992.

Gieryn, T. Boundaries of Science. In *Handbook of STS*, S. Jasonoff, G. Markle, J. Petersen, and T. Pinch, eds. Thousand Oaks, CA: Sage, 1995.

Guha, R. The Alternative Science Movement: An Interim Assessment. *Lokayan Bulletin* 6(3): 7–25, 1988.

Haraway, D., *Simians, Cyborgs, and Women: The Reinvention of Nature*. New York: Routledge, 1991.

Harding, S. Is Science Multicultural? Challenges, Resources, Opportunities, Uncertainties. *Configurations* 2: 301–30, 1994.

———. Science is 'Good to Think With.' *Social Text* 46/44: 15–26, 1996.

Hoodbhoy, P., *Islam and Science: Religious Orthodoxy and the Battle for Rationality*. London: Zed, 1991.

Horton, R., *Patterns of Thought in Africa and the West: Essays on Magic, Religion and Science*. London: Cambridge University Press, 1993.

Huntington, S. The West Unique, Not Universal. *Foreign Affairs*, Nov./Dec., 1996.

Jarvie, I. C., *Rationality and Relativism: In Search of a Philosophy and History of Anthropology.* London: Routledge and Kegan Paul, 1984.

Keller, E. F., *Secrets of Life. Secrets of Death: Essays on Language, Gender and Science.* New York: Columbia University Press, 1991.

Laudan, L., *Science and Relativism.* Chicago: University of Chicago Press, 1990.

Lewis, M. and K. E. Wigen. *The Myth of Continents: A Critique of Metageography.* Berkeley: University of California Press, 1997.

Lewontin, R. Billions and Billions of Demons. *New York Review of Books*, 9 Jan. 1997.

Lizhi, F. and P. Link. The Hope for China. *New York Review of Books*, 17 Oct. 1996.

Marglin, F. A. Smallpox in two Systems of Knowledge. In *Dominating Knowledge: Development, Culture and Resistance*, Frederique A. Marglin and Stephen A. Marglin, eds. Oxford: Clarendon, 1990.

Marglin, S. A. Losing Touch: The Cultural Conditions of Worker Accommodation and Resistance. In *Dominating Knowledge: Development, Culture and Resistance*, ed. Frederique Apfell Marglin and Stephen A. Marglin. Clarendon Press, 1990, pp. 217–282.

———. Towards the Decolonization of the Mind. In *Dominating Knowledge: Development, Culture and Resistance*, Frederique A. Marglin and Stephen A. Marglin, eds. Oxford: Clarendon, 1990.

Martin, E. Meeting Polemics with Irenics in the Science Wars. *Social Text* 46–47, 1996.

Miller, H. L., *Science and Dissent in Post-Mao China: The Politics of Knowledge.* Seattle: University of Washington Press, 1996.

Murphy, N. Scientific Realism and Postmodern Philosophy. *Brit. Jour. Phil.* 41: 291–303, 1990.

Nanda, M. Against Social De(con)struction of Science: Cautionary Tales from the Third World. In *In Defense of History: Marxism and the Postmodern Agenda*, E. M. Wood and J. B. Foster, eds. New York: Monthly Review Press, 1997a.

———. Restoring the Real: Rethinking Social Constructivist Theories of Science. In *Socialist Register*, L. Panitch, ed. London: Merlin Press, 1997b.

———. Science Wars in India. *Dissent*, Winter, 1997c.

Nandy, A. The Human Factor. *The Illustrated Weekly of India*, 17 Jan., 1988a.

———, ed. *Science, Hegemony and Violence: A Requiem for Modernity.* New Delhi: Oxford University Press, 1988b.

———. Culture, Violence and Development: A Primer for the Unsuspecting. *Thesis Eleven* 39: 1–18, 1994.

Needham, J., *The Grand Titration: Science and Society in East and West.* London: Allen & Unwin, 1969.

Nicholson, L. On the Postmodern Barricades: Feminism, Politics and Theory. In *Postmodernism and Social Theory*, Steven Seidman and David Wagner, eds. Cambridge, MA: Blackwell, 1992.

Nussbaum, M. Non Relative Virtues: An Aristotelian Approach. In *The Quality of Life*, M. Nussbaum and A. Sen, eds. Oxford: Clarendon Press, 1993.

———. Patriotism and Cosmopolitanism. In *For Love of Country: Debating the Limits of Patriotism*, Joshua Cohen, ed. Boston: Beacon Press, 1996.

Nussbaum, M. and A. Sen. Internal Criticism and Indian Rationalist Traditions. In *Relativism: Interpretations and Confrontations*, Michael Krausz, ed. Indiana: University of Notre Dame Press, 1989.

Pinnick, C. Cognitive Commitment and the Strong Program. *Social Epistemology* 6: 289–98, 1992.

Rentlen, A. D. Relativism and the Search for Human Rights. *Human Rights Quarterly* 90: 56–72, 1988.

Rorty, R. Human Rights, Rationality and Sentimentality. In *Human Rights: The Oxford Amnesty Lectures 1993*, Stephen Shute and Susan Hurley, eds. New York: Basic Books, 1993.

Sagan, C., *The Demon Haunted World: Science as a Candle in the Dark*. New York: Random House, 1995.

Sardar, Z., ed., *The Revenge of Athena: Science Exploitation and The Third World*. London, Mansell, 1988.

Schwarcz, V., *The Chinese Enlightenment: Intellectuals and the Legacy of the May Fourth Movement of 1919*. Berkeley, University of California Press, 1986.

Shiva, V. *Staying Alive*. New Delhi: Kali for Women, 1988.

Shweder, R. Post-Nietzschian Anthropology: The Idea of Multiple Objective Worlds. In *Relativism: Interpretation and Confrontation*, Michael Krausz, ed. Notre Dame, IN: University of Notre Dame Press, 1989.

Toulmin, S., *Cosmopolis: The Hidden Agenda of Modernity*. Chicago: University of Chicago Press, 1990.

Turner, B., *Orientalism, Postmodernism and Globalism*. New York: Routledge, 1994.

Watson-Verran, H. and D. Turnbull. Science and Other Indigenous Knowledge Dystems. In *Handbook of Science and Technology Studies*, S. Jasonoff, G. Markle, J. C. Peterson and T. Pinch, eds. Thousand Oaks, CA: Sage, 1995.

Zachariah, M., *Science for Social Revolution? Achievements and Dilemmas of a Development Movement*. London: Zed Books, 1994.

Index